New Paradigms in Cellular Physiology and Pathology

New Paradigms in Cellular Physiology and Pathology

Editor: Christopher Shelley

AMERICAN
MEDICAL PUBLISHERS
www.americanmedicalpublishers.com

AMERICAN
MEDICAL PUBLISHERS
www.americanmedicalpublishers.com

Cataloging-in-Publication Data

New paradigms in cellular physiology and pathology / edited by Christopher Shelley.
p. cm.
Includes bibliographical references and index.
ISBN 979-8-88740-333-5
1. Cell physiology. 2. Pathology, Cellular. 3. Cytology. 4. Physiology. I. Shelley, Christopher.
QH631 .N49 2023
574.876--dc23

American Medical Publishers,
41 Flatbush Avenue,
1st Floor, New York,
NY 11217, USA

ISBN 979-8-88740-333-5 (Hardback)

Contents

Preface

In my initial years as a student, I used to run to the library at every possible instance to grab a book and learn something new. Books were my primary source of knowledge and I would not have come such a long way without all that I learnt from them. Thus, when I was approached to edit this book; I became understandably nostalgic. It was an absolute honor to be considered worthy of guiding the current generation as well as those to come. I put all my knowledge and hard work into making this book most beneficial for its readers.

The cell is the basic structural and functional unit of life. Cell physiology refers to the biological study of the various processes that take place in the cell, which keep it alive. In living systems, free radicals such as reactive oxygen species (ROS) and reactive nitrogen species (RNS) play a crucial role in regulating a wide range of cell activities such as autophagy, metabolism, growth and development, solute transport, programmed cell death, and stress responses. Furthermore, ROS and RNS can damage tissues through oxidative stress and tissue dysfunction, and they can also act as molecular messengers for stress responses that are advantageous to the organism. Cellular pathology on the other hand, studies the diseases that occur in the cells, tissues and organs. The primary diagnostic tests that are utilized in cell pathology include cytopathology and histopathology. This book provides an insight into the new paradigms in the research on cellular physiology and pathology. It will serve as a valuable source of reference for graduate and post graduate students.

I wish to thank my publisher for supporting me at every step. I would also like to thank all the authors who have contributed their researches in this book. I hope this book will be a valuable contribution to the progress of the field.

Editor

1

Can Serum Nitrosoproteome Predict Longevity of Aged Women?

Daniele Capitanio [1], Pietro Barbacini [1], Beatrice Arosio [2,3], Franca Rosa Guerini [4], Enrica Torretta [5], Fabio Trecate [4], Matteo Cesari [3,6], Daniela Mari [7], Mario Clerici [4,8] and Cecilia Gelfi [1,5,*]

1. Department of Biomedical Sciences for Health, University of Milan, 20090 Segrate (MI), Italy; daniele.capitanio@unimi.it (D.C.); pietro.barbacini@unimi.it (P.B.)
2. Geriatric Unit, Fondazione IRCCS Ca' Granda Ospedale Maggiore Policlinico, 20122 Milan, Italy; beatrice.arosio@unimi.it
3. Department of Clinical Sciences and Community Health, University of Milan, 20122 Milan, Italy; matteo.cesari@unimi.it
4. IRCCS Fondazione Don Carlo Gnocchi ONLUS, 20148 Milan, Italy; fguerini@dongnocchi.it (F.R.G.); ftrecate@dongnocchi.it (F.T.); mario.clerici@unimi.it (M.C.)
5. IRCCS Istituto Ortopedico Galeazzi, 20161 Milan, Italy; enrica.torretta@grupposandonato.it
6. Geriatric Unit, IRCCS Istituti Clinici Scientifici Maugeri, 20138 Milan, Italy
7. Laboratorio Sperimentale di Ricerche di Neuroendocrinologia Geriatrica ed Oncologica, IRCCS Istituto Auxologico Italiano, 20145 Milan, Italy; daniela.mari@unimi.it
8. Department of Pathophysiology and Transplantation, University of Milan, 20122 Milan, Italy

* Correspondence: cecilia.gelfi@unimi.it

Abstract: Aging is characterized by increase in reactive oxygen (ROS) and nitrogen (RNS) species, key factors of cardiac failure and disuse-induced muscle atrophy. This study focused on serum nitroproteome as a trait of longevity by adopting two complementary gel-based techniques: two-dimensional differential in gel electrophoresis (2-D DIGE) and Nitro-DIGE coupled with mass spectrometry of albumin-depleted serum of aged (A, $n = 15$) and centenarian (C, $n = 15$) versus young females (Y, $n = 15$). Results indicate spots differently expressed in A and C compared to Y and spots changed in A vs. C. Nitro-DIGE revealed nitrosated protein spots in A and C compared to Y and spots changed in A vs. C only (p-value < 0.01). Nitro-proteoforms of alpha-1-antitripsin (SERPINA1), alpha-1-antichimotripsin (SERPINA3), ceruloplasmin (CP), 13 proteoforms of haptoglobin (HP), and inactive glycosyltransferase 25 family member 3 (CERCAM) increased in A vs. Y and C. Conversely, nitrosation levels decreased in C vs. Y and A, for immunoglobulin light chain 1 (IGLC1), serotransferrin (TF), transthyretin (TTR), and vitamin D-binding protein (VDBP). Immunoblottings of alcohol dehydrogenase 5/S-nitrosoglutathione reductase (ADH5/GSNOR) and thioredoxin reductase 1 (TRXR1) indicated lower levels of ADH5 in A vs. Y and C, whereas TRXR1 decreased in A and C in comparison to Y. In conclusion, the study identified putative markers in C of healthy aging and high levels of ADH5/GSNOR that can sustain the denitrosylase activity, promoting longevity.

Keywords: proteomics; nitrosative stress; aging; cardiovascular disease; muscle atrophy

1. Introduction

Longevity is the result of multiple factors including genetic, epigenetic, and lifestyle; however, circulating molecules able to predict the evolution from healthy toward stressor events overwhelming the antioxidant capacity of the entire organism are still absent. Aging is characterized by a progressive decline of the muscle functional performance (i.e., sarcopenia), which contributes to frailty, defined as the vulnerability of the body to exogenous and endogenous stressors [1]. One of the theories proposed to explain the aging process is based on the free radical theory proposed by Denhan Harman in 1956, suggesting that accumulation of oxygen species overcomes the cellular anti-oxidative capacity [2]. Following this theory, post-translational modifications of proteins such as nitrosation and peroxidation can be associated with the aging process including sarcopenia.

Recent results indicate that the increase in reactive oxygen (ROS) and nitrogen (RNS) species are key factors of cardiac and vasculature homeostasis and disuse-induced muscle atrophy [3,4]. The effects and the molecular mechanisms that regulate oxidative stress, in particular those related to nitrosative stress and the role of antioxidant defense in human skeletal muscle and at the whole-body level, are still poorly understood. Nitric oxide (NO) is an important signaling molecule produced endogenously from NO synthase (NOS). NO acts on proteins by means of selective covalent modification of cysteine residues through a redox reaction to form nitroso thiols (SNOs) [5]. This mechanism regulates a wide range of cellular functions and signal processes such as neurotransmission, muscle relaxation, and apoptosis. Physiologically, NO in cells is maintained at a minimum level to allow the carrying out of its regulatory role in various signal pathways. Conversely, the excessive production of NO triggers nitrosative stress and induces changes in S-nitrosated protein levels. It has been described that a prolonged muscle disuse, such as that observed in bedridden subjects, compromises the Ca^{2+} homeostasis that increases muscle deficit with alteration of its functional capabilities. This dysregulation can be improved with exercise as a countermeasure [6–8]. On muscle, the ryanodine receptor 1 (RYR1) nitrosation is directly related to the loss of muscle mass and function [9,10], and its regulation is a critical condition for the muscular homeostasis maintenance. In cardiac tissue, nitrosation of the ryanodine receptor 2 (RYR2) regulates peripheral vascular tone and β-adrenergic agonist-stimulated cardiac contractility [11]. It has also been demonstrated that the denitrosylase alcohol dehydrogenase 5/S-nitrosoglutathione reductase (alcohol dehydrogenase 5/S-nitrosoglutathione reductase (ADH5/GSNOR)) is downregulated in cells and animal models of aging and that mRNA levels from circulating peripheral blood mononuclear cells (PBMC) are retained in centenarians [12,13]. We expect that nitrosation of proteins could be a more general event and we assume that markers of nitrosative stress can be found not only in cardiac and skeletal muscle tissues but also in bodily fluids.

In this context, we would like to understand if there is a relationship between circulating biomarkers and loss of muscle mass and function or if other mechanisms contribute to longevity. Furthermore, it has been recently demonstrated that biomarkers of successful aging are sex-related [14,15], and thus gender has to be considered. In a previous study conducted in male aged subjects, the loss of muscle function was connected to the loss of phosphorylatable myosin light chain, which regulates the velocity of fiber contraction, and of changes in muscle metabolism, resulting in fat accumulation [16,17]. In astronauts undergoing 6 months of space flight, characterized by a loss of muscle mass and function at landing, deregulated miRNA were associated with microgravity exposure, pointing to a strong stress response caused by space flight that involved muscle tissue and proinflammatory molecules. In particular, the muscle biopsy of one of the astronauts showed a 80% increase of alpha-1-antitrypsin, an acute phase protein, and of its target miR-126-3p, suggesting that serum can tell us something about the muscle function decline and the effect at the whole body-level of a stressful condition such as a prolonged space flight [18]. More recently, sarcopenia in men was found to be associated with high cholesterol and lower creatinine levels. Conversely, in women, the maintenance of normal gait speed was found to be associated with a lower percentage of body fat and lower lactate dehydrogenase levels [15].

Several important studies have recently addressed the issue of a differential protein expression associated with aging, identifying a set of increased proteins as putative biomarkers of aging [19–22]. However, these studies on circulating proteome did not suggest any mechanism that can be targeted to counteract the aging decline affecting the entire organism.

In an attempt to contribute, we addressed the issue of nitrosignalling following levels of nitrosated proteins in serum of aged individuals and centenarians to find molecules that can be targeted to promote the evolution toward longevity and, indirectly, to expand the healthy age interval that have allowed centenarians to reach their exceptional age. It has been recently described that S-nitrosation increases the rate of mitochondrial fragmentation and inhibits mitophagy, which is tightly associated with the aging phenotype and decline of muscle mass and function [23]. The study indicates that excessive S-nitrosation of key mitochondrial proteins contributes to aging and related disorders [24–27], and this process is regulated by denitrosylases such as thioredoxin/thoredoxin reductase (Trx/TrxR) and ADH5/GSNOR [28,29]. Thioredoxin reductase, together with the reduced form of nicotinamide adenine dinucleotide phosphate (NADPH) and thioredoxin, is a component of the thioredoxin system that, with the glutathione–glutaredoxin system, controls the redox environment of mammalian cells. The Trx system protects cells from oxidative and nitrosative stress through its disulfide reductase activity regulating protein dithiol/disulfide balance. ADH5/GSNOR regulates intracellular concentration of reactive NO by catalyzing the breakdown of S-nitrosoglutathione (GSNO), a NO donor for cysteine thiols, thus indirectly modulating SNO formation through GSNO-mediated protein S-nitrosation. This suggests that these two systems target different molecules, since GSNOR specifically denitrosates GSNO, whereas the cytoplasmic and mitochondrial thioredoxins mediate denitrosation of multiple S-nitrosated proteins in relation to stimulus, substrate, and localization [30].

In an attempt to find modified molecules involved into the aging process and in the loss of muscle mass characterizing aging, we adopted a two-dimensional difference in gel electrophoresis (2-D DIGE) gel-based analysis of the serum proteome in conjunction with Nitro-DIGE and mass spectrometry of aged (A) women versus young (Y) women and centenarians (C) in order to identify putative markers associated with aging.

2. Results

2.1. Assessement of Muscle Performances

The loss of muscle force and function associated with sarcopenia was determined by hand grip test and short physical performance battery (SPPB) in aged women, and by hand grip test in centenarian women. SPPB indicated in five subjects a score of 7 or higher, while for the remaining subjects, the values ranged from 0 to 3, indicating a functional decline, as shown in Table S1. Figure 1 shows the muscle force decline determined by hand grip test in aged and centenarian women. The mean maximum hand grip strength observed among young subjects was 27.6 kg, whereas it decreased to 17.2 kg in aged women and to 8.6 kg in centenarians. The decline was increased in centenarians, although four subjects had a hand grip test value comparable with some of the aged women, and only three of them (aged 106, 107, and 113 years old, respectively) showed a significant decline of force levels (below 10 kg). Nonetheless, these three subjects had body mass index (BMI) values comparable to subjects of the same cohort. The Katz index of independence in activities of daily living did not show significant variations (Student's t-test, $p < 0.05$) among aged and centenarians, due to the great variability in both groups (2.2 ± 1.1 and 1.7 ± 1.7 for aged women and centenarians, respectively). Furthermore, 12 out of 15 aged subjects and 11 out of 15 centenarians had hypertension and were pharmacologically treated.

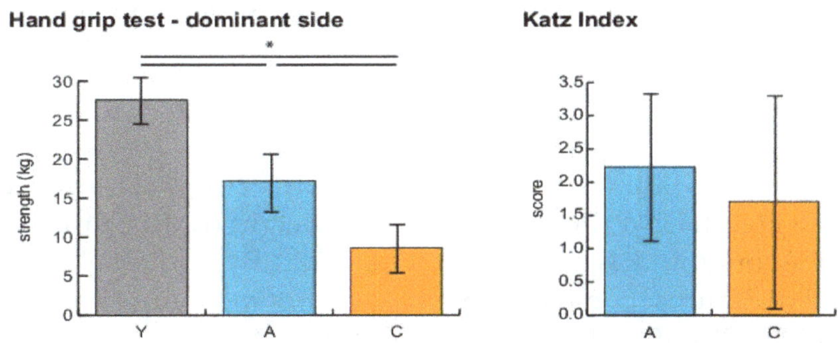

Figure 1. Histograms showing the mean (± SD) maximum hand grip strength (kg) and the Katz Index scores observed among the young (Y), aged (A), and centenarian (C) female subjects enrolled in this study (* = significant difference, ANOVA and Tukey, p-value < 0.05).

2.2. 2-D DIGE and Nitro-DIGE Proteomics

Albumin-depleted serum samples were analyzed by 2-D DIGE and Nitro-DIGE to evaluate changes in protein abundance and in their protein S-nitrosation levels among A, C, and Y subjects. Overall, 2-D DIGE revealed 120 changed over 978 matched spots among all gels. Among them, 114 spots were differently expressed (ANOVA and Tukey, $n = 6$, p-value < 0.01) in A and C compared to Y subjects (32 significantly changed in A, 12 in C, 70 changed with the same trend both in A and C vs. Y) (Figure 2A). Identification data are shown in Tables S2 and S3, and in Figure S1. Nitro-DIGE analysis revealed 206 nitrated protein spots as changed (ANOVA and Tukey, $n = 9$, p-value < 0.01) in A and C compared to Y (39 changed in A, 43 in C, whereas 119 changed with the same trend both in A and C vs. Y, and 5 counter-regulated in A vs. Y compared to C vs. Y) (Figure 2B).

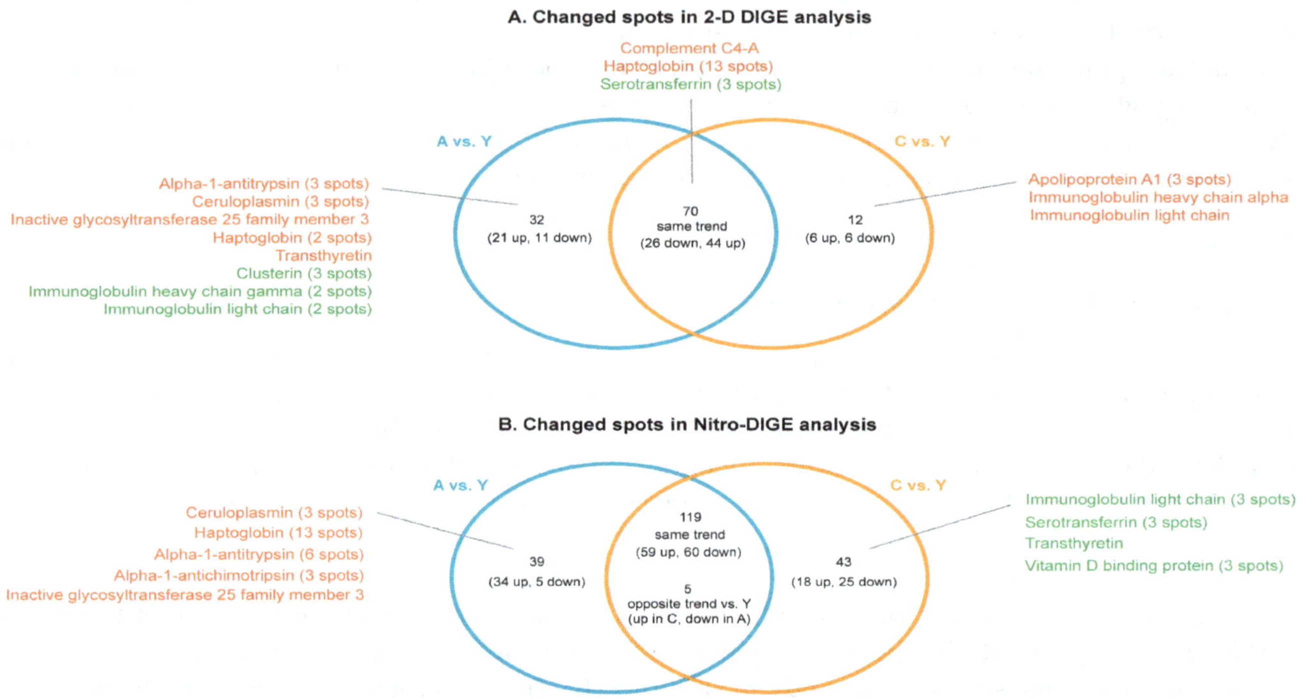

Figure 2. Schematic diagrams resuming findings obtained from two-dimensional differential in gel electrophoresis (2-D DIGE) (**A**) and Nitro-DIGE (**B**) proteomic analyses conducted on albumin-depleted sera from young (Y), aged (A), and centenarian (C) women. Protein names and number of identified spots (corresponding to different proteoforms) were reported (red = increased; green = decreased levels in A and/or C vs. Y, ANOVA and Tukey, $n = 6$ (2-D DIGE) or $n = 9$ (Nitro-DIGE), p-value < 0.01).

Mass spectrometry analysis allowed the identification of alpha-1-antitripsin (SERPINA1). SERPINA1 is composed by high (HMW) and low molecular weight (LMW) proteoforms. Three LMW spots were increased in A only and were not S-nitrosated. By contrast, six HMW spots were unchanged and their level of S-nitrosation increased in A but not in C. Another identified protein was alpha-1-antichimotripsin (SERPINA3). Three spots of SERPINA3 were unchanged in aging, and also in this case a higher degree of S-nitrosation was observed in A but not in C (Figure 3).

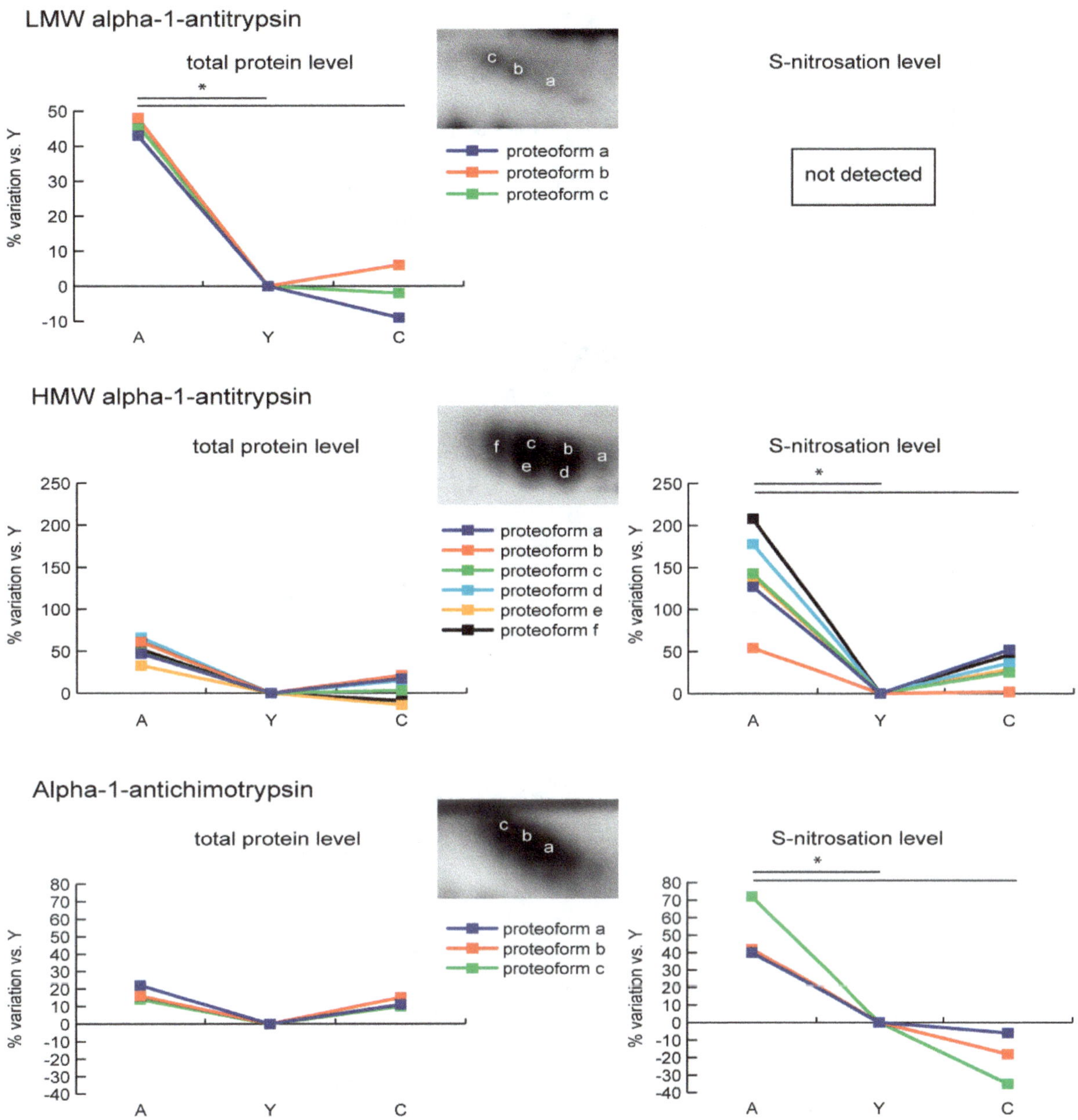

Figure 3. Line charts illustrating alpha-1-antitripsin (SERPINA1) and alpha-1-antichimotripsin (SERPINA3) proteoform abundance and S-nitrosation level variations (%) in sera of aged (A) and centenarian (C) compared to young (Y) women. Proteins were divided according to the statistical significance of the test (* = significant difference, ANOVA and Tukey, $n = 6$ (2-D DIGE) or $n = 9$ (Nitro-DIGE), p-value < 0.01).

As regards ceruloplasmin (CP), all the identified proteoforms were more abundant and nitrosated in A compared to Y and C. Haptoglobin (HP) is characterized by the presence of 15 proteoforms. Among them, two spots were increased in A not in C, whereas 13 increased both in A and C compared to Y. Of note, three were also significantly increased in A vs. C. Moreover, in this case, S-nitrosation appeared unchanged in proteoforms with increased protein levels in A compared to Y and C. Conversely, spot S-nitrosation of decreased proteoforms increased in A and C compared to Y, but particularly in A (Figure 4).

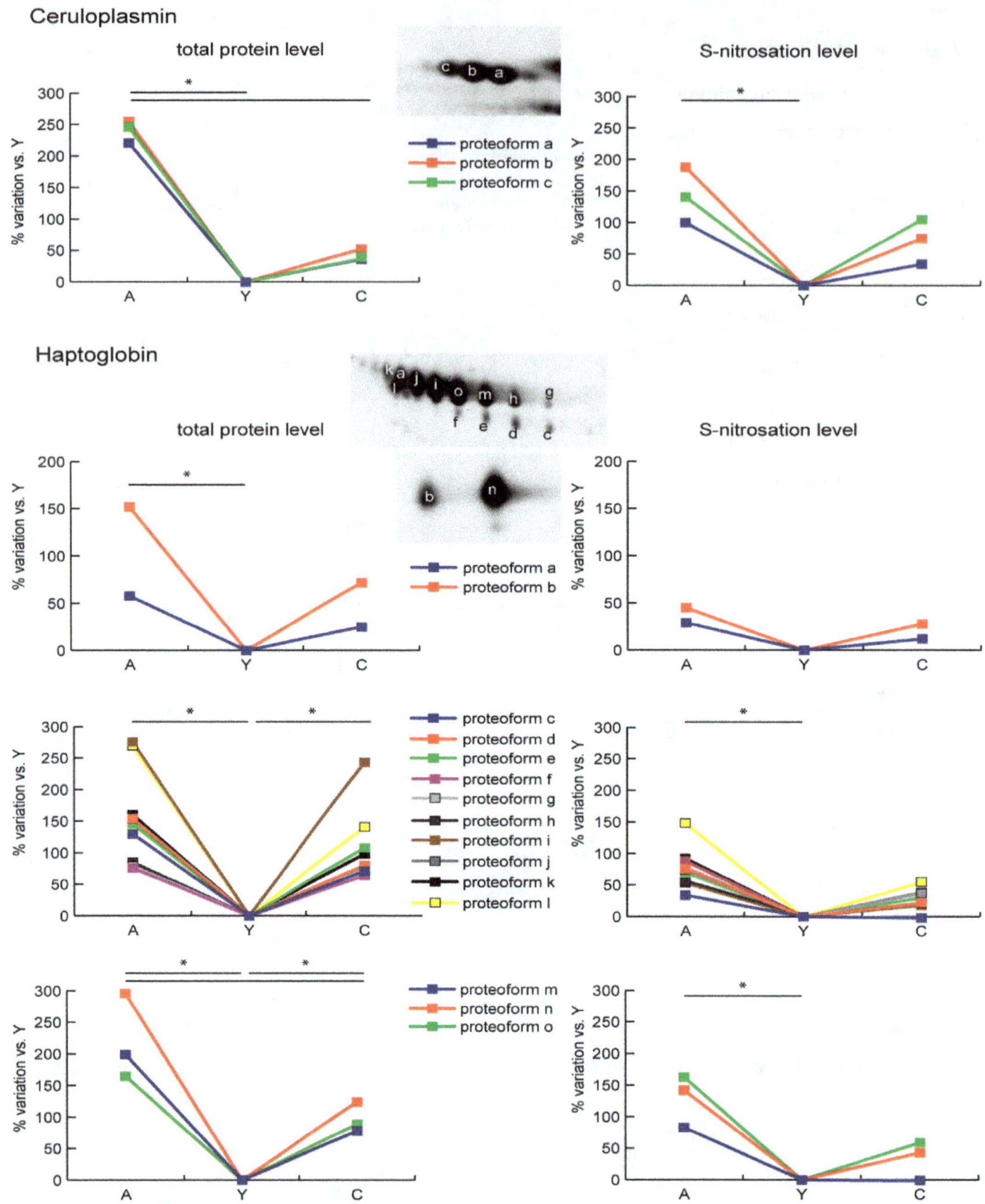

Figure 4. Line charts illustrating ceruloplasmin (CP) and haptoglobin (HP) proteoform abundance and S-nitrosation level variations (%) in sera of aged (A) and centenarian (C) compared to young (Y) women. Proteins were divided according to the statistical significance of the test (* = significant difference, ANOVA and Tukey, $n = 6$ (2-D DIGE) or $n = 9$ (Nitro-DIGE), p-value < 0.01).

Clusterin (CLU) and immunoglobulin heavy chain gamma (IGHG3) decreased in A vs. Y and C, and were not differentially S-nitrosated. Likewise, low molecular weight proteoforms (LMW) of immunoglobulin light chain 1 (IGLC1) decreased in A compared to Y and C. Nitrosation levels did not change significantly with age. Conversely, upper proteoforms (HMW), despite unchanged levels of total protein, showed decreased levels of S-nitrosation in C (Figure 5).

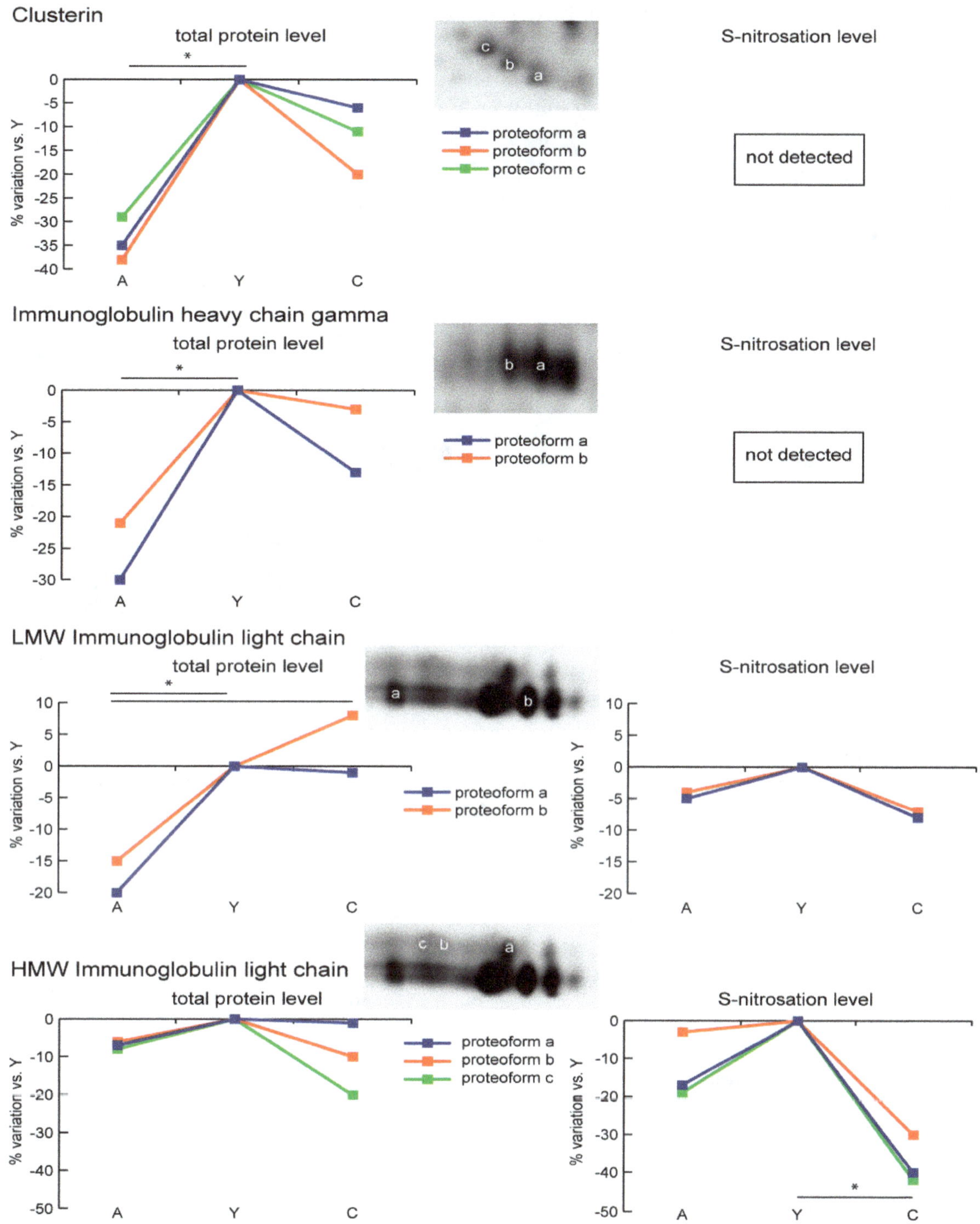

Figure 5. Line charts illustrating clusterin (CLU), immunoglobulin heavy chain gamma (IGHG3), and immunoglobulin light chain (IGLC1) proteoform abundance and S-nitrosation level variations (%) in sera of aged (A) and centenarian (C) compared to young (Y) women. Proteins were divided according to the statistical significance of the test (* = significant difference, ANOVA and Tukey, $n = 6$ (2-D DIGE) or $n = 9$ (Nitro-DIGE), p-value < 0.01).

Inactive glycosyltransferase 25 family member 3 (CERCAM) showed increased levels of total protein, whereas S-nitrosated CERCAM was observed in A only. Three proteoforms of serotransferrin (TF) decreased in A and C vs. Y; however, their nitrosation levels were decreased in C vs. Y only. Transthyretin (TTR) showed increased levels of total protein in A compared to Y and C. Furthermore, S-nitrosation levels decreased in C compared to Y and A. The three proteoforms of vitamin D-binding protein (VDBP) were unchanged with age; however, S-nitrosation level was lower in C in comparison to Y and A (Figure 6).

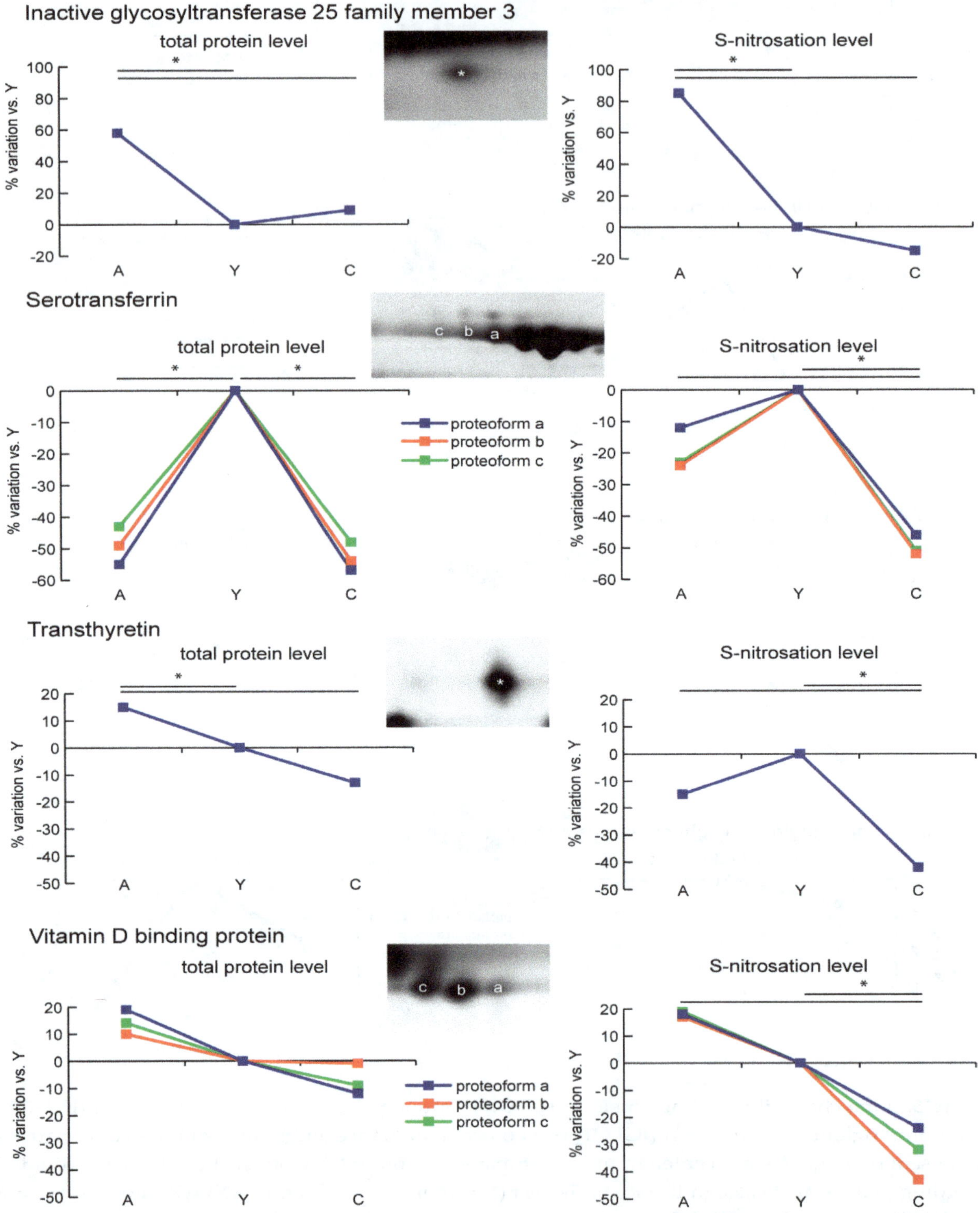

Figure 6. Line charts illustrating inactive glycosyltransferase 25 family member 3 (CERCAM), serotransferrin (TF), transthyretin (TTR), and vitamin D-binding protein (VDBP) proteoform abundance and S-nitrosation level variations (%) in sera of aged (A) and centenarian (C) compared to young (Y) women. Proteins were divided according to the statistical significance of the test (* = significant difference, ANOVA and Tukey, $n = 6$ (2-D DIGE) or $n = 9$ (Nitro-DIGE), p-value < 0.01).

Three proteoforms of apolipoprotein A1 (APOA1) and immunoglobulin heavy chain alpha (IGHA1) increased in C in comparison to Y and A. The complement C4-A (C4A) resulted in an increase both in A and C vs. Y, but increased more in C vs. A. No changes in S-nitrosation were detected (Figure 7).

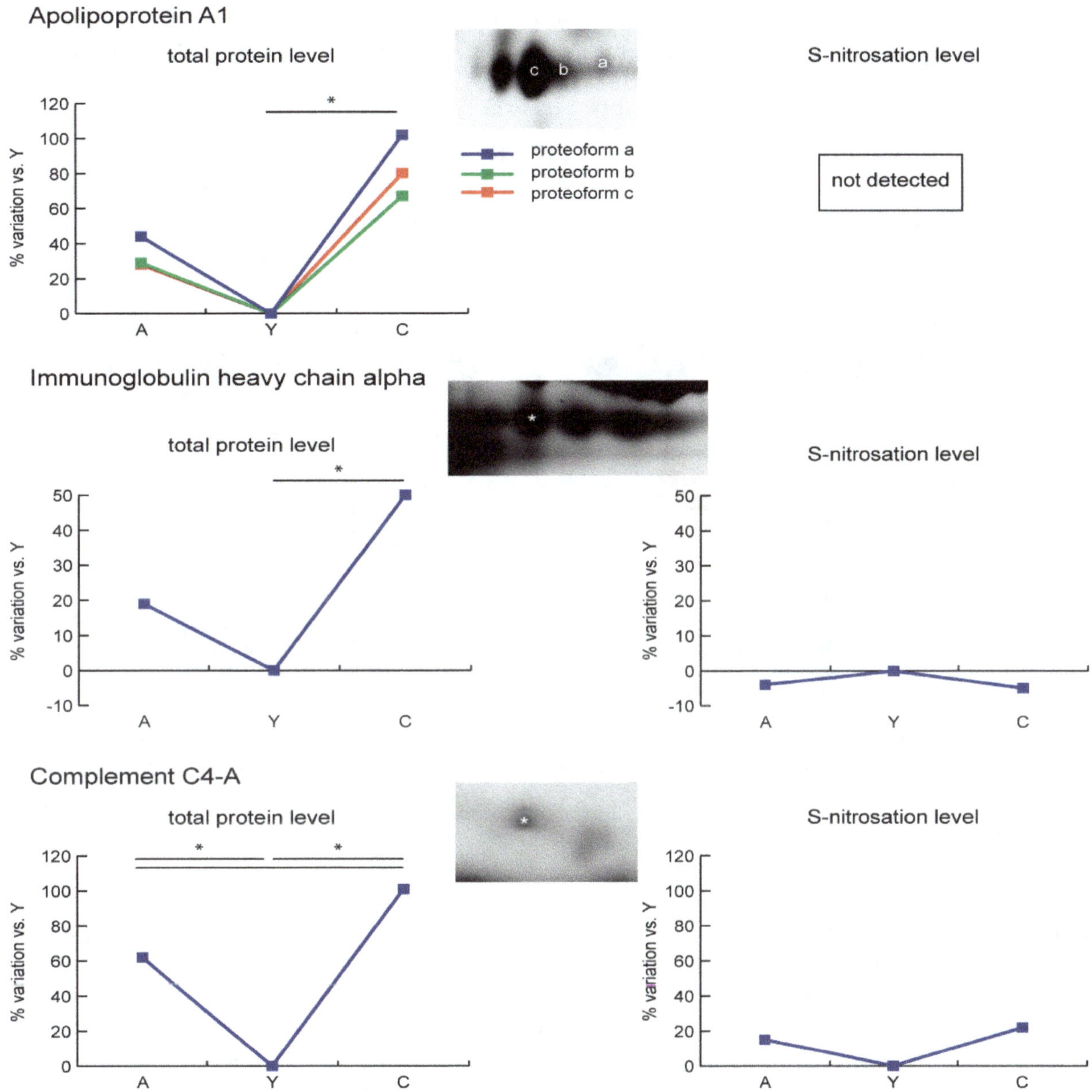

Figure 7. Line charts illustrating apolipoprotein A1 (APOA1), immunoglobulin heavy chain alpha (IGHA1), and complement C4-A (C4A) proteoform abundance and S-nitrosation level variations (%) in sera of aged (A) and centenarian (C) compared to young (Y) women. Proteins were divided according to the statistical significance of the test (* = significant difference, ANOVA and Tukey, $n = 6$ (2-D DIGE) or $n = 9$ (Nitro-DIGE), p-value < 0.01).

Concerning nitrosated proteins, albeit few exceptions (i.e., HP proteoforms c-o, CP, CERCAM and TF), we observed two general behaviors—proteoforms changed in total protein content generally did not appear to be differentially nitrosated with age (i.e., LMW SERPINA1, a and b proteoforms of HP, CLU, IGHG3, LMW IGLC1, APOA1, IGHA1, TTR in group A, C4A), whereas proteoforms that did not significantly change in total abundance were found to have different levels of S-nitrosation in aging (i.e., HMW SERPINA1, SERPINA3, HMW IGLC1, TTR in group C, VDBP).

In sum, protein S-nitrosation levels were found to be increased in A vs. Y and C, such as in HMW proteoforms of SERPINA1, SERPINA3, CP, 13 proteoforms of HP, and CERCAM. Conversely, nitrosation levels decreased in C vs. Y and A in HMW proteoforms of IGLC1, TF, TTR, and VDBP.

2.3. Tyrosine Nitration Assessment

Nitrated tyrosines were detected by immunoblotting. A slight increment of nitrated proteins was observed in A compared to Y and C, although this was not supported by statistical analysis (Figure S2).

2.4. Levels of Molecules Controlling the Nitrosative Stress in A, C, and Y Subjects

ADH5/GSNOR and thioredoxin reductase 1 (TRXR1) were assessed by immunoblotting. Figure 8 indicates that aged subjects displayed significantly lower levels of ADH5 compared to both Y and C, whereas TRXR1 levels were decreased in serum of A and C in comparison to Y.

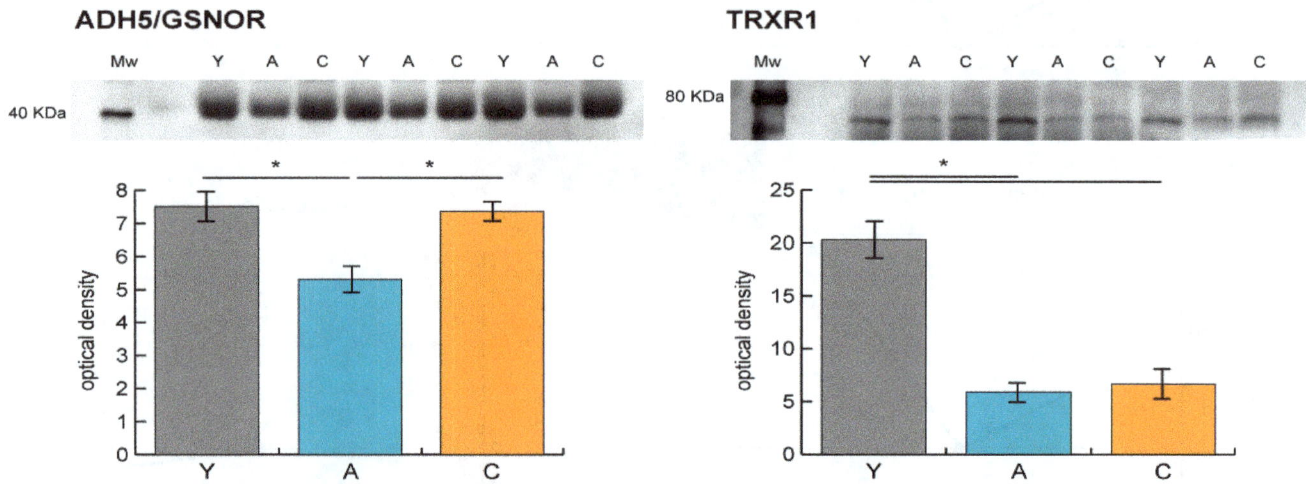

Figure 8. Representative histograms and immunoblot images of alcohol dehydrogenase 5/S-nitrosoglutathione reductase (ADH5/GSNOR) and thioredoxin reductase 1 (TRXR1) (mean ± SD; * = significant difference, ANOVA and Tukey's test, $n = 3$, p-value < 0.05) in aged (A), young (Y), and centenarian (C) samples.

3. Discussion

This pilot study was based on the serum proteome analysis of aged women in comparison with centenarian women. Results of physical performances were restricted to SPPB and hand grip test for aged women, and hand grip test for centenarians. It should be taken into account that the physical evaluation of centenarians was hampered by the unavailability of validated tools for this segment of the population, and often such procedures do not take into account the peculiar characteristics of centenarians (e.g., fatigue and sensory impairment). Nevertheless, the experience of centenarians is crucial for understanding mechanisms regulating aging and age-related conditions [31]. Sarcopenia initiates around the fourth decade of life; evidence suggests that skeletal muscle mass and skeletal muscle strength decline progressively, more in men than women, with up to 50% of mass being lost by the eighth decade of life [32,33]. Collectively, our data indicated that the decline of muscle performance was present in aged subjects, and centenarians under 105 years of age were characterized by a slower decline of muscle force and strength that became severe in very senescent subjects (106, 107, and 113 years old), where it was also difficult to perform measurements. It must also be highlighted that the majority of subjects had hypertension (12 out of 15 aged women and 11 out of 15 centenarians); however, all of them were pharmacologically treated. We were conscious about the contribution of hypertension in triggering cardiovascular disease and chronic kidney disorders (CKD); thus, due to the restricted number of subjects, particularly centenarian subjects, we decided to consider hypertension under control sharing non-hypertensive subjects homogeneously to balance their contribution inside each group.

Taking into account these preliminary observations, we based our investigation on the differential abundance of intact serum proteins and on their nitration status. The study adopted two gel-based complementary techniques (2-D DIGE and Nitro-DIGE) and unravelled specific proteoforms influenced by aging, providing hints on their level of nitrosation and describing, for the first time, levels of nitrated plasma proteins in aged subjects and of enzymes involved in the denitrosylation processes that were at variance between A and C in comparison to Y.

Overall, results indicate that for a group of proteins, nitrosation increased in A compared to Y, and, importantly, that C were characterized by lower levels of nitrosation in comparison to aged subjects, and in some cases their levels were also lower than Y.

Our data indicate that some proteins increased and had an unchanged degree of nitrosation compared to Y and C, also indicating that acidic proteoforms were preferential targets of nitrosation. Conversely, proteins with decreased levels in aging showed an increased nitrosation, and their nitrosation state was at a variance among A and C vs. Y, suggesting that this process can be associated with unbalanced homeostasis of the redox system and that these molecules can be selected as potential markers of longevity. Another underlined point is the level of SERPINA1. LMW proteoforms increased in aged subjects in comparison to Y, whereas HMW proteoforms were more nitrosated in A compared to C and Y. It could be of note that this molecule accumulates in muscle exposed to prolonged microgravity, suggesting a link between levels of this molecule in muscle tissue and its release into the bloodstream that is associated with muscle decline. Further study will be required to better clarify this issue.

Why are proteoforms with lower isoelectric point (pI) preferentially nitrosated? It can be postulated that the lower pI of proteoforms can be related to the location of the cysteine residues in the protein sequence that act as preferential targets of the sulfhydryl group for nitrosation due to their increased nucleophilicity. Thus, protein nitration is a specific process targeting only a restricted number of proteins and limited to key cysteine residues within these proteins [34].

Recent relevant studies observed that most age-associated proteins showed increased abundance with age, with the reason potentially being due to a loss of renal function in aging or a bias of the aptamer technology occurring on the basis of proteins that are targeted by the slow off-rate modified aptamers (SOMAmers) utilized to investigate the proteome [19–22,35,36]. Other proteomic studies in aged subjects using technologies such as two-dimensional gel electrophoresis [37,38] or quantitative mass spectrometry [35,39,40] showed a number of age-associated proteins that decreased as well as increased with age. In our study, the 2-D DIGE indicated more or less the same ratio of up- and downregulated, proteoforms in agreement with previous studies based on intact protein labelling and LC–MS/MS approaches. In our study, we observed increment of SERPINA1, CP, HP, CERCAM, and TTR, with some of them having already been described [18]. Furthermore, the study also indicates that specific proteoforms with lower pI were nitrosated. At variance, we observed a decrement of CLU, IGHG3, and LMW IGLC1 in A in comparison to Y and C. Among these proteins, CLU is known to enhance amyloid clearance and modulate neuroinflammation in mild cognitive impairment [41]. Its decrement can be associated with the loss of protective mechanisms in A.

Unexpectedly, proteins characterized by altered abundance in aging were also characterized by unchanged nitrosation levels and nitrosation-targeted specific proteoforms. The proteoform is a protein product of a single gene and it is a result of alternative splicing or post-translational modifications that affect proteolytic cleavage or terminal degradation [42]. Proteoforms may be depleted or enriched as they are released into the circulation. In light of this, the fact that specific plasma proteoforms are nitrosated could have implications on the origin of circulating proteins and/or in the level of NO and of molecules controlling nitrosation. The measurement of proteoforms in blood is a difficult task [43] that could in principle be overcome by mass spectrometry-based technologies, such as selected reaction monitoring (SRM) and multiple reaction monitoring (MRM); however, a strong effort on this direction is needed for it to be routinely implemented.

It has been recently described that higher levels of the denitrosylase ADH5/GSNOR enzymes promote longevity [12,44,45]. From the present study, it can be expected that the denitrosylase activity will decrease with age, as appears from the serum nitrosoprofile of A vs. Y. Results from immunoblotting indicate that a precise mechanism was targeted and that it was associated with ADH5/GSNOR, while the nitro/denitro processes under TRXR1 control were decreased both in A and C in comparison to Y.

The decrement of ADH5 confirmed our results and indicated that increased levels of nitrosated proteins both in cysteine and tyrosine (although in this case not statistically significant) in A can be due to the inhibition of ADH5/GSNOR denitrosylase activity, with C subjects appearing to be more protected from this negative signaling, sustaining the assumption that being able to denitrosylate well leads to living longer [12,13,26]. Interestingly, the level of the Trx/TrxR system was decreased both in A and C in comparison to Y, suggesting that denitrosylation in aging requires a tight regulation of the ADH5/GSNOR system while the general mechanism protecting cells from oxidative and nitrosative stress and regulating the dithiol/disulfide balance are not involved in the prolonged healthy aging status typical of centenarians.

It should be of note that specific proteins showed lower level of nitrosation in C vs. A, and they can be preferential targets of ADH5/GSNOR and therefore selected as possible targets for a precise therapeutic intervention to counteract frailty and promote healthy aging. Specifically, the HMW proteoforms of SERPINA1, SERPINA3, CP, 13 proteoforms of HP, and CERCAM being involved in tissues homeostasis can play a role on the fine-tuning of redox balance. Serpins regulate a wide range of biological processes, including inflammatory responses, and are key factors in the pathophysiology of cardiovascular diseases [46]. It could be of note that the increase of SERPINA1 and SERPINA3 in muscle and in serum was also demonstrated in various murine models of muscle atrophy [47,48], suggesting that a higher nitrosation state of these molecules could represent a trait of muscle decline. Furthermore, SERPINA3 has also been identified as a specific biomarker of delirium and Alzheimer's disease [49,50]. CP is an acute phase reactant, regulating NO homeostasis and working as a NO oxidase, suggesting that low levels may promote NO bioavailability, protecting C from endovascular dysfunctions [51]. Another protein of interest is HP, which has been recognized as an inflammatory indicator in cardiovascular disease. It is an acute-phase protein with a role in the neutralization and clearance of free heme [51].

Of note, a set of proteins also appeared less nitrosated in comparison to younger women (i.e., IGLC1, TF, TTR, VDBP). Of major interest are the lower levels of nitrosated IGLC1. These immunoglobulin light chains contribute to antigen recognition and accumulate in chronic kidney disorders (CKDs), where their concentration progressively increases according to CKD stage, leading to tubule-interstitial lesions. IGLC1, by interfering with essential functions and apoptotic cell death of neutrophils, may contribute to infectious and inflammatory complications, which are common in CKD patients [52]. It could be speculated that constitutive lower nitration levels could be protective for renal function, which represent a hallmark of the aging functional decline. Concerning VDBP, the free hormone hypothesis postulates that only hormones bound to a binding protein are released and enter in cells, suggesting that the lower nitration state of VDBP in centenarians may promote the binding of vitamin D as a protective factor for bone formation and neuromuscular function [53,54]. Regarding TTR, it is a precursor protein of senile systemic amyloidosis and S-nitrosated transthyretins, exhibiting higher amyloidogenicity than unmodified transthyretins. Nitrosated TTR was less abundant in C, suggesting a marked ability of C to retain the healthy status [55,56]. Further specific investigations should be addressed to precisely contextualize these proteins in aging, and therein the scope of this pilot study was to provide hints for studies on this direction.

In summary, we can conclude that the proteins indicated above can be putative markers of longevity, and, in particular, they are involved in the protection of cardiovascular and renal function decline and, indirectly, of muscle mass and function decline. Furthermore, the maintenance of high-level ADH5/GSNOR can sustain denitrosylase activity, promoting longevity. A number of limitations of this study exist, since we analyzed a restricted number of samples, even though the adoption of sub-pooling allowed us to reduce the variance among biological groups, increasing the power of the study to detect changes despite the restricted number of samples [57,58]. In comparison with recent larger studies based on differential protein abundance, this study utilized complex and labor-intensive methodologies not currently available in all labs. However, this is the first study suggesting nitrosation of circulating proteins as a possible marker to predict and monitor longevity, indicating specific molecules to be targeted to prevent frailty in aging. Therefore, it has to be considered as a starting point for more focused studies on larger cohorts.

4. Materials and Methods

4.1. Participants and Ethical Statement

Serum samples were collected from female subjects grouped according to the "age at time of sampling" into young (Y; n = 15, age range: 32–44 years old), aged (A; n = 15, 76–83 years old), and centenarian (C; n = 15, 105–114 years old). The general characteristics of enrolled subjects are summarized in Table S1. In particular, the functionality of the lower limbs was assessed in aged subjects using the SPPB, a group of measures that combines the results of the gait speed, chair stand, and balance tests [59]. For aged subjects and centenarians, the hand grip test by manual dynamometry was utilized to easily determine musculoskeletal function, weakness, and disability [60]. The functional status was further assessed as a measurement of the subject's ability to perform activities of daily living (i.e., bathing, dressing, toileting, transferring, continence, and feeding). The Katz index scores subjects for independence in each of these 6 functions. A score of 6 indicates full function, 4 indicates moderate impairment, and 2 or less indicates severe functional impairment.

All subjects gave their informed consent for inclusion before they participated in the study. The study was conducted in accordance with the Declaration of Helsinki, and the protocol was approved by the Ethics Committee of the Fondazione istituto di ricovero e cura a carattere scientifico (IRCCS) Ca' Granda Ospedale Maggiore Policlinico, Milan (Protocol identification code No. 2035, amendment 30/11/2011) and of the IRCCS Fondazione Don Carlo Gnocchi, Milan (Project identification code No. 2017-0622, amendement 04/2018).

4.2. Serum Albumin Depletion

Before serum albumin depletion, protein concentration was carefully assessed in all samples by a Pierce bicinchoninic acid assay (BCA) Protein Assay Kit (ThermoFisher Scientific, Rodano, Italy). Albumin depletion was performed according to the manufacturer's instructions using the Pierce Albumin Depletion Kit (ThermoFisher Scientific). Briefly, 400 μL of Pierce depletion resin (corresponding to 200 μL of settled resin) was transferred to a spin column, and the column was centrifuged and washed with 200 μL of binding/wash solution. After centrifuging at 12,000× g for 1 min, 50 μL of serum sample was added to the tube. The tube was incubated for 2 min at room temperature and centrifuged at 12,000× g for 1 min. To release unbound proteins, we washed spin tubes 3 times with 50 μL of binding/wash solution and then centrifuged them at 12,000× g for 1 min.

4.3. Sample Preparation

Serum protein concentration was quantified by BCA protein assay (ThermoFisher Scientific), and, for each group, samples were randomly selected and pooled into 3 sub-pools (5 subjects each). Proteins were selectively precipitated by acetone and resuspended in 7 M urea, 2 M thiourea, 4% 3-[(3-cholamidopropyl)dimethyl-ammonio]-1-propane sulfonate (CHAPS), 30 mM Tris, 1 mM phenylmethanesulfonyl fluoride (PMSF) (pH 8.5) (lysis buffer) for 2-D DIGE or in 300 mM sucrose, 250 mM 2-[4-(2-hydroxyethyl)-1-piperazinyl]-ethanesulfonic acid (HEPES) (pH 7.7), 1 mM Ethylenediaminetetraacetic acid (EDTA), and 0.1 mM neocuproine (HEN buffer) for Nitro-DIGE. After resuspension, protein concentration of sub-pools was determined by 2-D quant Kit (GE Healthcare) before 2-D DIGE. For Nitro-DIGE, sample preparation was carried out in the dark to prevent SNO decomposition. Protein concentration was determined using the BCA Protein Assay Kit and adjusted to 1 mg/mL with HEN buffer.

4.4. Two-Dimensional Difference in Gel Electrophoresis (2-D DIGE)

2-D DIGE was conducted on albumin-depleted serum extracts from 9 different sub-pools, each one constituted of 5 different age- and sex-matched subjects. All experimental procedures such as protein labeling, 2-D separation parameters, and protein analyses were performed as previously described [61]. For protein minimal labeling, cyanine dyes (Cy3 and Cy5) were adopted according to the manufacturer's instructions (GE healthcare, Little Chalfont, Buckinghamshire, UK). Briefly, 50 µg of each sub-pool extract was mixed with 400 pmol of CyDye (Cy5); the reaction was performed on ice in the dark for 30 min. The reaction was blocked by adding 1 mL L-lysine 10 mM for 10 min on ice. Internal standard (Cy3-labelled) was prepared by mixing 100 µg of each sub-pool (Y, A, and C). After protein labeling, 40 µg of each sample and 40 µg of internal standard were mixed. Samples were run in duplicate using 24 cm, 3–10 non-linear pH-gradient immobilized pH gradient (IPG) strips. Isoelectric focusing was performed on an IPGphor electrophoresis unit (GE Healthcare) using a gradient ranging from 200 to 8000 V, reaching a total of 75,000 Vh. Focused proteins on IPG strips were prepared for second dimension by reduction and alkylation. Second dimension was carried out using the Ettan Dalt II system (GE Healthcare) on 20×25 cm^2, 12% T, 2.5% C constant concentration polyacrylamide gels at 20 °C and 15 mA.

4.5. Identification of S-Nitrosated Proteins by 2-D CyDye-Maleimide DIGE (Nitro-DIGE)

A modified biotin switch method [62] with CyDye maleimide monoreactive sulfhydryl-reactive fluorescent dyes (GE Healthcare) to identify SNO proteins was used. Free thiols were blocked with 4X volume of 50 mM iodoacetamide (IAA) in HEN buffer containing 2.5% SDS for 30 min. Excess IAA was removed by cold acetone precipitation. Protein pellets were washed; dissolved in HEN buffer containing 1% SDS, 5 mM sodium ascorbate, and 1 µM copper sulfate [63]; and incubated at room temperature for 1 h in order to reduce S-nitrosothiols. After acetone precipitation, proteins were dissolved in labeling buffer (30 mM Tris-HCl (pH 8), 8 M urea, 4% CHAPS) at 2.5 mg/mL. CyDye DIGE Fluor reagent (10 µM Cy3 or Cy5) was added to each sample and incubated at room temperature for 1 h to label NO-released thiols. Each group consisted of at least three biological replicates; each replicate was labeled with Cy5, and a mixture containing an equal amount of all samples was labeled with Cy3 as the internal standard. After quenching with 50 mM dithiothreitol (DTT), labeled samples (internal standard versus each replicate) were mixed and separated by two-dimensional electrophoresis, as previously described.

4.6. Image Acquisition and Statistical Analysis

Images from CyDye-labeled gels were acquired by Typhoon 9200 Imager (GE Healthcare), and image analysis was performed by DeCyder software (version 6.5, GE Healthcare). Three groups of subjects were analyzed, and for each group (Y, A, and C), we only considered spots present in at least 90% of the samples. 2-D DIGE and Nitro-DIGE statistically significant differences were computed by analysis of variance (ANOVA) and corrected for Tukey's test (p-value < 0.01); when the use of ANOVA was not possible, the non-parametric Kruskal–Wallis (p-value < 0.01) test was adopted. False discovery rate was applied to correct for multiple tests to reduce the overall error. Statistically changed proteins underwent the power analysis, and only spots reaching a sensitivity cut-off > 0.8 were considered as differentially expressed.

4.7. Protein Identification

Protein identification was carried out by comparison of gel images to a human plasma reference 2-D map publicly available (https://world-2dpage.expasy.org/swiss-2dpage/viewer). Proteins not annotated in this map were identified by matrix-assisted laser desorption/ionization time-of-flight (MALDI-TOF) mass spectrometry (MS). For protein identification, semi-preparative gels were loaded with unlabelled sample (400 µg per strip); electrophoretic conditions were the same as 2-D DIGE, and gels were stained with a total-protein fluorescent stain (Krypton, ThermoFisher Scientific). Image acquisition was performed using a Typhoon 9200 laser scanner. Spots of interest were excised from gel using the Ettan spot picker robotic system (GE Healthcare), destained in 50% methanol/50 mM ammonium bicarbonate, and incubated with 30 µL of 6 ng/mL trypsin (Promega, Madison, Wisconsin, USA) dissolved in 10 mM ammonium bicarbonate for 16 h at 37 °C. Released peptides were subjected to reverse-phase chromatography (Zip-Tip C18 micro, Merck Millipore, Milano, Italy), eluted with 50% acetonitrile (ACN)/0.1% trifluoroacetic acid. Peptide mixture (1 µL) was diluted in an equal volume of 10 mg/mL alpha-cyano-4- hydroxycinnamic acid matrix dissolved in 70% ACN/30% citric acid and processed on an Ultraflex III MALDI-TOF/TOF (Bruker Daltonics, Bremen, Germany) mass spectrometer. MS was performed at an accelerating voltage of 20 kV, and spectra were externally calibrated using Peptide Mix calibration mixture (Bruker Daltonics); 1000 laser shots were taken per spectrum. Spectra were processed by FlexAnalysis software v. 3.0 (Bruker Daltonics), setting the signal to noise threshold value to 6, and search was carried out by correlation of uninterpreted spectra to *Homo sapiens* entries in Uniprot Proteomes UP5640 20200812 (97,065 sequences; 38,762,114 residues) using BioTools v. 3.2 (Bruker Daltonics) interfaced to the on-line MASCOT software (Matrix Science, London, UK) which utilizes a robust probabilistic scoring algorithm. The significance threshold was set at a p-value < 0.05. No mass and pI constraints were applied, and trypsin was set as enzyme. One missed cleavage per peptide was allowed, and carbamidomethylation was set as fixed modification while methionine oxidation was set as variable modification. Mass tolerance was set at 30 ppm for MS spectra.

In cases where this approach was unsuccessful, we performed additional searches using electrospray ionization–MS/MS, as previously described [64].

4.8. *Immunoblotting*

Protein extracts (50 µg) from pooled Y, A, and C serum samples were loaded in triplicate and resolved on 12–18% gradient polyacrylamide gels. Blots were incubated with rabbit or mouse primary antibodies as follows: anti-nitrotyrosine (Cayman Chemicals, Ann Arbor, Michigan, USA, no. 10189540, 1:200), anti-ADH5 (Santa Cruz Biotechnology, Dallas, Texas, USA, sc-293460, 1:500), anti-TRXR1 (Novus Biologicals, Bio-Techne, Milano, Italy, NBP1-81791, 1:500). After washing, membranes were incubated with anti-rabbit (GE Healthcare, 1:10,000) or anti-mouse (KPL, Seracare, Milford, Massachusetts, USA, 1:5000) secondary antibodies conjugated with horseradish peroxidase. Signals were visualized by chemiluminescence using the ECL Prime Detection Kit and the Image Quant LAS 4000 (GE Healthcare) analysis system. Band quantification was performed using the Image Quant TL v. 8.1(GE Healthcare) software followed by statistical analysis (ANOVA + Tukey, $n = 3$, p-value < 0.05). Band intensities were normalized against the total amount of proteins stained by Sypro ruby total-protein stain.

Author Contributions: Conceptualization, C.G.; methodology, D.C.; formal analysis, D.C. and P.B.; investigation, D.C., P.B. and E.T.; resources, D.M., F.R.G., B.A. and F.T.; project administration, C.G. and D.C.; writing—original draft, C.G. and D.C.; writing—review and editing, C.G., M.C. (Mario Clerici), D.C. and M.C. (Matteo Cesari); funding acquisition, M.C. (Mario Clerici), F.R.G. and C.G. All authors have read and agreed to the published version of the manuscript.

Acknowledgments: The authors acknowledge Manuela Garbellini and Rossella Miglioli for technical support.

Abbreviations

2-D DIGE	Two-dimensional difference in gel electrophoresis
ADH5/GSNOR	Alcohol dehydrogenase 5/S-nitrosoglutathione reductase
ANOVA	Analysis of variance
APOA1	Apolipoprotein A1
BCA	Bicinchoninic acid assay
BMI	Body mass index
C4A	Complement C4-A
CERCAM	Inactive glycosyltransferase 25 family member 3
CHAPS	3-[(3-cholamidopropyl)dimethyl-ammonio]-1-propane sulfonate
CKD	Chronic kidney disorder
CLU	Clusterin
CP	Ceruloplasmin
DTT	Dithiothreitol
EDTA	Ethylenediaminetetraacetic acid
GSNO	S-nitrosoglutathione
HEPES	2-[4-(2-hydroxyethyl)-1-piperazinyl]-ethanesulfonic acid
HMW	High molecular weight
HP	Haptoglobin
IAA	Iodoacetamide
IGHA1	Immunoglobulin heavy chain alpha
IGHG3	Immunoglobulin heavy chain gamma
IGLC1	Immunoglobulin light chain 1
IPG	Immobilized pH gradient
IRCCS	Istituto di ricovero e cura a carattere scientifico (scientific institute for research, hospitalization and healthcare)
LC–MS/MS	Liquid chromatography tandem mass spectrometry
LMW	Low molecular weight
MALDI-TOF	Matrix-assisted laser desorption/ionization time-of-flight
MRM	Multiple reaction monitoring
MS	Mass spectrometry
NADPH	Nicotinamide adenine dinucleotide phosphate, reduced form
Nitro-DIGE	2-D CyDye-maleimide difference in gel electrophoresis
NO	Nitric oxide
NOS	NO synthase
PBMC	Peripheral blood mononuclear cells
pI	Isoelectric point
PMSF	Phenylmethanesulfonyl fluoride
RNS	Reactive nitrogen species
ROS	Reactive oxygen species
RYR1	Ryanodine receptor 1 (skeletal muscle type)
RYR2	Ryanodine receptor 2 (cardiac muscle type)
SERPINA1	Alpha-1-antitripsin
SERPINA3	Alpha-1-antichimotripsin
SNOs	Nitroso thiols
SOMAmers	Slow off-rate modified aptamers
SPPB	Short physical performance battery
SRM	Selected reaction monitoring
TF	Serotransferrin
Trx/TrxR	Thioredoxin/thoredoxin reductase
TRXR1	Thioredoxin reductase 1
TTR	Transthyretin
VDBP	Vitamin D-binding protein

References

1. Cesari, M.; Calvani, R.; Marzetti, E. Frailty in older persons. *Clin. Geriatr. Med.* **2017**, *33*, 293–303. [CrossRef] [PubMed]
2. Harman, D. Aging: A theory based on free radical and radiation chemistry. *J. Gerontol.* **1956**, *11*, 298–300. [CrossRef] [PubMed]

3. Vasilaki, A.; Richardson, A.; Van Remmen, H.; Brooks, S.V.; Larkin, L.; McArdle, A.; Jackson, M.J. Role of nerve-muscle interactions and reactive oxygen species in regulation of muscle proteostasis with ageing. *J. Physiol.* **2017**, *595*, 6409–6415. [CrossRef] [PubMed]
4. Schulman, I.H.; Hare, J.M. Regulation of cardiovascular cellular processes by S-nitrosylation. *Biochim. Biophys. Acta (BBA)-Gen. Subj.* **2012**, *1820*, 752–762. [CrossRef] [PubMed]
5. Seth, D.; Hess, D.T.; Hausladen, A.; Wang, L.; Wang, Y.J.; Stamler, J.S. A multiplex enzymatic machinery for cellular protein s-nitrosylation. *Mol. Cell* **2018**, *69*, 451–464.e6. [CrossRef] [PubMed]
6. Moriggi, M.; Vasso, M.; Fania, C.; Capitanio, D.; Bonifacio, G.; Salanova, M.; Blottner, D.; Rittweger, J.; Felsenberg, D.; Cerretelli, P.; et al. Long term bed rest with and without vibration exercise countermeasures: Effects on human muscle protein dysregulation. *Proteomics* **2010**, *10*, 3756–3774. [CrossRef]
7. Salanova, M.; Gambara, G.; Moriggi, M.; Vasso, M.; Ungethuem, U.; Belavy, D.L.; Felsenberg, D.; Cerretelli, P.; Gelfi, C.; Blottner, D. Vibration mechanosignals superimposed to resistive exercise result in baseline skeletal muscle transcriptome profiles following chronic disuse in bed rest. *Sci. Rep.* **2015**, *5*, 17027. [CrossRef]
8. Salanova, M.; Gelfi, C.; Moriggi, M.; Vasso, M.; Vigano, A.; Minafra, L.; Bonifacio, G.; Schiffl, G.; Gutsmann, M.; Felsenberg, D.; et al. Disuse deterioration of human skeletal muscle challenged by resistive exercise superimposed with vibration: Evidence from structural and proteomic analysis. *FASEB J.* **2014**, *28*, 4748–4763. [CrossRef]
9. Bellinger, A.M.; Reiken, S.; Carlson, C.; Mongillo, M.; Liu, X.; Rothman, L.; Matecki, S.; Lacampagne, A.; Marks, A.R. Hypernitrosylated ryanodine receptor calcium release channels are leaky in dystrophic muscle. *Nat. Med.* **2009**, *15*, 325–330. [CrossRef]
10. Bellinger, A.M.; Reiken, S.; Dura, M.; Murphy, P.W.; Deng, S.X.; Landry, D.W.; Nieman, D.; Lehnart, S.E.; Samaru, M.; LaCampagne, A.; et al. Remodeling of ryanodine receptor complex causes "leaky" channels: A molecular mechanism for decreased exercise capacity. *Proc. Natl. Acad. Sci. USA* **2008**, *105*, 2198–2202. [CrossRef] [PubMed]
11. Beigi, F.; Gonzalez, D.R.; Minhas, K.M.; Sun, Q.A.; Foster, M.W.; Khan, S.A.; Treuer, A.V.; Dulce, R.A.; Harrison, R.W.; Saraiva, R.M.; et al. Dynamic denitrosylation via S-nitrosoglutathione reductase regulates cardiovascular function. *Proc. Natl. Acad. Sci. USA* **2012**, *109*, 4314–4319. [CrossRef] [PubMed]
12. Rizza, S.; Filomeni, G. Denitrosylate and live longer: How ADH5/GSNOR links mitophagy to aging. *Autophagy* **2018**, *14*, 1285–1287. [CrossRef] [PubMed]
13. Montagna, C.; Rizza, S.; Cirotti, C.; Maiani, E.; Muscaritoli, M.; Musaro, A.; Carri, M.T.; Ferraro, E.; Cecconi, F.; Filomeni, G. nNOS/GSNOR interaction contributes to skeletal muscle differentiation and homeostasis. *Cell Death Dis.* **2019**, *10*, 354. [CrossRef]
14. Kane, A.E.; Sinclair, D.A. Frailty biomarkers in humans and rodents: Current approaches and future advances. *Mech. Ageing Dev.* **2019**, *180*, 117–128. [CrossRef] [PubMed]
15. Waters, D.L.; Vlietstra, L.; Qualls, C.; Morley, J.E.; Vellas, B. Sex-specific muscle and metabolic biomarkers associated with gait speed and cognitive transitions in older adults: A 9-year follow-up. *GeroScience* **2020**, *42*, 585–593. [CrossRef]
16. Gelfi, C.; Vigano, A.; Ripamonti, M.; Pontoglio, A.; Begum, S.; Pellegrino, M.A.; Grassi, B.; Bottinelli, R.; Wait, R.; Cerretelli, P. The human muscle proteome in aging. *J. Proteome Res.* **2006**, *5*, 1344–1353. [CrossRef]
17. Theron, L.; Gueugneau, M.; Coudy, C.; Viala, D.; Bijlsma, A.; Butler-Browne, G.; Maier, A.; Bechet, D.; Chambon, C. Label-free quantitative protein profiling of vastus lateralis muscle during human aging. *Mol. Cell. Proteom.* **2014**, *13*, 283–294. [CrossRef]
18. Capri, M.; Morsiani, C.; Santoro, A.; Moriggi, M.; Conte, M.; Martucci, M.; Bellavista, E.; Fabbri, C.; Giampieri, E.; Albracht, K.; et al. Recovery from 6-month spaceflight at the International Space Station: Muscle-related stress into a proinflammatory setting. *FASEB J.* **2019**, *33*, 5168–5180. [CrossRef]
19. Menni, C.; Kiddle, S.J.; Mangino, M.; Vinuela, A.; Psatha, M.; Steves, C.; Sattlecker, M.; Buil, A.; Newhouse, S.; Nelson, S.; et al. Circulating proteomic signatures of chronological age. *J. Gerontol. Ser. A Biol. Sci. Med. Sci.* **2015**, *70*, 809–816. [CrossRef]
20. Tanaka, T.; Biancotto, A.; Moaddel, R.; Moore, A.Z.; Gonzalez-Freire, M.; Aon, M.A.; Candia, J.; Zhang, P.; Cheung, F.; Fantoni, G.; et al. Plasma proteomic signature of age in healthy humans. *Aging Cell* **2018**, *17*, e12799. [CrossRef]

21. Lind, L.; Sundstrom, J.; Larsson, A.; Lampa, E.; Arnlov, J.; Ingelsson, E. Longitudinal effects of aging on plasma proteins levels in older adults-associations with kidney function and hemoglobin levels. *PLoS ONE* **2019**, *14*, e0212060. [CrossRef] [PubMed]
22. Lehallier, B.; Gate, D.; Schaum, N.; Nanasi, T.; Lee, S.E.; Yousef, H.; Moran Losada, P.; Berdnik, D.; Keller, A.; Verghese, J.; et al. Undulating changes in human plasma proteome profiles across the lifespan. *Nat. Med.* **2019**, *25*, 1843–1850. [CrossRef] [PubMed]
23. Rizza, S.; Cardaci, S.; Montagna, C.; Di Giacomo, G.; De Zio, D.; Bordi, M.; Maiani, E.; Campello, S.; Borreca, A.; Puca, A.A.; et al. S-nitrosylation drives cell senescence and aging in mammals by controlling mitochondrial dynamics and mitophagy. *Proc. Natl. Acad. Sci. USA* **2018**, *115*, E3388–E3397. [CrossRef] [PubMed]
24. Liu, L.; Yan, Y.; Zeng, M.; Zhang, J.; Hanes, M.A.; Ahearn, G.; McMahon, T.J.; Dickfeld, T.; Marshall, H.E.; Que, L.G.; et al. Essential roles of S-nitrosothiols in vascular homeostasis and endotoxic shock. *Cell* **2004**, *116*, 617–628. [CrossRef]
25. Wei, W.; Li, B.; Hanes, M.A.; Kakar, S.; Chen, X.; Liu, L. S-nitrosylation from GSNOR deficiency impairs DNA repair and promotes hepatocarcinogenesis. *Sci. Transl. Med.* **2010**, *2*, 19ra13. [CrossRef]
26. Montagna, C.; Cirotti, C.; Rizza, S.; Filomeni, G. When S-nitrosylation gets to mitochondria: From signaling to age-related diseases. *Antioxid. Redox Signal.* **2020**, *32*, 884–905. [CrossRef]
27. Rizza, S.; Montagna, C.; Di Giacomo, G.; Cirotti, C.; Filomeni, G. S-nitrosation and ubiquitin-proteasome system interplay in neuromuscular disorders. *Int. J. Cell Biol.* **2014**, *2014*, 428764. [CrossRef]
28. Sengupta, R.; Holmgren, A. Thioredoxin and thioredoxin reductase in relation to reversible S-nitrosylation. *Antioxid. Redox Signal.* **2013**, *18*, 259–269. [CrossRef]
29. Haun, F.; Nakamura, T.; Shiu, A.D.; Cho, D.H.; Tsunemi, T.; Holland, E.A.; La Spada, A.R.; Lipton, S.A. S-nitrosylation of dynamin-related protein 1 mediates mutant huntingtin-induced mitochondrial fragmentation and neuronal injury in Huntington's disease. *Antioxid. Redox Signal.* **2013**, *19*, 1173–1184. [CrossRef]
30. Benhar, M.; Forrester, M.T.; Hess, D.T.; Stamler, J.S. Regulated protein denitrosylation by cytosolic and mitochondrial thioredoxins. *Science* **2008**, *320*, 1050–1054. [CrossRef]
31. Franceschi, C.; Passarino, G.; Mari, D.; Monti, D. Centenarians as a 21st century healthy aging model: A legacy of humanity and the need for a world-wide consortium (WWC100+). *Mech. Ageing Dev.* **2017**, *165*, 55–58. [CrossRef] [PubMed]
32. Walston, J.D. Sarcopenia in older adults. *Curr. Opin. Rheumatol.* **2012**, *24*, 623–627. [CrossRef] [PubMed]
33. Narici, M.V.; Maffulli, N. Sarcopenia: Characteristics, mechanisms and functional significance. *Br. Med. Bull.* **2010**, *95*, 139–159. [CrossRef]
34. Guikema, B.; Lu, Q.; Jourd'heuil, D. Chemical considerations and biological selectivity of protein nitrosation: Implications for NO-mediated signal transduction. *Antioxid. Redox Signal.* **2005**, *7*, 593–606. [CrossRef] [PubMed]
35. Santos-Lozano, A.; Valenzuela, P.L.; Llavero, F.; Lista, S.; Carrera-Bastos, P.; Hampel, H.; Pareja-Galeano, H.; Galvez, B.G.; Lopez, J.A.; Vazquez, J.; et al. Successful aging: Insights from proteome analyses of healthy centenarians. *Aging* **2020**, *12*, 3502–3515. [CrossRef]
36. Schafer, M.J.; Miller, J.D.; LeBrasseur, N.K. Cellular senescence: Implications for metabolic disease. *Mol. Cell. Endocrinol.* **2017**, *455*, 93–102. [CrossRef] [PubMed]
37. Misek, D.E.; Kuick, R.; Wang, H.; Galchev, V.; Deng, B.; Zhao, R.; Tra, J.; Pisano, M.R.; Amunugama, R.; Allen, D.; et al. A wide range of protein isoforms in serum and plasma uncovered by a quantitative intact protein analysis system. *Proteomics* **2005**, *5*, 3343–3352. [CrossRef]
38. Byerley, L.O.; Leamy, L.; Tam, S.W.; Chou, C.W.; Ravussin, E. Development of a serum profile for healthy aging. *Age* **2010**, *32*, 497–507. [CrossRef]
39. Waldera-Lupa, D.M.; Kalfalah, F.; Florea, A.M.; Sass, S.; Kruse, F.; Rieder, V.; Tigges, J.; Fritsche, E.; Krutmann, J.; Busch, H.; et al. Proteome-wide analysis reveals an age-associated cellular phenotype of in situ aged human fibroblasts. *Aging* **2014**, *6*, 856–878. [CrossRef]
40. Wang, Z.; Zhang, R.; Liu, F.; Jiang, P.; Xu, J.; Cao, H.; Du, X.; Ma, L.; Lin, F.; Cheng, L.; et al. TMT-Based quantitative proteomic analysis reveals proteomic changes involved in longevity. *Proteom. Clin. Appl.* **2019**, *13*, e1800024. [CrossRef]

41. Thambisetty, M.; An, Y.; Kinsey, A.; Koka, D.; Saleem, M.; Guntert, A.; Kraut, M.; Ferrucci, L.; Davatzikos, C.; Lovestone, S.; et al. Plasma clusterin concentration is associated with longitudinal brain atrophy in mild cognitive impairment. *NeuroImage* **2012**, *59*, 212–217. [CrossRef] [PubMed]
42. Smith, L.M.; Kelleher, N.L. Proteoform: A single term describing protein complexity. *Nat. Methods* **2013**, *10*, 186–187. [CrossRef] [PubMed]
43. Semba, R.D.; Zhang, P.; Zhu, M.; Fabbri, E.; Gonzalez-Freire, M.; Moaddel, R.; Geng-Spyropoulos, M.; Ferrucci, L. A targeted proteomic assay for the measurement of plasma proteoforms related to human aging phenotypes. *Proteomics* **2017**, *17*. [CrossRef] [PubMed]
44. Barnett, S.D.; Buxton, I.L.O. The role of S-nitrosoglutathione reductase (GSNOR) in human disease and therapy. *Crit. Rev. Biochem Mol. Biol.* **2017**, *52*, 340–354. [CrossRef]
45. Choi, M.S. Pathophysiological role of s-nitrosylation and transnitrosylation depending on s-nitrosoglutathione levels regulated by s-nitrosoglutathione reductase. *Biomol. Ther.* **2018**, *26*, 533–538. [CrossRef]
46. Mkaouar, H.; Akermi, N.; Kriaa, A.; Abraham, A.L.; Jablaoui, A.; Soussou, S.; Mokdad-Gargouri, R.; Maguin, E.; Rhimi, M. Serine protease inhibitors and human wellbeing interplay: New insights for old friends. *Peer J.* **2019**, *7*, e7224. [CrossRef]
47. Gueugneau, M.; d'Hose, D.; Barbe, C.; de Barsy, M.; Lause, P.; Maiter, D.; Bindels, L.B.; Delzenne, N.M.; Schaeffer, L.; Gangloff, Y.G.; et al. Increased Serpina3n release into circulation during glucocorticoid-mediated muscle atrophy. *J. Cachex. Sarcopenia Muscle* **2018**, *9*, 929–946. [CrossRef]
48. Naderi, J.; Bernreuther, C.; Grabinski, N.; Putman, C.T.; Henkel, B.; Bell, G.; Glatzel, M.; Sultan, K.R. Plasminogen activator inhibitor type 1 up-regulation is associated with skeletal muscle atrophy and associated fibrosis. *Am. J. Pathol.* **2009**, *175*, 763–771. [CrossRef]
49. Padmanabhan, J.; Levy, M.; Dickson, D.W.; Potter, H. Alpha1-antichymotrypsin, an inflammatory protein overexpressed in Alzheimer's disease brain, induces tau phosphorylation in neurons. *Brain* **2006**, *129*, 3020–3034. [CrossRef]
50. Poljak, A.; Hill, M.; Hall, R.J.; MacLullich, A.M.; Raftery, M.J.; Tai, J.; Yan, S.; Caplan, G.A. Quantitative proteomics of delirium cerebrospinal fluid. *Transl. Psychiatry* **2014**, *4*, e477. [CrossRef]
51. Shiva, S.; Wang, X.; Ringwood, L.A.; Xu, X.; Yuditskaya, S.; Annavajjhala, V.; Miyajima, H.; Hogg, N.; Harris, Z.L.; Gladwin, M.T. Ceruloplasmin is a NO oxidase and nitrite synthase that determines endocrine NO homeostasis. *Nat. Chem. Biol.* **2006**, *2*, 486–493. [CrossRef] [PubMed]
52. Chmielewski, M.; Heimbürger, O.; Stenvinkel, P.; Lindholm, B. Uremic toxicity. In *Nutritional Management of Renal Disease*; Kopple, J.D., Massry, S.G., Kalantar-Zadeh, K., Eds.; Academic Press: Cambridge, MA, USA, 2013; pp. 49–77. [CrossRef]
53. Ensrud, K.E.; Ewing, S.K.; Fredman, L.; Hochberg, M.C.; Cauley, J.A.; Hillier, T.A.; Cummings, S.R.; Yaffe, K.; Cawthon, P.M. Circulating 25-hydroxyvitamin D levels and frailty status in older women. *J. Clin. Endocrinol. Metab.* **2010**, *95*, 5266–5273. [CrossRef] [PubMed]
54. Wang, Y.; Wang, Y.J.; Zhan, J.K.; Tang, Z.Y.; Huang, W.; Tan, P.; Gao, S.; Ma, C.L.; Jian, Z.J.; Liu, Y.S. Vitamin d binding protein affects the correlation of 25(oh)d and frailty in the older men. *Int. J. Endocrinol.* **2014**, *2014*, 543783. [CrossRef] [PubMed]
55. Ruberg, F.L.; Berk, J.L. Transthyretin (TTR) cardiac amyloidosis. *Circulation* **2012**, *126*, 1286–1300. [CrossRef] [PubMed]
56. Saito, S.; Ando, Y.; Nakamura, M.; Ueda, M.; Kim, J.; Ishima, Y.; Akaike, T.; Otagiri, M. Effect of nitric oxide in amyloid fibril formation on transthyretin-related amyloidosis. *Biochemistry* **2005**, *44*, 11122–11129. [CrossRef] [PubMed]
57. Karp, N.A.; Lilley, K.S. Design and analysis issues in quantitative proteomics studies. *Proteomics* **2007**, *7*, 42–50. [CrossRef]
58. Karp, N.A.; Spencer, M.; Lindsay, H.; O'Dell, K.; Lilley, K.S. Impact of replicate types on proteomic expression analysis. *J. Proteome Res.* **2005**, *4*, 1867–1871. [CrossRef] [PubMed]
59. Guralnik, J.M.; Ferrucci, L.; Pieper, C.F.; Leveille, S.G.; Markides, K.S.; Ostir, G.V.; Studenski, S.; Berkman, L.F.; Wallace, R.B. Lower extremity function and subsequent disability: Consistency across studies, predictive models, and value of gait speed alone compared with the short physical performance battery. *J. Gerontol. Ser. A Biol. Sci. Med. Sci.* **2000**, *55*, M221–M231. [CrossRef]

60. Amaral, C.A.; Amaral, T.L.M.; Monteiro, G.T.R.; Vasconcellos, M.T.L.; Portela, M.C. Hand grip strength: Reference values for adults and elderly people of Rio Branco, Acre, Brazil. *PLoS ONE* **2019**, *14*, e0211452. [CrossRef]
61. Capitanio, D.; Moriggi, M.; Torretta, E.; Barbacini, P.; De Palma, S.; Vigano, A.; Lochmuller, H.; Muntoni, F.; Ferlini, A.; Mora, M.; et al. Comparative proteomic analyses of Duchenne muscular dystrophy and Becker muscular dystrophy muscles: Changes contributing to preserve muscle function in Becker muscular dystrophy patients. *J. Cachex. Sarcopenia Muscle* **2020**, *11*, 547–563. [CrossRef]
62. Jaffrey, S.R.; Snyder, S.H. The biotin switch method for the detection of S-nitrosylated proteins. *Sci. STKE* **2001**, *2001*, pl1. [CrossRef] [PubMed]
63. Kettenhofen, N.J.; Wang, X.; Gladwin, M.T.; Hogg, N. In-gel detection of S-nitrosated proteins using fluorescence methods. *Methods Enzymol.* **2008**, *441*, 53–71. [CrossRef] [PubMed]
64. Capitanio, D.; Vasso, M.; Fania, C.; Moriggi, M.; Vigano, A.; Procacci, P.; Magnaghi, V.; Gelfi, C. Comparative proteomic profile of rat sciatic nerve and gastrocnemius muscle tissues in ageing by 2-D DIGE. *Proteomics* **2009**, *9*, 2004–2020. [CrossRef] [PubMed]

2

Free Radicals as a Double-Edged Sword: The Cancer Preventive and Therapeutic Roles of Curcumin

Nehal Gupta [1], Kshitij Verma [2], Sarath Nalla [3], Alok Kulshreshtha [3], Rajiv Lall [3] and Sahdeo Prasad [3,*]

1 Department of Medicine, University of California San Francisco, 1600 Divisadero Street, San Francisco, CA 94115, USA; nehal.gupta@ucsf.edu
2 Discovery Chemistry, Genentech, Inc., 1 DNA Way, South San Francisco, CA 94080, USA; verma.kshitij@gene.com
3 Noble Pharma, LLC, 4602 Domain Drive, Menomonie, WI 54751, USA; sarathn@noblepharmallc.com (S.N.); Alokk@noblepharmallc.com (A.K.); lallr123@yahoo.com (R.L.)
* Correspondence: spbiotech@gmail.com or sahdeop@noblepharmallc.com

Academic Editor: Alejandro Samhan-Arias

Abstract: Free radicals, generally composed of reactive oxygen species (ROS) and reactive nitrogen species (RNS), are generated in the body by various endogenous and exogenous systems. The overproduction of free radicals is known to cause several chronic diseases including cancer. However, increased production of free radicals by chemotherapeutic drugs is also associated with apoptosis in cancer cells, indicating the dual nature of free radicals. Among various natural compounds, curcumin manifests as an antioxidant in normal cells that helps in the prevention of carcinogenesis. It also acts as a prooxidant in cancer cells and is associated with inducing apoptosis. Curcumin quenches free radicals, induces antioxidant enzymes (catalase, superoxide dismutase, glutathione peroxidase), and upregulates antioxidative protein markers–Nrf2 and HO-1 that lead to the suppression of cellular oxidative stress. In cancer cells, curcumin aggressively increases ROS that results in DNA damage and subsequently cancer cell death. It also sensitizes drug-resistant cancer cells and increases the anticancer effects of chemotherapeutic drugs. Thus, curcumin shows beneficial effects in prevention, treatment and chemosensitization of cancer cells. In this review, we will discuss the dual role of free radicals as well as the chemopreventive and chemotherapeutic effects of curcumin and its analogues against cancer.

Keywords: free radicals; ROS; cancer; curcumin; chemoprevention; therapy

---

1. Introduction

Free radicals are generally reactive oxygen species (ROS) and reactive nitrogen species (RNS), which are capable of independent existence. ROS and RNS can broadly be categorized into two groups—radicals and non-radicals. Superoxide ($O_2^{-\bullet}$), oxygen radical ($O_2^{\bullet\bullet}$), hydroxyl ($OH^\bullet$), alkoxyradical ($RO^\bullet$), peroxyl radical ($ROO^{\bullet\bullet}$), nitric oxide ($NO^\bullet$) and nitrogen dioxide ($NO_2^\bullet$) are various examples of radical species [1]. The non-radical species are comprised of hydrogen peroxide (H_2O_2), hypochlorous acid (HOCl), hypobromous acid (HOBr), ozone (O_3), singlet oxygen ($^1O_2$), nitrous acid (HNO_2), nitrosyl cation (NO^+), nitroxyl anion (NO^-), dinitrogen trioxide (N_2O_3), dinitrogen tetraoxide (N_2O_4), nitronium cation (NO_2^+), organic peroxides (ROOH), aldehydes (HCOR) and peroxynitrite (ONOOH). Radical species are comparatively highly reactive, due to the presence of an unpaired electron that imparts a high degree of electrophilicity. These free radicals act as oxidants or reductants by either donating an electron to or accepting an electron from other reactive molecules [2].

Free radicals are generated in the body by various endogenous and exogenous systems including pathological states and exposure to different physiochemical conditions. The formation of free radicals in the body is a continuous process through the enzymatic and non-enzymatic reactions. Mitochondria, peroxisomes, and phagocytic organelles participate in the enzymatic production of free radicals while non-enzymatic free radicals are produced by ionizing radiation and non-enzymatic reactions of oxygen with organic compounds [3]. Accumulated evidences suggest that free radicals cause progressive adverse effects in the body by increasing oxidative stress. Although the body has an antioxidant defense mechanism to balance the redox system, excessive production of ROS and RNS results in oxidative and nitrosative stress respectively. This chronic oxidative or nitrosative stress manifests in the form of a number of diseases including cancer. This is evident by a notion that the overproduction of free radicals has a close relation with increased incidences of cancer. It has also been found that a higher consumption of fats and oils leads to the increased production of free radicals correlated with the increased death rate from different types of cancer [4].

To combat the detrimental effects of free radical species ROS and RNS, a system of antioxidant defense is expressed in all living cells. The antioxidant system can be classified in multiple ways. It may be based on the source as exogenous (derived from dietary food intake) or endogenous (produced in the body), solubility in water (e.g., vitamin C) or lipids (e.g., vitamin E), or the size and nature of the antioxidant molecule, i.e., enzymatic (e.g., catalase) or non-enzymatic (e.g., flavonoids). Studies have shown that antioxidants reduce the occurrence of carcinogenesis. It has been observed that the consumption of fruits and vegetables, rich in antioxidants, in daily diet causes a reduction in the risk for cancer incidence [5]. In contrast, several studies have also shown that antioxidant from dietary supplements such as β-carotene and retinol promote tumor growth and metastasis in cancer patients [6–8]. However, various anticancer drugs kill cancer cells by producing free radicals [9]. Thus, these studies indicate that free radicals serve a dual purpose—inducing carcinogenesis and imparting in cancer cell death.

2. Free Radicals as a Double Edge Sword

Reactive oxygen species (ROS), produced either exogenously (e.g., radiation, chemicals, hyperoxia) or endogenously (normal cellular metabolism), are related to a wide variety of human disorders. Excessive ROS causes damage to proteins, DNA, and RNA leading to genetic alterations in cells. On the other hand, low levels of ROS are essential for various biological functions, including cell survival, cell growth, proliferation and differentiation, and immune response [10]. Lethal production of ROS by certain agents also causes death of cancer cells. These observations indicate the dual role of ROS in terms of cancer promoting and cancer suppressing activity. It is evident that ROS participates simultaneously in both the Ras-Raf-MEK1/2-ERK1/2 oncogenic signaling and the p38 mitogen-activated protein kinases (p38MAPK) tumor suppressing pathways [11]. Thus, depending on the cell type, ROS may function as cytoprotective or oncogenic.

2.1. Carcinogenesis

The high accumulation of ROS and/or low level of antioxidants in cells causes an imbalance in redox status, which is known as oxidative stress. Oxidative stress is implicated in a variety of pathologies including cancer. ROS are short-lived and highly electrophilic molecules generated by the partial reduction of oxygen to form superoxide ($O_2^{-\bullet}$), hydrogen peroxide (H_2O_2), and hydroxyl radical ($HO^{\bullet}$) as well as secondary metabolites including lipid peroxides, peroxynitrite ($ONOO^-$), and hypochlorous acid (HOCl). Because of their high reactivity, they cause damage to the macromolecules and disrupt biochemical function [12].

It has been known that ROS causes DNA strand breaks and oxidative damage to the nucleotides, subsequently resulting in mutagenesis and eventually cancer. The susceptible target of ROS in DNA is guanine that causes G→T transversions [13]. Additionally, ROS can also cause mutations by oxidative damage to a range of target sites in genetic materials including purines and pyrimidines, alkali labile sites, single strand breaks and disruption of DNA repair processes, leading to genetic instability [14,15]. ROS-induced carcinogenesis is supported by a study describing that elevated levels of ROS leads to modification of nucleobases in cancerous and precancerous tissues [16]. The initiation of cancer in humans caused by ROS is further supported by the presence of oxidative DNA modifications in cancer tissues [17]. Oxidative DNA damage leading to the development of breast cancer has also been reported. For instance, in inflammatory breast cancer, an increase in DNA base damage and 8-oxo-dG adducts leads to malignant cancer progression [18]. The role of oxidative stress in the development of hepatocellular carcinoma has also been reported, since ROS caused accumulation of 8-OHdG by oxidative DNA damage in the cells and further development of hepatocellular carcinoma [19]. The association of oxidative DNA damage and carcinogenesis have been found in a variety of other cancers. However, a comparative measurement of distinctively modified DNA bases in tumor tissue and their respective normal tissues is required to provide further insights into the involvement of ROS in carcinogenesis.

ROS also induces modifications of redox-sensitive amino acid residues in regulatory proteins including cysteine oxidation. Such modifications can modulate the regulatory effects of proteins and enzymes. The regulatory proteins such as kinases (MAPK and PI3K/Akt), transcription factors (Nrf2, AP-1, NF-κB, STAT3, and p53), components of ubiquitin/proteasome system and autophagy/lysosomal protein degradation systems, molecular chaperones, and cytoskeletal proteins are susceptible to altered physiological function by ROS [20]. These modifications in the regulatory proteins of non-cancerous cells dysregulate cellular homeostasis and as a consequence lead to initiation of carcinogenesis. Stimulation of these signaling pathways by ROS is also found to be involved in proliferation, migration, and invasion of human breast, liver, prostate, lung, skin, pancreatic, and many other cancer cells. Lipids are the other cellular target of ROS. ROS reacts with polyunsaturated or polydesaturated fatty acids to initiate lipid peroxidation [21]. The peroxidation of lipids generate numerous genotoxic molecules such as malondialdehyde, 2-alkenals and 4-hydroxy-2-alkenals [22]. ROS-induced lipid peroxidation may result in alteration of membrane structure, disruption in membrane permeability and interruption in immune system recognition, which cumulatively promotes cancer initiation and progression. The involvement of lipid peroxidation in carcinogenesis was made evident by the presence of thiobarbituric acid-reactive substances (a byproduct of lipid peroxidation) in the serum of patients with colorectal cancer [23].

2.2. Therapeutic

Cancer cells have a hypermetabolism that produces a large amount of ROS compared to normal cells. At the same time, cancer cells have marked antioxidant capacity that help in maintaining the redox balance. Recently, anticancer therapies that induce oxidative stress by increasing ROS and/or inhibiting antioxidant level have received great attention [24]. The increased ROS disrupts redox homeostasis and causes damage to the cancer cells and ultimately cell death. Numerous anticancer drugs utilize the principle of oxidative stress-induced chemotherapy.

Since cancer cells possess comparatively high ROS levels, these cells have a higher sensitivity towards increased prooxidant and decreased antioxidant levels [25,26]. This increased sensitivity to prooxidants induces cancer cell death using the process of apoptosis, necroptosis, and autophagy [27], which mediates either by direct induction of ROS generation and/or suppression of antioxidant levels [28]. Some anticancer drugs induce ROS generation and cause cancer cell death. For example, motexafin gadolinium, an electron acceptor, suppresses the irradiation-induced DNA repair and increases the effect of radiotherapy in many cancers [29–31]. Anthracycline-based anticancer drugs—such as doxorubicin—also accumulate hydroxyl radicals and cause the death of cancer cells [32].

Another drug, 2-methoxyestradiol, induces mitochondrial production of hydrogen peroxide [33] and subsequently activates c-Jun N-terminal kinase (JNK), resulting in initiation of apoptosis [34,35]; additionally, it can also potentiate the therapeutic action of other anticancer agents [36,37]

Besides these, several natural compounds also exhibit anticancer effects by inducing ROS. Resveratrol, a dietary product present in grapes, vegetables and berries, has been shown to exert an anti-cancer effects on melanoma cells through the induction of ROS. It has also been shown to induce ROS-p38-p53 pathway by increasing the gene expression of phosphorylated p38 MAPK and activating the p53 and ER stress pathway [38]. Ursolic acid, another natural compound, has been shown to induce apoptosis and enhance oxaliplatin-induced inhibition of colorectal cancer cell proliferation in both in vitro and in xenograft nude mouse models. The enhanced anti-proliferative effect of ursolic acid was found to be correlated with increased ROS production and decreased expression of drug resistance genes [39]. A combination of gambogic acid, a natural compound derived from the gamboge, (*Garcinia* species), also caused synergistic reduction of cell viability in SKOV-3 cells with doxorubicin and this effect was found to be correlated with increased cellular ROS accumulation [40]. Similarly, there are several other bioactive dietary polyphenols including quercetin, curcumin, capsaicin, epigallocatechin-3-gallate, piperine, phenethyl isothiocyanate, benzyl isothiocyanate, and others that exert antitumor effects by inducing ROS-mediated cytotoxicity in cancer cells [41].

Other types of anticancer drugs cause cancer cell death by depleting antioxidant levels. For instance, buthionine sulfoximine synergistically affects the chlorin e6 (Ce6)-based photodynamic therapy (PDT) of colorectal cancer cells [42] and several other cancer types including prostate, breast, lung, colon, cervix, bladder, and kidney cancers by inhibiting glutathione [43]. Similarly, imexon—another small-molecule chemotherapeutic agent—disrupts glutathione activity and induces cancer cell death. This agent is widely used in the treatment of advanced cancers of the breast, lung, and prostate [44,45]. Although such anticancer drugs disrupt the redox homeostasis in cancer cells, the inhibition of antioxidant enzymes may also introduce the liability of causing deleterious side effects in normal tissues and organs.

3. Cancer Preventive Role of Curcumin through Suppression of Free Radicals

As described above, ROS plays a key role in carcinogenesis. Therefore, reducing ROS levels can reserve the initiation of tumorigenesis. Curcumin has shown to be a well-established antioxidant compound by innumerable researchers. Various studies have also shown that curcumin blocks the tumorigenesis process in many rodent carcinogenesis models including skin, duodenal, forestomach, and colon carcinogenesis and therefore works as a chemopreventive agent [46–48]. It regulates multiple cell signaling pathways including oxidative stress and inhibits the initiation of carcinogenesis (Figure 1). The chemopreventive effect of curcumin was found to be enhanced when combined with other chemopreventive agents.

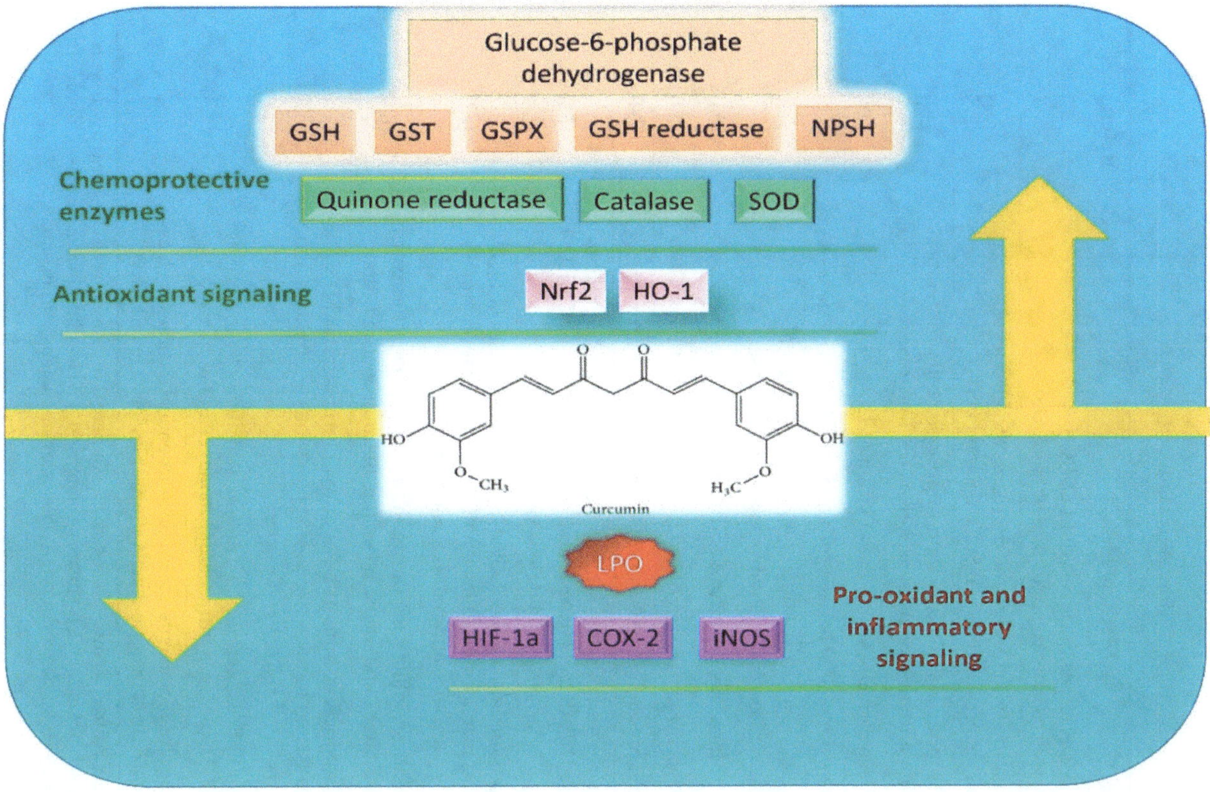

Figure 1. Chemopreventive role of curcumin by suppressing reactive oxygen species (ROS).

3.1. Curcumin Mediates Chemopreventive Effect through Its Antioxidant Property

As curcumin has antioxidant activity, it exhibits a chemopreventive effect by reducing ROS levels and by inducing antioxidant enzymes (Table 1). In a study, curcumin has shown chemopreventive property by suppressing colon carcinogenesis in azoxymethane-dextran sulfate sodium (AOM-DSS) treated mice. The chemopreventive effect of curcumin was found to be associated with the downregulation of multiple regulatory pathways including oxidative stress, as analyzed by RNA sequencing of tumors [49]. In another study, curcumin was found to prevent amino-1-methyl-6-phenylimidazo [4,5-b]pyridine (PhIP)-induced cytotoxicity in normal breast epithelial cells. It has been shown that curcumin prevents PhIP-induced DNA adduct formation and DNA double stand breaks through induction of various antioxidant enzyme systems [50]. Curcumin has also been shown to have protective effect against lung carcinogenesis through its antioxidant property. Curcumin at a dose of 60 mg/kg in drinking water suppressed benzopyrene induced ROS and lipid peroxidation in mice. Moreover, curcumin restored benzopyrene-induced reduced level of glutathione in mice, indicating the protective role of curcumin as an antioxidant [51]. Curcumin also inhibited methylglyoxal (MG)-induced apoptotic events in human hepatoma G2 cells. It has been found that curcumin abolishes MG-stimulated intracellular ROS formation, and subsequent apoptotic biochemical changes in hepatoma G2 cells [52].

Curcumin also exhibits potential benefit against arsenic-induced oxidative stress in humans. A field trial was conducted in volunteers consuming arsenic contaminated ground water. Among 286 volunteers, half were treated with curcumin (500 mg twice daily) and remaining half were assigned with placebo. After three months, DNA damage and oxidative stress were analyzed in blood. The blood samples from endemic regions showed severe DNA damage with increased levels of ROS and lipid peroxidation. The antioxidant enzymes showed depleted activity in the placebo treated group. However, curcumin intervention reduced the DNA damage, retarded ROS generation and lipid peroxidation and raised the level of antioxidant activity [53]. Thus, curcumin exhibits a protective role in humans.

Table 1. Cancer preventive properties of curcumin mediated through inhibition of oxidative stress.

Properties	Models	Mechanisms	References
Anti-carcinogenesis	BP-induced lung tumor in mice	Decreases the levels of LPO, ROS, as well as increases activities of SOD, GST	[51]
		In combination with resveratrol, decreases the LPO level and restores activities of SOD, GR, and GST	[54]
	CoCl2-induced hypoxia in HCC	decreased hypoxia-induced HIF-1α protein, suppressed cell proliferation, migration and invasiveness, as well as EMT changes	[55]
	AOM-DSS-induced colon cancer in mice	Decreases DNA CpG methylation of Tnf	[49]
	DEN induced HCC in rats	Combats oxidative damage of hepatic cells and inhibits carcinogenesis	[56]
Chemopreventive	ddY mice	Increases the activity of antioxidant enzymes GPx, GR, glucose-6-phosphate dehydrogenase and catalase	[57]
	Sprague-Dawley rats.	Increases activity of GST enzyme	[58]
	Renal epithelial cells	Stimulates the expression of Nrf2, increases in HO-1	[59]
	Bovine aortic endothelial cells	Increases the expression of HO-1 mRNA, protein and its activity	[60]
	Spontaneous ovarian cancer in hen	Reduces tumor sizes and number, inhibits NF-κB and STAT3 signaling pathways, decreases KRAS and HRAS mutations, and induces NRF2/HO-1 antioxidant pathway	[61]
Chemoprotective	Hemin-induced cytotoxicity in rat neurons.	Attenuates ROS production, reduces GSH/GSSG ratio, increases GR, GST and SOD enzymes, increases HO-1 level and Nrf2 translocation into the nucleus, and reduces cell death	[62]
	TAA-induced liver inflammation and fibrosis in rats	Reduces oxidative stress, inhibits apoptosis, induced autophagy, decreases fetoprotein AST activity, and increased serum albumin concentration.	[63]
Anti-cytotoxic	PhIP-induced cytotoxicity in breast epithelial cells	Decreases ROS production, inhibits DNA adduct formation and DNA double stand breaks, and induces expression of various antioxidant and DNA repair genes	[50]
	Dox-induced cytotoxicity in 3T3 normal cells	With resveratrol and EEAC increases cell antioxidant ability by improving the activity of SOD, prevents intracellular damage, and reduces ROS	[64]
	MG-induced cell death in human hepatoma G2 cells	Abolishes oxidative stress, prevents apoptotic biochemical changes such as release of cytochrome c, caspase-3 activation, and cleavage of PARP	[52]

TAA—Thioacetamide, AST—Aspartate aminotransferase, MG-Methylglyoxal, EEAC—Ethanolic extract of Antrodia cinnamomea, SOD—Superoxide dismutase, ROS—Reactive oxygen species, PhIP-Amino-1-methyl-6-phenylimidazo [4,5-b]pyridine, Nrf2—nuclear factor erythroid 2–related factor 2, HO-1—heme oxygenase-1, GSH—glutathione, GSSG—glutathione disulfide, GR—glutathione reductase, GPx—glutathione peroxidase, GST—glutathione-S-transferase, LPO—lipid peroxidation, AOM—azoxymethane-dextran sulfate.

Curcumin also exerts chemopreventive efficacy by increasing the activity of antioxidant enzymes and phase II-metabolic enzymes in the liver and kidneys. Curcumin (2%) when fed to mice for 30 days showed a significant enhancement in the activities of glutathione peroxidase, glutathione reductase, glucose-6-phosphate dehydrogenase, and catalase. In addition, curcumin enhances the activities of glutathione *S*-transferase and quinone reductase. This suggests that curcumin may prevent liver and kidney carcinogenesis by inducing antioxidant enzymes [57]. Moreover, in an in vivo rat model, curcumin decreased the oxidative stress levels and showed a dose dependent increase in the NPSH levels in rat liver, which corresponds roughly to tissue glutathione content. Glutathione peroxidase activity was also found to be elevated with increasing curcumin dosing [58], while it reduced lipid peroxidation and salvaged hepatic glutathione antioxidant defense in rats [65]. Such observations clearly point towards the chemopreventive actions of curcumin by inhibiting ROS and inducing antioxidant activities.

The transcription factor Nrf2 is believed to be one of the crucial players in regulating the various antioxidant genes including hemeoxygenase (HO-1). Activation of the Nrf2 antioxidant pathway signaling is reported to play an important role in cancer prevention [66]. Interestingly, curcumin has been shown to activate Nrf2 and HO-1 signaling. In one study, curcumin activated Nrf2 by promoting its nuclear translocation in a time-dependent manner in neuronal cells, and also induced HO-1 protein levels in a concentration-dependent manner [62]. Similarly, curcumin treatment also promoted translocation of Nrf2 protein into the nucleus in renal epithelial cells. This translocation was associated with elevated HO-1 activity [59]. In addition, curcumin treatment increased the expression of HO-1 in vascular endothelial cells leading to the alleviation of oxidative damage [60]. It has also been shown that dietary curcumin consumption at a dose of 53 mg/day reduced ovarian cancer incidence in hens. Curcumin was mixed with a basal diet and were mixed homogenously. This curcumin mixed diet was administered to hens. Interestingly, this chemopreventive effect of curcumin was found to be mediated through reduction in oxidative stress and lipid peroxidation in serum and ovarian tissues. The mechanism behind the chemopreventive effect of curcumin in hen ovarian tissue was an upregulation of antioxidant related proteins such as Nrf2 and HO-1 [61]. Another transcription factor, hypoxia-inducible factor 1 (HIF-1) is known to be associated with increased ROS levels in hypoxic conditions [67,68]. In vitro studies have indicated that curcumin can inhibit the expression of HIF-1 in hepatic carcinoma cells [55,69], possibly suggesting the inhibition of hypoxia-induced ROS by curcumin.

Curcumin is a known potent anti-inflammatory agent that prevents tumor progression and exerts a chemopreventive effect on carcinogenesis via suppressing inflammation. Increasing evidences suggest the involvement of ROS as a mechanism by which chronic inflammation drives cancer initiation and progression. Chronic inflammation stimulates ROS generation which leads to tissue injury and promotes carcinogenesis [70–72]. TNFα is an inflammatory cytokine that activates NF-κB and increases expression of other inflammatory enzymes such as COX-2 and inducible nitric oxide synthase (iNOS). Importantly, studies support that a positive feedback loop between TNFα and ROS promote inflammation-induced carcinogenesis [73–76]. Curcumin has shown to prevent colon carcinogenesis in mice and cholangiocarcinogenesis in hamsters by suppressing the expression of pro-inflammatory enzymes iNOS and COX-2 [77,78]. Curcumin also reduced the iNOS mRNA expression in ex vivo cultured macrophages in a concentration-dependent manner [79]. Moreover, it inhibits thioacetamide (TAA)-induced liver inflammation and fibrosis in rats. It further reduced oxidative stress in liver, inhibited apoptosis, and induced autophagy, and thus protects the first dysplastic stage of hepatocellular carcinoma (HCC) in rats [63]. These studies suggest the potential benefit of curcumin as a chemopreventive agent by suppressing inflammation induced ROS.

Considering the poor absorption and bioavailability of curcumin, several analogues of curcumin have been synthesized and optimized for better antioxidant properties [80]. Dimethoxy curcumin, when tested in human peripheral blood mononuclear cells showed a decrease in lipid peroxidation status and an increase in the catalase activity [81]. In another study, efforts have been directed towards synthesizing curcumin derivatives as Nrf2 activators. Curcumin nanoparticles as polylactide co-glycolide (PLGA) nanocapsulated curcumin have shown chemopreventive activity against diethylnitrosamine (DEN) induced HCC in rat. Oral administration of curcumin nanoparticles (20 mg/kg body weight for 16 weeks) in DEN induced HCC rats exerted a significant protection against HCC by restoring redox homeostasis in liver cells [56]. In order to enhance solubility of curcumin, it is encapsulated with biodegradable nanoparticulate formulation based on PLGA and a stabilizer polyethylene glycol (PEG)-5000. In line with the previous studies, the synthesized compounds were found to be cytoprotective against oxidative stress [82].

3.2. Chemopreventive Potential of Curcumin in Combination with Other Compounds

The chemopreventive efficacy of curcumin has been evaluated in combination with several other natural and synthetic compounds. Interestingly, the chemopreventive potential of curcumin in combination with resveratrol has been shown in multiple studies. Curcumin in combination with resveratrol prevented lung carcinogenesis in rats [83]. In a study, supplementation with curcumin and resveratrol to benzo(a)pyrene-treated mice resulted in the diminution of the molecular events during the promotional phase of lung carcinogenesis. Curcumin (60 mg/kg orally in drinking water) and resveratrol (5.7 μg/mL orally in drinking water) were administered individually or in combination—3 times a week for a total duration of 22 weeks—to benzo(a)pyrene-treated mice. Supplementation of curcumin and resveratrol alone resulted in improvement in cellular integrity, nuclear deformation and premature mitochondrial senescence, however the effects were potentiated with the combined supplementation of phytochemicals [84]. Benzo(a)pyrene treatment also caused significant increase in the levels of lipid peroxidation along with reduction in glutathione levels and the decreased activities of SOD in lungs of mice. Administration of curcumin and resveratrol to benzo(a)pyrene-treated mice decreased the levels of lipid peroxidation and elevated the activities of SOD respectively. Further, the combination also showed an increase in the levels of both reduced glutathione and SOD activity along with decreased level of lipid peroxidation when compared with the monotherapy [85]. Another study showed that treatment of resveratrol and curcumin in combination significantly decreased the benzo(a)pyrene-induced lipid peroxidation, and restored several antioxidant enzymes including SOD, GR, and GST, which eventually may have contributed to lung cancer prevention [54]. Interestingly, curcumin has shown to reduce doxorubicin-induced cardiotoxicity by inhibiting production of ROS. Additionally, curcumin has synergistic effects with doxorubicin against MCF-7 breast cancer cells [64]. Thus, use of curcumin as an adjuvant, alongside other chemopreventive agents could have substantial clinical implications to further develop novel modalities for chemoprevention.

4. Cancer Therapeutic Role of Curcumin through Induction of ROS

The cancer therapeutic benefits of curcumin have been demonstrated in multiple types of cancers including, breast, prostate, skin, lung, liver, brain, stomach, cervical, ovary, multiple myeloma, leukemia and several other types of cancer [86]. Prior experimental evidences have shown that curcumin is effective in potentiating and enhancing anticancer effects of other cancer therapeutic drugs. Moreover, it sensitizes the drug resistant cells and increases therapeutic potential of anticancer drugs [87]. At the molecular level, curcumin exerts anticancer and chemosensitizing effects by targeting multiple pathways including ROS mediated cell signaling (Figure 2).

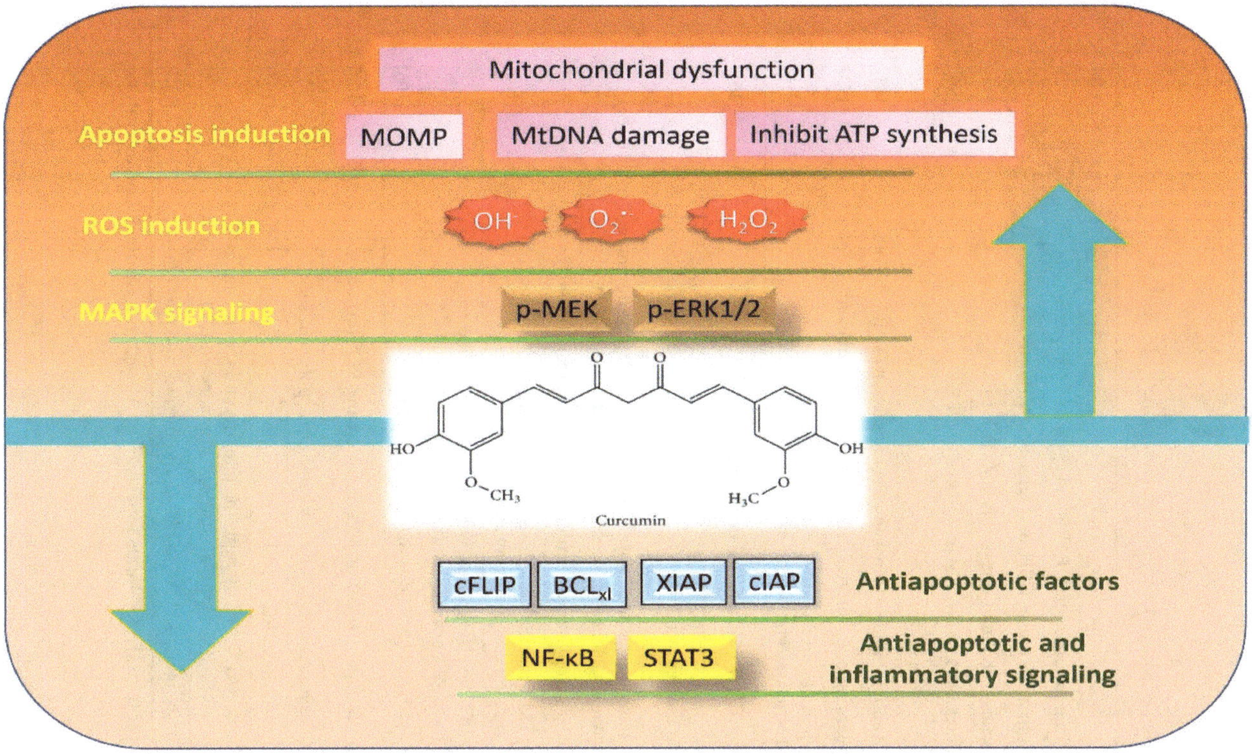

Figure 2. Chemotherapeutic role of curcumin mediated through induction of ROS.

4.1. ROS Mediated Anticancer Effect of Curcumin

Along with chemopreventive properties, curcumin has also chemotherapeutic effects against different types of cancers through multiple mechanisms including oxidative stress pathway (Table 2). Curcumin regulates cellular redox balance by disrupting mitochondrial homeostasis and enhancing cellular oxidative stress. It oxidizes thiols in the mitochondrial membrane, triggers opening of mitochondrial permeability transition pore, mitochondrial swelling, loss of mitochondrial membrane potential, and inhibition of ATP synthesis, leading to initiation of cellular apoptosis in cancer cells [88]. Curcumin further increases the generation of variety of ROS, including hydroxy radicals, superoxides, and H_2O_2 [89,90]. Indeed, in gastric cancer cells curcumin induced the generation of excessive ROS that resulted in depletion of mtDNA and POLG (polymerase γ that reduce mitochondrial oxygen consumption) and subsequent occurrence of cell death [91]. Furthermore, curcumin induces ROS production and decreases mitochondrial transmembrane potential thereby activating DNA damage/repair pathway and mitochondrial apoptosis. Similarly, curcumin exhibited cytotoxicity in NSCLC through the accumulation of cellular ROS. The involvement of ROS was confirmed by using ROS scavengers like catalase and N-acetyl-l-cysteine (NAC) that reversed curcumin-induced cell death in NSCLC [92].

Table 2. Cancer therapeutic properties of curcumin and its analogues mediated through generation.

Properties	Models	Mechanism	Reference
Apoptosis	Myeloid leukemia K562 cells	Releases cytochrome c from mitochondria, PARP and caspase-9 cleavages	[93]
	Melanoma A375 cells	Induces ROS burst, decreases GSH, and wrecks MMP	[94]
	Gastric cancer BGC-823 cells	Induces ROS, activates ASK1, and phosphorylates JNK protein	[95]
	Leukemic Jurkat and K562 cells	Downregulates IAPs, pAkt, c-Myc, and cyclin D1	[96]
	Breast cancer MCF-7, MDAMB, HepG2 cells	Generates ROS	[97]
Cell cycle arrest	Breast cancer MCF-7 cells	Downregulates cyclin B1, Cdc2 and NF-κB by decreasing the interaction of pIκB-NF-κB	[98]
Cell cycle arrest and apoptosis	HT-29 colon cancer cells	Induced ROS generation, DNA fragmentation, chromatin condensation, and nuclear shrinkage	[99]
	K562 cells and xenograft mouse	Derivative **PGV-1** induces prometaphase arrest in the M phase and induces cell senescence and death by increasing ROS.	[100]
	Prostate carcinoma PC-3 and DU145 and xenograft mice	Analogue Ca 37 induces ROS production	[101]
	Prostate cancer RM-1 and DU145 cell lines and xenograft mice	Analog WZ35 induces ROS overproduction, intracellular calcium surge, and mitochondrial disruption	[102]
	NCI-H460 cells	Analogues hexamethoxy-diarylpentadienones (1 and 2) upregulate p53 and p21 and downregulate Cdc25C, cyclin B1/Cdk1 in a Michael acceptor- and ROS-dependent fashion	[103]
	NSCLC A549 and SPC-A1 cell lines	Causes ROS production, DNA damage, endoplasmic reticulum stress and mitochondrial instability.	[92]
Chemosensitization	Glioblastoma	DMC synergistically increases TMZ-induced apoptosis by increasing ROS production, inactivating JAK/STAT3 signaling pathway and caspase-3 cleavage	[104]
Anti-tumorigenesis	CML-derived K562 cells, xenograft mouse	Derivatives upregulate ROS levels, compete with co-enzymes to bind to the respective ROS metabolic enzymes and inhibit their activities	[105]
Anti-angiogenesis	HUVECs, CAMs	Analog **A2** induces NADH/NADPH oxidase-derived ROS	[106]
Tumor re-incidence and metastasis inhibition	B16F10 cells, syngeneic mice	Nanoformulation increases intracellular curcumin accumulation and ROS formation	[107]
Anti-tumorigenesis	Gastric cancer BGC-823 cells, xenograft mice	Enhances oxidative stress, decreases mtDNA content and DNA polymerase γ	[91]
	Leukemic K562 cells, xenograft mice	Induces ROS level	[108]
Autophagy and apoptosis	lymphoma HuT-78 cells	Produces ROS, inhibits c-FLIP, Bcl-xL, cIAP, XIAP, disrupts the integrity of IKK and beclin-1 by degrading Hsp90, inhibits NF-κB, accumulates autophagy marker LC3-I	[109]
Autophagy	Colon cancer HCT116 cells	Generates ROS, converts autophagic marker LC3-I to LC3-II and degrades sequestome-1	[110]

MMP—matrix metallopeptidase, ASK1—apoptosis signal-regulating kinase 1, IAPs—inhibitors of apoptosis proteins, JAK—Janus kinase, STAT3—signal transducer and activator of transcription 3, mtDNA—mitochondrial DNA, GSH—glutathione, ROS—reactive oxygen species.

It has also been observed that the treatment of leukemic cells with curcumin increases intracellular ROS levels over a threshold which was found to be linked with the induction of anti-tumorigenic activity [108]. In melanoma A375 cells, curcumin-induced cell death was also found to be associated with increased oxidative stress. It has been shown that curcumin causes oxidative stress through inducing ROS burst, decreasing glutathione, and wrecking mitochondria membrane potential (MMP), which were reversed by ROS inhibitor NAC [94]. Curcumin also affects the signaling pathways regulated by ROS. In human gastric cancer BGC-823 cells, curcumin induced apoptosis through the generation of ROS as evidenced by the inhibition of curcumin-mediated apoptosis upon antioxidant (NAC or trion) application. Notably, it has been observed that curcumin activated oxidative stress-related kinase ASK1, up-regulated an upstream effector of JNK, MKK4, and phosphorylated JNK protein expression in BGC-823 cells [95]. The modulation of these proteins by curcumin resulted in apoptosis in BGC-823 cells. Similarly, in colon cancer cells, curcumin treatment markedly decreased its cell viability and proliferation potential through generation of ROS. However, the pretreatment with NAC suppressed the growth inhibitory effect of curcumin on HT-29 cells [99].

Along with ROS production, depletion of antioxidant enzymes enhances the anticancer property of curcumin. In a study, curcumin suppressed the growth of human leukemic cells via ROS-independent glutathione depletion, which leads to caspase activation and further apoptosis in leukemic cells. Curcumin treatment to leukemic cells also downregulates the expression of the inhibitor of apoptosis proteins (IAPs), phospho-Akt, c-Myc, and cyclin D1, probably through the suppression of glutathione [96]. In another study, curcumin induced apoptosis in MCF-7, MDAMB-231, and HepG2 cells through generation of ROS. However, depletion of glutathione by buthionine sulfoximine resulted in increased generation of ROS and that enhanced curcumin-mediated apoptosis [97]. It has been reported that curcumin causes rapid generation of ROS in human cutaneous T-cell lymphoma (HuT-78) cells. This increased ROS further modulated different cell survival and cell death pathways and subsequently induced apoptosis. Curcumin also downregulates the expression of antiapoptotic proteins c-FLIP, Bcl-xL, cellular inhibitor of apoptosis protein (cIAP), and X-linked IAP (XIAP) in a ROS-dependent manner, which further induces cancer cell death [109]. Besides apoptosis, curcumin also induces autophagic cell death in HCT116 human colon cancer cell by inducing ROS production [110]. Curcumin also hinders cancer stemness through production of ROS. It inhibits proliferation, sphere forming and colony forming abilities of glioblastoma [111] and liver cancer [112] stem cells by ROS mediated activation of MAPK pathway, suppression of NF-κB signaling, downregulation of STAT3 activity and IAP family members.

Besides curcumin, its analogues (Figure 3) also exhibit ROS mediated chemotherapeutic activity against various cancers. In a study, curcumin analogue 1,5-bis(3-hydroxyphenyl)-1,4-pentadiene-3-one (Ca 37), which is a monocarbonyl analogue of curcumin, exhibited antitumor activity in both in vitro and in prostate xenografted tumor models. Furthermore, a combination of Ca 37 and curcumin resulted in enhanced antitumor activity in prostate cancer cells. However, administration of ROS quencher NAC abrogated Ca 37 mediated tumor growth inhibition; indicating that induction of ROS plays a vital role in the growth inhibitory activity of Ca 37 in prostate cancer cells [101]. In another study, two symmetrical hexamethoxy-diarylpentadienones (1 and 2) monocarbonyl analogues of curcumin have been reported to possess significantly enhanced cytotoxicity as compared to the parent molecule. In comparison to curcumin, these analogues induce a more potent burst of ROS [103]. WZ35, another structural monocarbonyl analogue of curcumin, was examined for anti-prostate cancer effects in both in vitro and in vivo models and showed a reduction in cancer cell viability, an increase in apoptosis, and G2/M cell cycle arrest through overproduction of ROS. Interestingly, WZ35-induced apoptosis in prostate cancer cells was found to be completely reversed by ROS inhibition. Additionally, in an animal study, WZ35 inhibited prostate homograft tumor growth with increased ROS accumulation, mitochondrial disruption, and cell apoptosis in tumor tissues [102], indicating ROS plays a key role in mediating WZ35 tumor growth inhibition.

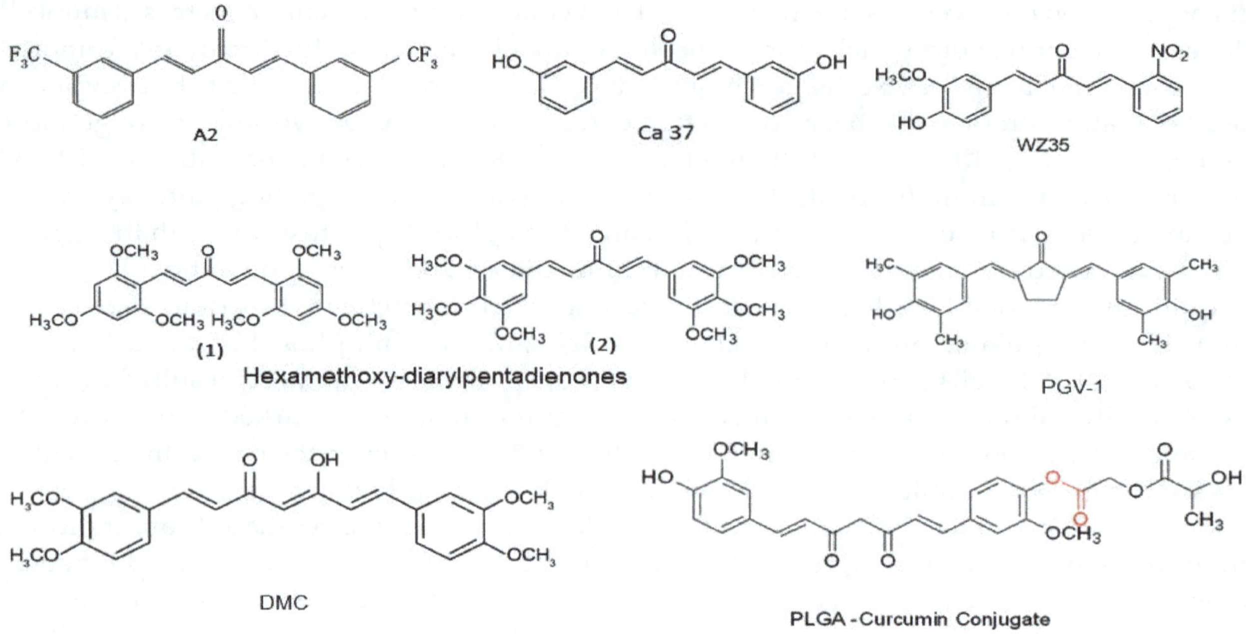

Figure 3. Chemical structure of curcumin analogues, which exhibits anticancer effects mediated through ROS.

Another monocarbonyl analogue of curcumin, **A2** without the β-diketone moiety, displayed antiangiogenic activity. It has been demonstrated that **A2** exerts its antiangiogenic activity mainly through inducing endothelial cell death via elevating NADH/NADPH oxidase-derived ROS [106]. Furthermore, curcumin analogue pentagamavunon-1 (**PGV-1**) has been studied for its inhibitory activity on tumor cells in vitro and in vivo. **PGV-1** (2,5-bis-(4-hydroxy, 3′,5′-dimethyl)-benzylidine-cyclopentanone) is a monocarbonyl analogue of curcumin and has stronger potency than the lead compound, curcumin. It has been found that **PGV-1** exhibits 60-fold higher efficacy compared to that of curcumin in inhibiting K562 cell growth. **PGV-1** also inhibited the proliferation of leukemia, breast adenocarcinoma, cervical cancer, uterine cancer, and pancreatic cancer cells. Interestingly, **PGV-1**-induced cell death is mediated by an increase in intracellular ROS levels through inhibition of ROS-metabolic enzymes [100]. Nakamae et al. (2019) synthesized 39 novel curcumin derivatives and examined their anti-proliferative and anti-tumorigenic properties. They found that these analogues exhibit anti-proliferative activity toward human cancer cell lines in a manner sensitive to antioxidant NAC. Some analogues markedly increased ROS levels and efficiently induced cell death as well as suppressed tumor formation in a xenograft mouse model. Finally, it was found that the anti-tumorigenic activity of these analogues was well-correlated with an increase in ROS levels [105]. Furthermore, nanoformulation of curcumin (curcumin-loaded nanoemulsion) increased intracellular curcumin accumulation and ROS formation, while preventing migration and invasion of melanoma cells [107].

4.2. Chemosensitizing Effect of Curcumin Mediated through ROS

Accumulated preclinical evidences suggest that the effectiveness of chemotherapy is being limited due to drug resistance, therapeutic selectivity, and undesirable side effects. In this regard, curcumin has proved its ability to sensitize drug-resistant cancer cells and enhance therapeutic efficacy of drugs. As ROS production by curcumin plays a pivotal role in cancer cell death, it can also mediate chemosensitization. Studies showed that curcumin improves the sensitivity of cancer cells to chemotherapy drugs by regulating a variety of signaling pathways [113]. An in vitro study demonstrated that cisplatin in combination with curcumin exhibit higher anti-tumor activity in HepG2 cells compared with mono-drug therapy. This synergistic effect was found to be the result of increased intracellular ROS levels in HCC cells [114]. Another study showed that curcumin also exhibits a pro-oxidant effect and exerts a chemosensitive property. It sensitizes cisplatin-resistant cells by targeting Nrf-2, NF-κB and STAT3 phosphorylation [115]. The apoptosis in bladder cancer (253J-Bv and T24) cells was found to be increased when curcumin was treated in combination with cisplatin as compared to single agent treatment. The synergistic induction of apoptosis by curcumin and cisplatin was associated with increased ROS production together with upregulation of p-MEK and p-ERK1/2 signaling [116].

The cancer therapeutic response of curcumin has been shown to be improved by tolfenamic acid (a non-steroidal anti-inflammatory drug). Tolfenamic acid has been reported to inhibit the growth of human cancer cells in vitro and in vivo. However, the combination of tolfenamic acid and curcumin showed higher growth inhibition when compared to either single agent. Further, another study showed that the combination treatment upregulated the ROS level that led to an increase in apoptosis in colorectal cancer cells [117]. Subtoxic concentrations of curcumin has also demonstrated to sensitize human renal cancer cells to the tumor necrosis factor-related apoptosis inducing ligand (TRAIL)-mediated apoptosis. Curcumin induces generation of ROS that led to the expression of death receptor 5 (DR5) and further enhancement in TRAIL mediated apoptosis [118]. An increase in ROS production and mitochondria depolarization by the combined treatment of curcumin and tamoxifen was also observed. This combination of drugs resulted in synergistic induction of autophagy along with apoptosis in chemoresistant melanoma cells, indicating the role of curcumin-induced ROS in combination chemotherapy [119]. ROS generation by curcumin has also been shown to enhance irinotecan's therapeutic effects on colorectal cancer cells by inhibiting cell viability and inducing cell cycle arrest and apoptosis [120]. Curcumin analogue also sensitizes drug-resistant cancer cells. In a study, curcumin analog ALZ003 inhibited the survival of TMZ-sensitive and -resistant glioblastoma in both in vitro and in vivo models through the accumulation of ROS, lipid peroxidation and suppression of GPX4 [121]. Demethoxycurcumin (DMC), a naturally occurring curcumin analogue has been shown to enhance antitumor effect of temozolomide (TMZ). In a study, treatment of DMC prior to TMZ in glioblastoma cells resulted in a significant increase in caspase-3 signaling, mitochondria-related apoptosis and a marked inhibition of cell growth in vitro through production of ROS as well as inactivation of JAK/STAT3 signaling pathway [104].

Combinations of drugs may enhance the therapeutic effect of individual drug treatment. Combinations of natural products are broadly explored in cancer therapy. In a study, curcumin was combined with arabinogalactan (found in Larch trees) and treated to the human breast cancer cells. This combination promoted cell growth inhibition and apoptosis induction in human breast cancer cells by increasing ROS level as well as loss of mitochondrial membrane potential and reduction of glutathione. In addition, this combination inhibited the progression of breast tumors in mouse model [122]. It has been observed that curcumin in combination with quercetin induced apoptosis by increasing ROS and decreasing glutathione levels, as well as by inducing loss of mitochondrial membrane potential. The combination of quercetin and curcumin potentiates individual's apoptotic effect and reduces effective dose of individual agent [93]. Nanoparticles loaded with curcumin and quercetin have also shown synergistic antitumor properties on a breast cancer cell line. Cell viability study showed a pronounced antitumor effect in combination compared to the individual drug on the MCF7 cell line, which was associated with an increase in intracellular ROS level [123]. Curcumin and resveratrol have also been found to exhibit a synergistic anticancer effect in colon cancer. The combination of curcumin and resveratrol has shown to upregulate intracellular ROS levels and further elicit a synergistic antiproliferative effect in Hepa1-6 cells [124]. Besides these, there are several natural compounds as well as therapeutic drugs that have shown to exhibit synergistic anticancer effect with curcumin mediated by ROS induction.

5. Conclusions

Oxidative stress—caused by an imbalance between free radical generation and the antioxidant defense system of the body—potentially results in the initiation and progression of cancer. However, free radicals, specially ROS, play an important role in therapeutic drug action. Thus, ROS has both pathological and pharmacological importance in the body. These findings suggest the use of strategies to either suppress or induce ROS. One of the strategies is to decrease oxidative stress by quenching ROS and/or increasing antioxidant enzymes that can subsequently inhibit carcinogenesis. Another strategy is to use ROS inducing agents in cancer patients that can induce apoptosis in cancer cells. Use of antioxidants in cancer patients have shown unfavorable results for the treatment efficacy of therapeutic drugs. Studies have shown that antioxidants cause increased cancer incidences, probably due to the inhibition of ROS mediated apoptosis of cancer cells. However, high consumption of fruits and vegetables rich in antioxidants are associated with decreased risk of cancer incidence. In view of this, curcumin is considered as an antioxidant as well as a prooxidant, since it reduces oxidative stress in normal cells and induces ROS production in cancer cells. Although several studies have investigated the beneficial role of curcumin, this review will help to better understand the ROS associated chemopreventive and anticancer mechanism of curcumin.

Author Contributions: S.P. contributed to the concept, design and to the writing of the manuscript; N.G., K.V. and S.N. contributed to the literature search and writing of manuscript; A.K. and R.L. reviewed and provided feedback in the manuscript designing and writing. All authors have read and agreed to the published version of the manuscript.

Acknowledgments: We thank Shyanne Page-Hefley from the Department of Pediatrics for carefully proofreading the manuscript.

References

1. Kurutas, E.B. The importance of antioxidants which play the role in cellular response against oxidative/nitrosative stress: Current state. *Nutr. J.* **2015**, *15*, 1–22. [CrossRef] [PubMed]
2. Cheeseman, K.; Slater, T. An introduction to free radical biochemistry. *Br. Med. Bull.* **1993**, *49*, 481–493. [CrossRef] [PubMed]
3. Liu, T.; Stern, A.; Roberts, L.J.; Morrow, J.D. The isoprostanes: Novel prostaglandin-like products of the free radical-catalyzed peroxidation of arachidonic acid. *J. Biomed. Sci.* **1999**, *6*, 226–235. [CrossRef] [PubMed]
4. Lea, A. Dietary factors associated with death-rates from certain neoplasms in man. *Lancet* **1966**, *2*, 332–333. [CrossRef]
5. Key, T.J.; Bradbury, K.E.; Perez-Cornago, A.; Sinha, R.; Tsilidis, K.K.; Tsugane, S. Diet, nutrition, and cancer risk: What do we know and what is the way forward? *BMJ* **2020**, *368*. [CrossRef]
6. Neuhouser, M.L.; Barnett, M.J.; Kristal, A.R.; Ambrosone, C.B.; King, I.B.; Thornquist, M.; Goodman, G.G. Dietary supplement use and prostate cancer risk in the Carotene and Retinol Efficacy Trial. *Cancer Epidemiol. Prev. Biomark.* **2009**, *18*, 2202–2206. [CrossRef]
7. Goodman, G.E.; Thornquist, M.D.; Balmes, J.; Cullen, M.R.; Meyskens, F.L., Jr.; Omenn, G.S.; Valanis, B.; Williams, J.H., Jr. The Beta-Carotene and Retinol Efficacy Trial: Incidence of lung cancer and cardiovascular disease mortality during 6-year follow-up after stopping β-carotene and retinol supplements. *J. Natl. Cancer Inst.* **2004**, *96*, 1743–1750. [CrossRef]
8. Wiel, C.; Le Gal, K.; Ibrahim, M.X.; Jahangir, C.A.; Kashif, M.; Yao, H.; Ziegler, D.V.; Xu, X.; Ghosh, T.; Mondal, T. BACH1 stabilization by antioxidants stimulates lung cancer metastasis. *Cell* **2019**, *178*, 330–345.e22. [CrossRef]
9. Sinha, B.K. Free radicals in anticancer drug pharmacology. *Chem. Biol. Interact.* **1989**, *69*, 293–317. [CrossRef]
10. Martin, K.; Barrett, J. Reactive oxygen species as double-edged swords in cellular processes: Low-dose cell signaling versus high-dose toxicity. *Hum. Exp. Toxicol.* **2002**, *21*, 71–75. [CrossRef]
11. Pan, J.-S.; Hong, M.-Z.; Ren, J.-L. Reactive oxygen species: A double-edged sword in oncogenesis. *World J. Gastroenterol. WJG* **2009**, *15*, 1702. [CrossRef] [PubMed]
12. Phaniendra, A.; Jestadi, D.B.; Periyasamy, L. Free radicals: Properties, sources, targets, and their implication in various diseases. *Indian J. Clin. Biochem.* **2015**, *30*, 11–26. [CrossRef] [PubMed]
13. Du, M.Q.; Carmichael, P.L.; Phillips, D.H. Induction of activating mutations in the human c-Ha-ras-1 proto-oncogene by oxygen free radicals. *Mol. Carcinog.* **1994**, *11*, 170–175. [CrossRef] [PubMed]

14. Wang, D.; Kreutzer, D.A.; Essigmann, J.M. Mutagenicity and repair of oxidative DNA damage: Insights from studies using defined lesions. *Mutat. Res. Fundam. Mol. Mech. Mutagenes.* **1998**, *400*, 99–115. [CrossRef]
15. Domínguez, A.G.; Ramírez, L.P.; Monroy, R.H.; Angeles-Angeles, A. Non-Hodgkin's Lymphoma Secondary to Hodgkin's Disease Treated With Chemo-And Radiotherapy. Report of a Case. *Rev. Investig. Clin. Organo Hosp. Enferm. Nutr.* **1992**, *44*, 393–398.
16. Olinski, R.; Jaruga, P.; Zastawny, T.H. Oxidative DNA base modifications as factors in carcinogenesis. *Acta Biochim. Pol.* **1998**, *45*, 561–572. [CrossRef]
17. Poulsen, H.E.; Prieme, H.; Loft, S. Role of oxidative DNA damage in cancer initiation and promotion. *Eur. J. Cancer Prev.* **1998**, *7*, 9–16.
18. Malins, D.C.; Haimanot, R. Major alterations in the nucleotide structure of DNA in cancer of the female breast. *Cancer Res.* **1991**, *51*, 5430–5432.
19. Ichiba, M.; Maeta, Y.; Mukoyama, T.; Saeki, T.; Yasui, S.; Kanbe, T.; Okano, J.I.; Tanabe, Y.; Hirooka, Y.; Yamada, S. Expression of 8-hydroxy-2′-deoxyguanosine in chronic liver disease and hepatocellular carcinoma. *Liver Int.* **2003**, *23*, 338–345. [CrossRef]
20. Moldogazieva, N.T.; Lutsenko, S.V.; Terentiev, A.A. Reactive Oxygen and Nitrogen Species–Induced Protein Modifications: Implication in Carcinogenesis and Anticancer Therapy. *Cancer Res.* **2018**, *78*, 6040–6047. [CrossRef]
21. Barrera, G. Oxidative stress and lipid peroxidation products in cancer progression and therapy. *ISRN Oncol.* **2012**, *2012*. [CrossRef]
22. Csala, M.; Kardon, T.; Legeza, B.; Lizák, B.; Mandl, J.; Margittai, É.; Puskás, F.; Száraz, P.; Szelényi, P.; Bánhegyi, G. On the role of 4-hydroxynonenal in health and disease. *Biochim. Biophys. Acta (BBA) Mol. Basis Dis.* **2015**, *1852*, 826–838. [CrossRef]
23. Lauschke, H.; Tolba, R.; Burger, B.; Minor, T.; Hirner, A. Lipid peroxidation as additional marker in patients with colorectal cancer. *Eur. Surg. Res.* **2002**, *34*, 346–350. [CrossRef]
24. Prasad, S.; Gupta, S.C.; Tyagi, A.K. Reactive oxygen species (ROS) and cancer: Role of antioxidative nutraceuticals. *Cancer Lett.* **2017**, *387*, 95–105. [CrossRef]
25. Aykin-Burns, N.; Ahmad, I.M.; Zhu, Y.; Oberley, L.W.; Spitz, D.R. Increased levels of superoxide and H2O2 mediate the differential susceptibility of cancer cells versus normal cells to glucose deprivation. *Biochem. J.* **2009**, *418*, 29–37. [CrossRef]
26. Singh, A.; Misra, V.; Thimmulappa, R.K.; Lee, H.; Ames, S.; Hoque, M.O.; Herman, J.G.; Baylin, S.B.; Sidransky, D.; Gabrielson, E. Dysfunctional KEAP1–NRF2 interaction in non-small-cell lung cancer. *PLoS Med.* **2006**, *3*, e420. [CrossRef]
27. Neumann, C.A.; Fang, Q. Are peroxiredoxins tumor suppressors? *Curr. Opin. Pharmacol.* **2007**, *7*, 375–380. [CrossRef]
28. Wang, J.; Yi, J. Cancer cell killing via ROS: To increase or decrease, that is the question. *Cancer Biol. Ther.* **2008**, *7*, 1875–1884. [CrossRef]
29. Khuntia, D.; Mehta, M. Motexafin gadolinium: A clinical review of a novel radioenhancer for brain tumors. *Expert Rev. Anticancer. Ther.* **2004**, *4*, 981–989. [CrossRef]
30. Mehta, M.P.; Rodrigus, P.; Terhaard, C.; Rao, A.; Suh, J.; Roa, W.; Souhami, L.; Bezjak, A.; Leibenhaut, M.; Komaki, R. Survival and neurologic outcomes in a randomized trial of motexafin gadolinium and whole-brain radiation therapy in brain metastases. *J. Clin. Oncol.* **2003**, *21*, 2529–2536. [CrossRef]
31. Rodrigus, P. Motexafin gadolinium: A possible new radiosensitiser. *Expert Opin. Investig. Drugs* **2003**, *12*, 1205–1210. [CrossRef] [PubMed]
32. Kotamraju, S.; Chitambar, C.R.; Kalivendi, S.V.; Joseph, J.; Kalyanaraman, B. Transferrin receptor-dependent iron uptake is responsible for doxorubicin-mediated apoptosis in endothelial cells role of oxidant-induced iron signaling in apoptosis. *J. Biol. Chem.* **2002**, *277*, 17179–17187. [CrossRef] [PubMed]
33. Mooberry, S.L. Mechanism of action of 2-methoxyestradiol: New developments. *Drug Resist. Updates* **2003**, *6*, 355–361. [CrossRef] [PubMed]
34. Djavaheri-Mergny, M.; Wietzerbin, J.; Besancon, F. 2-Methoxyestradiol induces apoptosis in Ewing sarcoma cells through mitochondrial hydrogen peroxide production. *Oncogene* **2003**, *22*, 2558–2567. [CrossRef]
35. Kachadourian, R.; Liochev, S.I.; Cabelli, D.E.; Patel, M.N.; Fridovich, I.; Day, B.J. 2-Methoxyestradiol does not inhibit superoxide dismutase. *Arch. Biochem. Biophys.* **2001**, *392*, 349–353. [CrossRef]

36. Lakhani, N.J.; Sarkar, M.A.; Venitz, J.; Figg, W.D. 2-Methoxyestradiol, a promising anticancer agent. *Pharmacother. J. Hum. Pharmacol. Drug Ther.* **2003**, *23*, 165–172. [CrossRef]
37. Mukhopadhyay, T.; Roth, J.A. Superinduction of wild-type p53 protein after 2-methoxyestradiol treatment of Ad5p53-transduced cells induces tumor cell apoptosis. *Oncogene* **1998**, *17*, 241–246. [CrossRef]
38. Heo, J.R.; Kim, S.M.; Hwang, K.A.; Kang, J.H.; Choi, K.C. Resveratrol induced reactive oxygen species and endoplasmic reticulum stress-mediated apoptosis, and cell cycle arrest in the A375SM malignant melanoma cell line. *Int. J. Mol. Med.* **2018**, *42*, 1427–1435. [CrossRef]
39. Zhang, Y.; Huang, L.; Shi, H.; Chen, H.; Tao, J.; Shen, R.; Wang, T. Ursolic acid enhances the therapeutic effects of oxaliplatin in colorectal cancer by inhibition of drug resistance. *Cancer Sci.* **2018**, *109*, 94–102. [CrossRef]
40. Wang, J.; Yuan, Z. Gambogic acid sensitizes ovarian cancer cells to doxorubicin through ROS-mediated apoptosis. *Cell Biochem. Biophys.* **2013**, *67*, 199–206. [CrossRef]
41. NavaneethaKrishnan, S.; Rosales, J.L.; Lee, K.-Y. ROS-mediated cancer cell killing through dietary phytochemicals. *Oxid. Med. Cell. Longev.* **2019**, *2019*. [CrossRef] [PubMed]
42. Lee, H.M.; Kim, D.H.; Lee, H.L.; Cha, B.; Kang, D.H.; Jeong, Y.-I. Synergistic effect of buthionine sulfoximine on the chlorin e6-based photodynamic treatment of cancer cells. *Arch. Pharmacal Res.* **2019**, *42*, 990–999. [CrossRef] [PubMed]
43. Maeda, H.; Hori, S.; Ohizumi, H.; Segawa, T.; Kakehi, Y.; Ogawa, O.; Kakizuka, A. Effective treatment of advanced solid tumors by the combination of arsenic trioxide and L-buthionine-sulfoximine. *Cell Death Differ.* **2004**, *11*, 737–746. [CrossRef] [PubMed]
44. Moulder, S.; Dhillon, N.; Ng, C.; Hong, D.; Wheler, J.; Naing, A.; Tse, S.; La Paglia, A.; Dorr, R.; Hersh, E. A phase I trial of imexon, a pro-oxidant, in combination with docetaxel for the treatment of patients with advanced breast, non-small cell lung and prostate cancer. *Investig. New Drugs* **2010**, *28*, 634–640. [CrossRef] [PubMed]
45. Sheveleva, E.V.; Landowski, T.H.; Samulitis, B.K.; Bartholomeusz, G.; Powis, G.; Dorr, R.T. Imexon induces an oxidative endoplasmic reticulum stress response in pancreatic cancer cells. *Mol. Cancer Res.* **2012**, *10*, 392–400. [CrossRef] [PubMed]
46. Huang, M.-T.; Ma, W.; Lu, Y.-P.; Chang, R.L.; Fisher, C.; Manchand, P.S.; Newmark, H.L.; Conney, A.H.; You, M. Effects of curcumin, demethoxycurcumin, bisdemethoxycurcumin and tetrahydrocurcumin on 12-O-tetradecanoylphorbol-13-acetateinduced tumor promotion. *Carcinogenesis* **1995**, *16*, 2493–2497. [CrossRef] [PubMed]
47. Huang, M.-T.; Wang, Z.; Georgiadis, C.; Laskin, J.; Conney, A. Inhibitory effects of curcumin on tumor initiation by benzo [a] pyrene and 7, 12-dimethylbenz [a] anthracene. *Carcinogenesis* **1992**, *13*, 2183–2186. [CrossRef]
48. Huang, M.-T.; Lou, Y.-R.; Ma, W.; Newmark, H.L.; Reuhl, K.R.; Conney, A.H. Inhibitory effects of dietary curcumin on forestomach, duodenal, and colon carcinogenesis in mice. *Cancer Res.* **1994**, *54*, 5841–5847.
49. Guo, Y.; Wu, R.; Gaspar, J.M.; Sargsyan, D.; Su, Z.-Y.; Zhang, C.; Gao, L.; Cheng, D.; Li, W.; Wang, C. DNA methylome and transcriptome alterations and cancer prevention by curcumin in colitis-accelerated colon cancer in mice. *Carcinogenesis* **2018**, *39*, 669–680. [CrossRef]
50. Jain, A.; Samykutty, A.; Jackson, C.; Browning, D.; Bollag, W.B.; Thangaraju, M.; Takahashi, S.; Singh, S.R. Curcumin inhibits PhIP induced cytotoxicity in breast epithelial cells through multiple molecular targets. *Cancer Lett.* **2015**, *365*, 122–131. [CrossRef]
51. Liu, Y.; Wu, Y.; Zhang, P. Protective effects of curcumin and quercetin during benzo (a) pyrene induced lung carcinogenesis in mice. *Eur. Rev. Med. Pharmacol. Sci.* **2015**, *19*, 1736–1743.
52. CHAN, W.H.; WU, H.J.; HSUUW, Y.D. Curcumin inhibits ROS formation and apoptosis in methylglyoxal-treated human hepatoma G2 cells. *Ann. N. Y. Acad. Sci.* **2005**, *1042*, 372–378. [CrossRef]
53. Biswas, J.; Sinha, D.; Mukherjee, S.; Roy, S.; Siddiqi, M.; Roy, M. Curcumin protects DNA damage in a chronically arsenic-exposed population of West Bengal. *Hum. Exp. Toxicol.* **2010**, *29*, 513–524. [CrossRef]
54. Liu, Y.; Wu, Y.; Yu, Y.; Cao, C.; Zhang, J.; Li, K.; Zhang, P. Curcumin and resveratrol in combination modulate drug-metabolizing enzymes as well as antioxidant indices during lung carcinogenesis in mice. *Hum. Exp. Toxicol.* **2015**, *34*, 620–627. [CrossRef]
55. Duan, W.; Chang, Y.; Li, R.; Xu, Q.; Lei, J.; Yin, C.; Li, T.; Wu, Y.; Ma, Q.; Li, X. Curcumin inhibits hypoxia inducible factor-1α-induced epithelial-mesenchymal transition in HepG2 hepatocellular carcinoma cells. *Mol. Med. Rep.* **2014**, *10*, 2505–2510. [CrossRef]

56. Ghosh, D.; Choudhury, S.T.; Ghosh, S.; Mandal, A.K.; Sarkar, S.; Ghosh, A.; Saha, K.D.; Das, N. Nanocapsulated curcumin: Oral chemopreventive formulation against diethylnitrosamine induced hepatocellular carcinoma in rat. *Chem. Biol. Interact.* **2012**, *195*, 206–214. [CrossRef]
57. Iqbal, M.; Sharma, S.D.; Okazaki, Y.; Fujisawa, M.; Okada, S. Dietary supplementation of curcumin enhances antioxidant and phase II metabolizing enzymes in ddY male mice: Possible role in protection against chemical carcinogenesis and toxicity. *Pharmacol. Toxicol.* **2003**, *92*, 33–38. [CrossRef]
58. Piper, J.T.; Singhal, S.S.; Salameh, M.S.; Torman, R.T.; Awasthi, Y.C.; Awasthi, S. Mechanisms of anticarcinogenic properties of curcumin: The effect of curcumin on glutathione linked detoxification enzymes in rat liver. *Int. J. Biochem. Cell Biol.* **1998**, *30*, 445–456. [CrossRef]
59. Balogun, E.; Hoque, M.; Gong, P.; Killeen, E.; Green, C.J.; Foresti, R.; Alam, J.; Motterlini, R. Curcumin activates the haem oxygenase-1 gene via regulation of Nrf2 and the antioxidant-responsive element. *Biochem. J.* **2003**, *371*, 887–895. [CrossRef]
60. Motterlini, R.; Foresti, R.; Bassi, R.; Green, C.J. Curcumin, an antioxidant and anti-inflammatory agent, induces heme oxygenase-1 and protects endothelial cells against oxidative stress. *Free Radic. Biol. Med.* **2000**, *28*, 1303–1312. [CrossRef]
61. Sahin, K.; Orhan, C.; Tuzcu, M.; Sahin, N.; Tastan, H.; Özercan, İ.H.; Güler, O.; Kahraman, N.; Kucuk, O.; Ozpolat, B. Chemopreventive and antitumor efficacy of curcumin in a spontaneously developing hen ovarian cancer model. *Cancer Prev. Res.* **2018**, *11*, 59–67. [CrossRef] [PubMed]
62. González-Reyes, S.; Guzmán-Beltrán, S.; Medina-Campos, O.N.; Pedraza-Chaverri, J. Curcumin pretreatment induces Nrf2 and an antioxidant response and prevents hemin-induced toxicity in primary cultures of cerebellar granule neurons of rats. *Oxid. Med. Cell. Longev.* **2013**, *2013*. [CrossRef] [PubMed]
63. Elmansi, A.M.; El-Karef, A.A.; El-Shishtawy, M.M.; Eissa, L.A. Hepatoprotective effect of curcumin on hepatocellular carcinoma through autophagic and apoptic pathways. *Ann. Hepatol.* **2017**, *16*, 607–618. [CrossRef]
64. Sheu, M.-T.; Jhan, H.-J.; Hsieh, C.-M.; Wang, C.-J.; Ho, H.-O. Efficacy of antioxidants as a complementary and alternative medicine (CAM) in combination with the chemotherapeutic agent doxorubicin. *Integr. Cancer Ther.* **2015**, *14*, 184–195. [CrossRef]
65. Sreepriya, M.; Bali, G. Effects of administration of Embelin and Curcumin on lipid peroxidation, hepatic glutathione antioxidant defense and hematopoietic system during N-nitrosodiethylamine/Phenobarbital-induced hepatocarcinogenesis in Wistar rats. *Mol. Cell. Biochem.* **2006**, *284*, 49–55. [CrossRef]
66. Zhang, Y.; Gordon, G.B. A strategy for cancer prevention: Stimulation of the Nrf2-ARE signaling pathway. *Mol. Cancer Ther.* **2004**, *3*, 885 893.
67. Qutub, A.A.; Popel, A.S. Reactive oxygen species regulate hypoxia-inducible factor 1α differentially in cancer and ischemia. *Mol. Cell. Biol.* **2008**, *28*, 5106–5119. [CrossRef]
68. Zhao, T.; Zhu, Y.; Morinibu, A.; Kobayashi, M.; Shinomiya, K.; Itasaka, S.; Yoshimura, M.; Guo, G.; Hiraoka, M.; Harada, H. HIF-1-mediated metabolic reprogramming reduces ROS levels and facilitates the metastatic colonization of cancers in lungs. *Sci. Rep.* **2014**, *4*, 3793. [CrossRef]
69. Chang, Y.; Jiang, M.; Liu, K.; Li, X. Curcumin inhibited hypoxia induced epithelial-mesenchymal transition in hepatic carcinoma cell line HepG2 in vitro. *Zhongguo Zhong Xi Yi Jie He Za Zhi Zhongguo Zhongxiyi Jiehe Zazhi Chin. J. Integr. Tradit. West. Med.* **2013**, *33*, 1102–1106.
70. El-Kenawi, A.; Ruffell, B. Inflammation, ROS, and mutagenesis. *Cancer Cell* **2017**, *32*, 727–729. [CrossRef]
71. Agita, A.; Alsagaff, M.T. Inflammation, immunity, and hypertension. *Acta Med. Indones* **2017**, *49*, 158–165.
72. Mittal, M.; Siddiqui, M.R.; Tran, K.; Reddy, S.P.; Malik, A.B. Reactive oxygen species in inflammation and tissue injury. *Antioxid. Redox Signal.* **2014**, *20*, 1126–1167. [CrossRef]
73. Sprague, A.H.; Khalil, R.A. Inflammatory cytokines in vascular dysfunction and vascular disease. *Biochem. Pharmacol.* **2009**, *78*, 539–552. [CrossRef]
74. Yan, Y.; Li, J.; Ouyang, W.; Ma, Q.; Hu, Y.; Zhang, D.; Ding, J.; Qu, Q.; Subbaramaiah, K.; Huang, C. NFAT3 is specifically required for TNF-α-induced cyclooxygenase-2 (COX-2) expression and transformation of Cl41 cells. *J. Cell Sci.* **2006**, *119*, 2985–2994. [CrossRef]
75. He, Y.; Xiao, Y.; Yang, X.; Li, Y.; Wang, B.; Yao, F.; Shang, C.; Jin, Z.; Wang, W.; Lin, R. SIRT6 inhibits TNF-α-induced inflammation of vascular adventitial fibroblasts through ROS and Akt signaling pathway. *Exp. Cell Res.* **2017**, *357*, 88–97. [CrossRef]

76. Lee, I.-T.; Lin, C.-C.; Lee, C.-Y.; Hsieh, P.-W.; Yang, C.-M. Protective effects of (−)-epigallocatechin-3-gallate against TNF-α-induced lung inflammation via ROS-dependent ICAM-1 inhibition. *J. Nutr. Biochem.* **2013**, *24*, 124–136. [CrossRef]
77. Murakami, A.; Furukawa, I.; Miyamoto, S.; Tanaka, T.; Ohigashi, H. Curcumin combined with turmerones, essential oil components of turmeric, abolishes inflammation-associated mouse colon carcinogenesis. *Biofactors* **2013**, *39*, 221–232. [CrossRef]
78. Prakobwong, S.; Khoontawad, J.; Yongvanit, P.; Pairojkul, C.; Hiraku, Y.; Sithithaworn, P.; Pinlaor, P.; Aggarwal, B.B.; Pinlaor, S. Curcumin decreases cholangiocarcinogenesis in hamsters by suppressing inflammation-mediated molecular events related to multistep carcinogenesis. *Int. J. Cancer* **2011**, *129*, 88–100. [CrossRef]
79. Chan, M.M.-Y.; Huang, H.-I.; Fenton, M.R.; Fong, D. In vivo inhibition of nitric oxide synthase gene expression by curcumin, a cancer preventive natural product with anti-inflammatory properties. *Biochem. Pharmacol.* **1998**, *55*, 1955–1962. [CrossRef]
80. Youssef, K.M.; El-Sherbeny, M.A.; El-Shafie, F.S.; Farag, H.A.; Al-Deeb, O.A.; Awadalla, S.A.A. Synthesis of curcumin analogues as potential antioxidant, cancer chemopreventive agents. *Arch. Pharm. Int. J. Pharm. Med. Chem.* **2004**, *337*, 42–54. [CrossRef]
81. Simon, E.; Aswini, P.; Sameer Kumar, V.; Mankadath, G. Curcumin and its synthetic analogue dimethoxycurcumin differentially modulates antioxidant status of normal human peripheral blood mononuclear cells. *Free Radic. Res.* **2018**, *52*, 583–591. [CrossRef]
82. Tu, Z.-S.; Wang, Q.; Sun, D.-D.; Dai, F.; Zhou, B. Design, synthesis, and evaluation of curcumin derivatives as Nrf2 activators and cytoprotectors against oxidative death. *Eur. J. Med. Chem.* **2017**, *134*, 72–85. [CrossRef]
83. Liu, D.; He, B.; Lin, L.; Malhotra, A.; Yuan, N. Potential of curcumin and resveratrol as biochemical and biophysical modulators during lung cancer in rats. *Drug Chem. Toxicol.* **2019**, *42*, 328–334. [CrossRef]
84. Malhotra, A.; Nair, P.; Dhawan, D. Premature mitochondrial senescence and related ultrastructural changes during lung carcinogenesis modulation by curcumin and resveratrol. *Ultrastruct. Pathol.* **2012**, *36*, 179–184. [CrossRef] [PubMed]
85. Malhotra, A.; Nair, P.; Dhawan, D. Modulatory effects of curcumin and resveratrol on lung carcinogenesis in mice. *Phytother. Res.* **2010**, *24*, 1271–1277. [CrossRef]
86. Giordano, A.; Tommonaro, G. Curcumin and cancer. *Nutrients* **2019**, *11*, 2376. [CrossRef]
87. Gupta, S.C.; Prasad, S.; Kim, J.H.; Patchva, S.; Webb, L.J.; Priyadarsini, I.K.; Aggarwal, B.B. Multitargeting by curcumin as revealed by molecular interaction studies. *Nat. Prod. Rep.* **2011**, *28*, 1937–1955. [CrossRef]
88. Morin, D.; Barthélémy, S.; Zini, R.; Labidalle, S.; Tillement, J.-P. Curcumin induces the mitochondrial permeability transition pore mediated by membrane protein thiol oxidation. *FEBS Lett.* **2001**, *495*, 131–136. [CrossRef]
89. Watson, J.L.; Hill, R.; Yaffe, P.B.; Greenshields, A.; Walsh, M.; Lee, P.W.; Giacomantonio, C.A.; Hoskin, D.W. Curcumin causes superoxide anion production and p53-independent apoptosis in human colon cancer cells. *Cancer Lett.* **2010**, *297*, 1–8. [CrossRef]
90. Thayyullathil, F.; Chathoth, S.; Hago, A.; Patel, M.; Galadari, S. Rapid reactive oxygen species (ROS) generation induced by curcumin leads to caspase-dependent and-independent apoptosis in L929 cells. *Free Radic. Biol. Med.* **2008**, *45*, 1403–1412. [CrossRef] [PubMed]
91. Wang, L.; Chen, X.; Du, Z.; Li, G.; Chen, M.; Chen, X.; Liang, G.; Chen, T. Curcumin suppresses gastric tumor cell growth via ROS-mediated DNA polymerase γ depletion disrupting cellular bioenergetics. *J. Exp. Clin. Cancer Res.* **2017**, *36*, 1–14. [CrossRef] [PubMed]
92. Wang, C.; Song, X.; Shang, M.; Zou, W.; Zhang, M.; Wei, H.; Shao, H. Curcumin exerts cytotoxicity dependent on reactive oxygen species accumulation in non-small-cell lung cancer cells. *Future Oncol.* **2019**, *15*, 1243–1253. [CrossRef] [PubMed]
93. Mutlu Altundağ, E.; Yılmaz, A.M.; Koçtürk, S.; Taga, Y.; Yalçın, A.S. Synergistic induction of apoptosis by quercetin and curcumin in chronic myeloid leukemia (K562) cells. *Nutr. Cancer* **2018**, *70*, 97–108. [CrossRef] [PubMed]
94. Liao, W.; Xiang, W.; Wang, F.-F.; Wang, R.; Ding, Y. Curcumin inhibited growth of human melanoma A375 cells via inciting oxidative stress. *Biomed. Pharmacother.* **2017**, *95*, 1177–1186. [CrossRef] [PubMed]

95. Liang, T.; Zhang, X.; Xue, W.; Zhao, S.; Zhang, X.; Pei, J. Curcumin induced human gastric cancer BGC-823 cells apoptosis by ROS-mediated ASK1-MKK4-JNK stress signaling pathway. *Int. J. Mol. Sci.* **2014**, *15*, 15754–15765. [CrossRef] [PubMed]

96. Kizhakkayil, J.; Thayyullathil, F.; Chathoth, S.; Hago, A.; Patel, M.; Galadari, S. Glutathione regulates caspase-dependent ceramide production and curcumin-induced apoptosis in human leukemic cells. *Free Radic. Biol. Med.* **2012**, *52*, 1854–1864. [CrossRef]

97. Syng-ai, C.; Kumari, A.L.; Khar, A. Effect of curcumin on normal and tumor cells: Role of glutathione and bcl-2. *Mol. Cancer Ther.* **2004**, *3*, 1101–1108.

98. Berrak, Ö.; Akkoç, Y.; Arısan, E.D.; Çoker-Gürkan, A.; Obakan-Yerlikaya, P.; Palavan-Ünsal, N. The inhibition of PI3K and NFκB promoted curcumin-induced cell cycle arrest at G2/M via altering polyamine metabolism in Bcl-2 overexpressing MCF-7 breast cancer cells. *Biomed. Pharmacother.* **2016**, *77*, 150–160. [CrossRef]

99. Agarwal, A.; Kasinathan, A.; Ganesan, R.; Balasubramanian, A.; Bhaskaran, J.; Suresh, S.; Srinivasan, R.; Aravind, K.; Sivalingam, N. Curcumin induces apoptosis and cell cycle arrest via the activation of reactive oxygen species–independent mitochondrial apoptotic pathway in Smad4 and p53 mutated colon adenocarcinoma HT29 cells. *Nutr. Res.* **2018**, *51*, 67–81. [CrossRef]

100. Lestari, B.; Nakamae, I.; Yoneda-Kato, N.; Morimoto, T.; Kanaya, S.; Yokoyama, T.; Shionyu, M.; Shirai, T.; Meiyanto, E.; Kato, J.-y. Pentagamavunon-1 (PGV-1) inhibits ROS metabolic enzymes and suppresses tumor cell growth by inducing M phase (prometaphase) arrest and cell senescence. *Sci. Rep.* **2019**, *9*, 1–12. [CrossRef]

101. Luo, C.; Li, Y.; Zhou, B.; Yang, L.; Li, H.; Feng, Z.; Li, Y.; Long, J.; Liu, J. A monocarbonyl analogue of curcumin, 1, 5-bis (3-hydroxyphenyl)-1, 4-pentadiene-3-one (Ca 37), exhibits potent growth suppressive activity and enhances the inhibitory effect of curcumin on human prostate cancer cells. *Apoptosis* **2014**, *19*, 542–553. [CrossRef] [PubMed]

102. Chen, M.; Zhou, B.; Zhong, P.; Rajamanickam, V.; Dai, X.; Karvannan, K.; Zhou, H.; Zhang, X.; Liang, G. Increased Intracellular Reactive Oxygen Species Mediates the Anti-Cancer Effects of WZ35 via Activating Mitochondrial Apoptosis Pathway in Prostate Cancer Cells. *Prostate* **2017**, *77*, 489–504. [CrossRef] [PubMed]

103. Li, Y.; Zhang, L.-P.; Dai, F.; Yan, W.-J.; Wang, H.-B.; Tu, Z.-S.; Zhou, B. Hexamethoxylated monocarbonyl analogues of curcumin cause G2/M cell cycle arrest in NCI-H460 cells via Michael acceptor-dependent redox intervention. *J. Agric. Food Chem.* **2015**, *63*, 7731–7742. [CrossRef] [PubMed]

104. Shi, L.; Fei, X.; Wang, Z. Demethoxycurcumin was prior to temozolomide on inhibiting proliferation and induced apoptosis of glioblastoma stem cells. *Tumor Biol.* **2015**, *36*, 7107–7119. [CrossRef]

105. Nakamae, I.; Morimoto, T.; Shima, H.; Shionyu, M.; Fujiki, H.; Yoneda-Kato, N.; Yokoyama, T.; Kanaya, S.; Kakiuchi, K.; Shirai, T. Curcumin derivatives verify the essentiality of ROS upregulation in tumor suppression. *Molecules* **2019**, *24*, 4067. [CrossRef]

106. Liu, B.; Cui, L.-s.; Zhou, B.; Zhang, L.-l.; Liu, Z.-h.; Zhang, L. Monocarbonyl curcumin analog A2 potently inhibits angiogenesis by inducing ROS-dependent endothelial cell death. *Acta Pharmacol. Sin.* **2019**, *40*, 1412–1423. [CrossRef]

107. Guerrero, S.; Inostroza-Riquelme, M.; Contreras-Orellana, P.; Diaz-Garcia, V.; Lara, P.; Vivanco-Palma, A.; Cárdenas, A.; Miranda, V.; Robert, P.; Leyton, L. Curcumin-loaded nanoemulsion: A new safe and effective formulation to prevent tumor reincidence and metastasis. *Nanoscale* **2018**, *10*, 22612–22622. [CrossRef]

108. Larasati, Y.A.; Yoneda-Kato, N.; Nakamae, I.; Yokoyama, T.; Meiyanto, E.; Kato, J.-y. Curcumin targets multiple enzymes involved in the ROS metabolic pathway to suppress tumor cell growth. *Sci. Rep.* **2018**, *8*, 1–13. [CrossRef]

109. Khan, M.A.; Gahlot, S.; Majumdar, S. Oxidative stress induced by curcumin promotes the death of cutaneous T-cell lymphoma (HuT-78) by disrupting the function of several molecular targets. *Mol. Cancer Ther.* **2012**, *11*, 1873–1883. [CrossRef]

110. Lee, Y.J.; Kim, N.-Y.; Suh, Y.-A.; Lee, C. Involvement of ROS in curcumin-induced autophagic cell death. *Korean J. Physiol. Pharmacol.* **2011**, *15*, 1–7. [CrossRef]

111. Gersey, Z.C.; Rodriguez, G.A.; Barbarite, E.; Sanchez, A.; Walters, W.M.; Ohaeto, K.C.; Komotar, R.J.; Graham, R.M. Curcumin decreases malignant characteristics of glioblastoma stem cells via induction of reactive oxygen species. *BMC Cancer* **2017**, *17*, 1–11. [CrossRef] [PubMed]

112. Marquardt, J.U.; Gomez-Quiroz, L.; Camacho, L.O.A.; Pinna, F.; Lee, Y.-H.; Kitade, M.; Domínguez, M.P.; Castven, D.; Breuhahn, K.; Conner, E.A. Curcumin effectively inhibits oncogenic NF-κB signaling and restrains stemness features in liver cancer. *J. Hepatol.* **2015**, *63*, 661–669. [CrossRef] [PubMed]
113. Tan, B.L.; Norhaizan, M.E. Curcumin combination chemotherapy: The implication and efficacy in cancer. *Molecules* **2019**, *24*, 2527. [CrossRef] [PubMed]
114. Cheng, Y.; Zhao, P.; Wu, S.; Yang, T.; Chen, Y.; Zhang, X.; He, C.; Zheng, C.; Li, K.; Ma, X. Cisplatin and curcumin co-loaded nano-liposomes for the treatment of hepatocellular carcinoma. *Int. J. Pharm.* **2018**, *545*, 261–273. [CrossRef] [PubMed]
115. Paciello, F.; Fetoni, A.R.; Mezzogori, D.; Rolesi, R.; Di Pino, A.; Paludetti, G.; Grassi, C.; Troiani, D. The dual role of curcumin and ferulic acid in counteracting chemoresistance and cisplatin-induced ototoxicity. *Sci. Rep.* **2020**, *10*, 1–17. [CrossRef]
116. Park, B.H.; Lim, J.E.; Jeon, H.G.; Seo, S.I.; Lee, H.M.; Choi, H.Y.; Jeon, S.S.; Jeong, B.C. Curcumin potentiates antitumor activity of cisplatin in bladder cancer cell lines via ROS-mediated activation of ERK1/2. *Oncotarget* **2016**, *7*, 63870. [CrossRef]
117. Sankpal, U.T.; Nagaraju, G.P.; Gottipolu, S.R.; Hurtado, M.; Jordan, C.G.; Simecka, J.W.; Shoji, M.; El-Rayes, B.; Basha, R. Combination of tolfenamic acid and curcumin induces colon cancer cell growth inhibition through modulating specific transcription factors and reactive oxygen species. *Oncotarget* **2016**, *7*, 3186. [CrossRef]
118. Jung, E.M.; Park, J.-W.; Choi, K.S.; Park, J.-W.; Lee, H.I.; Lee, K.-S.; Kwon, T.K. Curcumin sensitizes tumor necrosis factor-related apoptosis-inducing ligand (TRAIL)-mediated apoptosis through CHOP-independent DR5 upregulation. *Carcinogenesis* **2006**, *27*, 2008–2017. [CrossRef]
119. Chatterjee, S.J.; Pandey, S. Chemo-resistant melanoma sensitized by tamoxifen to low dose curcumin treatment through induction of apoptosis and autophagy. *Cancer Biol. Ther.* **2011**, *11*, 216–228. [CrossRef]
120. Huang, Y.-F.; Zhu, D.-J.; Chen, X.-W.; Chen, Q.-K.; Luo, Z.-T.; Liu, C.-C.; Wang, G.-X.; Zhang, W.-J.; Liao, N.-Z. Curcumin enhances the effects of irinotecan on colorectal cancer cells through the generation of reactive oxygen species and activation of the endoplasmic reticulum stress pathway. *Oncotarget* **2017**, *8*, 40264. [CrossRef]
121. Chen, T.-C.; Chuang, J.-Y.; Ko, C.-Y.; Kao, T.-J.; Yang, P.-Y.; Yu, C.-H.; Liu, M.-S.; Hu, S.-L.; Tsai, Y.-T.; Chan, H. AR ubiquitination induced by the curcumin analog suppresses growth of temozolomide-resistant glioblastoma through disrupting GPX4-Mediated redox homeostasis. *Redox Biol.* **2020**, *30*, 101413. [CrossRef] [PubMed]
122. Moghtaderi, H.; Sepehri, H.; Attari, F. Combination of arabinogalactan and curcumin induces apoptosis in breast cancer cells in vitro and inhibits tumor growth via overexpression of p53 level in vivo. *Biomed. Pharmacother.* **2017**, *88*, 582–594. [CrossRef] [PubMed]
123. Mansourizadeh, F.; Alberti, D.; Bitonto, V.; Tripepi, M.; Sepehri, H.; Khoee, S.; Crich, S.G. Efficient synergistic combination effect of Quercetin with Curcumin on breast cancer cell apoptosis through their loading into Apo ferritin cavity. *Colloids Surf. B Biointerfaces* **2020**, *191*, 110982. [CrossRef] [PubMed]
124. Du, Q.; Hu, B.; An, H.-M.; Shen, K.-P.; Xu, L.; Deng, S.; Wei, M.-M. Synergistic anticancer effects of curcumin and resveratrol in Hepa1-6 hepatocellular carcinoma cells. *Oncol. Rep.* **2013**, *29*, 1851–1858. [CrossRef]

3

Cannabidiol Regulates the Expression of Keratinocyte Proteins Involved in the Inflammation Process through Transcriptional Regulation

Anna Jastrząb, Agnieszka Gęgotek and Elżbieta Skrzydlewska *

Department of Analytical Chemistry, Medical University of Bialystok, 15-089 Bialystok, Poland
* Correspondence: elzbieta.skrzydlewska@umb.edu.pl

Abstract: Cannabidiol (CBD), a natural phytocannabinoid without psychoactive effect, is a well-known anti-inflammatory and antioxidant compound. The possibility of its use in cytoprotection of cells from harmful factors, including ultraviolet (UV) radiation, is an area of ongoing investigation. Therefore, the aim of this study was to evaluate the effect of CBD on the regulatory mechanisms associated with the redox balance and inflammation in keratinocytes irradiated with UVA [30 J/cm^2] and UVB [60 mJ/cm^2]. Spectrophotometric results show that CBD significantly enhances the activity of antioxidant enzymes such as superoxide dismutase and thioredoxin reductase in UV irradiated keratinocytes. Furthermore, despite decreased glutathione peroxidase and reductase activities, CBD prevents lipid peroxidation, which was observed as a decreased level of 4-HNE and 15d-PGJ_2 (measured using GC/MS and LC/MS). Moreover, Western blot analysis of protein levels shows that, under stress conditions, CBD influences interactions of transcription factors Nrf2- NFκB by inhibiting the NFκB pathway, increasing the expression of Nrf2 activators and stimulating the transcription activity of Nrf2. In conclusion, the antioxidant activity of CBD through Nrf2 activation as well as its anti-inflammatory properties as an inhibitor of NFκB should be considered during design of new protective treatments for the skin.

Keywords: cannabidiol; UV radiation; keratinocytes; antioxidants; inflammation; intracellular signaling; Nrf2; NFκB

---

1. Introduction

It is believed that various disorders of human skin are the result of oxidative stress occurring in skin cells. Therefore, maintaining the redox balance is very important, especially because the skin remains in constant contact with the environment, plays an important role in the body's response to environmental factors, and also mediates the transmission of environmental signals to the organism [1]. Exposure to solar UV radiation is a causative factor in acute skin photodamage, chronic photoaging, and photocarcinogenesis [2,3]. UVA and UVB radiation have different biological effects on skin cells, but the common feature is to cause a redox imbalance with a shift towards oxidation. This is a result of increased reactive oxygen species (ROS) generation, which under physiological conditions are involved in cell signaling, but in the case of overproduction can also lead to cell damage. Cell damage is favored in the setting of UV radiation due to the reduction of effective endogenous antioxidant protection [4].

Oxidative stress caused by both endogenous and environmental factors reflects an imbalance between the number of oxidants produced in skin cells and the amount of antioxidant gene products (superoxide dismutase [SOD], catalase [CAT], glutathione [GSH], peroxidase [Px], etc.) maintaining the redox balance [5]. From this point of view, the expression of Nrf2, a transcription factor ubiquitously expressed in all tissues, including skin, which modulates the levels of cytoprotective protein expression, seems very important. Nrf2 protein levels and its activity are tightly regulated [6]. Activation is significantly related to the oxidative modification of the cysteine reactant Keap1, its cytosol inhibitor, which is responsible for targeting Nrf2 to ubiquitination and proteasomal degradation [7]. In contrast, the activation mechanisms of Nrf2 involve kinase signaling pathways, with Nrf2 phosphorylation being ultimately responsible for its translocation into the cell nucleus [8]. Natural compounds, such as sulforaphane and its derivatives, as well as various drugs, also play a role as activators or inhibitors of Nrf2 [9,10].

One of the natural compounds that can support the skin's antioxidant system is the main component of the Cannabis sativa extract, cannabidiol (CBD). CBD is a phytocannabinoid, which has no psychoactive effect, but aroused interest due to the therapeutic potential for many disease states that have been studied in animal models [11–14]. It has antioxidant, anti-inflammatory, anxiolytic, antitumor and anti-cancer properties [15,16]. Due to the chemical structure (Figure 1), CBD, like other antioxidants, interferes with the free radical chain reactions mainly in the propagation and the final phase, capturing free radicals or transforming them into less reactive forms [17]. In addition, CBD leads to a decrease in oxidative conditions in cells, which is associated with chelation of transition metals and reduction of prooxidative enzymes, which leads to prevention of ROS [18]. Notwithstanding the reduction of levels of oxidants, CBD modifies the redox balance also by changing the level/activity of antioxidants [19].

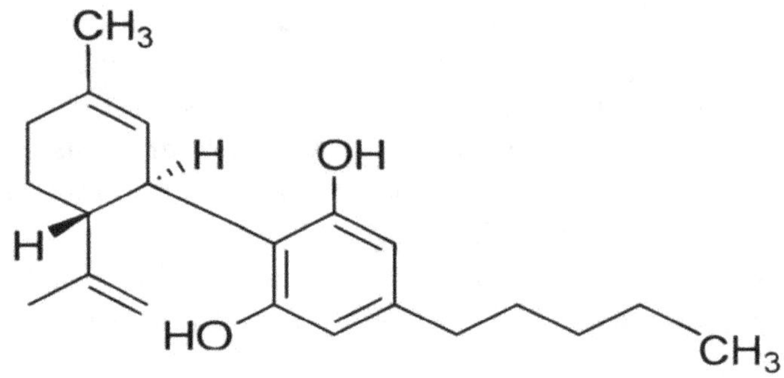

Figure 1. Cannabidiol chemical structure.

CBD, like endocannabinoids, acts not only directly, but also indirectly through receptors. By activating A2A adenosine receptors, it inhibits the NFκB inflammatory pathway and reduces the level of proinflammatory cytokines, including TNFα [20]. This phytocannabinoid has a low affinity for cannabinoid receptor 1 (CB1) and behaves like an incompetent antagonist; therefore, it can increase stimulation of the cannabinoid system. It inhibits the anandamide transport and its degradation of by fatty acid amide hydrolase [21]. In addition, CBD exhibits weak inverse agonism towards the cannabinoid receptor 2 (CB2). It can act as an "indirect" CB1/CB2 agonist by weak inhibition of anandamide enzymatic hydrolysis [22]. CBD may also have effects at the level of other endocannabinoids, including 2-arachidonylglicerol (2-AG) [23]. Moreover, it modulates the activity of G protein-coupled receptors (GPCRs) that are activated by endocannabinoids and related compounds [24].

The broad action of CBD does not remain indifferent to the genetic material. It has been found that CBD by inhibition of the DNA repair enzyme PARP-1 can induce cytotoxicity [25] and acts destructively in relation of DNA of cancer-transformed cells therefore CBD has been suggested as a promising antitumor factor [26]. Moreover, anti-inflammatory and modulatory effects of CBD promote the attenuation of various autoimmune conditions in animal models, including skin diseases such as psoriasis [27]. In addition, by affecting skin cells, it inhibits the proliferation and differentiation of human keratinocytes via CB1 and CB2 independent epigenetic mechanisms [28].

Therefore, the aim of this study was to evaluate the effect of CBD on regulatory mechanisms related to redox balance and inflammation in keratinocytes after UVA and UVB irradiation.

2. Materials and Methods

2.1. Cell Culture and Treatment

Human keratinocytes (CDD 1102 KERTr) were obtained from American Type Culture Collection (ATCC, Virginia, USA). Keratinocytes were cultured in Keratinocyte Serum-Free Medium (Gibco, Grand Island, NY) that contained fetal bovine serum (10%), epidermal growth factor (EGF 1–53; 5 μg/L), 50 U/mL penicillin and 50 μg/mL streptomycin. Cells were cultured in a humidified atmosphere with 5% CO_2 at 37 °C. When the cells (passage 9–11) reached 70% confluency, they were washed with warm PBS (37 °C). Cells were exposed to UV radiation in cold PBS (4 °C) to avoid heat stress and oxidation of the medium components. The exposure dose that corresponded to 70% cell viability measured by the MTT assay was used [29]. The cells were irradiated on ice at a distance of 15 cm from the assembly of 6 lamps (Bio-Link Crosslinker BLX 312/365; Vilber Lourmat, Germany), 6W each, which corresponds to 4.2 mW/cm^2 and 4.08 mW/cm^2, respectively for UVA (365 nm) and UVB (312 nm). Total radiation doses were 30 J/cm^2 UVA and 60 mJ/cm^2 UVB. After radiation, cells were incubated for 24 h under standard conditions without rinsing; control cells that did not receive irradiation were incubated in parallel.

To examine the effect of CBD on keratinocytes all cell groups, control cells and cells after UVA and UVB irradiation were cultured in medium containing 1 μM CBD (THC-Pharm, Frankfurt, Germany). After 24 h incubation, all cells were washed with PBS, collected by scraping into cold PBS and centrifuged. Cells were then resuspended in PBS and subjected to sonication. Total protein content in cell lysate was measured using a Bradford assay [30].

2.2. Cell Viability

To examine the effect of CBD on UV irradiated keratinocytes, cells were incubated after irradiation for 24 h under standard conditions in medium containing 0, 0.1, 0.5, 1, 2, 4, 10, 25, 50, 100 nmoles/mL CBD in ethanol. The MTT assay was used to examine the effect of CBD concentration on keratinocyte viability compared to control cells [29].

2.3. Determination of Superoxide Anion Generation

The generation of superoxide anions was detected using an electron spin resonance (ESR) spectrometer e-scan (Noxygen GmbH/Bruker Biospin GmbH, Berlin, Germany), where selective interaction of the spin probes CMH (1-hydroxy-3-methoxy-carbonyl-2,2,5,5-tetrame-thylpyrrolidine, 200 μM) with ROS formed a stable nitroxide CM-radical with a half-life of 4 h. Thus, ROS formation was measured by assessing the kinetics of nitroxide accumulation based on the electron spin resonance amplitude of the low field component of ESR spectra. The rate of superoxide radical formation was determined by measuring superoxide dismutase (SOD)-inhibited nitroxide generation [31].

2.4. Determination of Antioxidant Enzyme Activity

SOD (Cu/Zn–SOD, EC.1.15.1.1) activity was determined according to the method of Misra and Fridovich [32] as modified by Sykes and co-authors [33], which measures the activity of cytosolic SOD. One unit of SOD was defined as the amount of the enzyme that inhibits epinephrine oxidation to adrenochrome by 50%. Enzyme specific activity was expressed in units per milligram of protein.

Glutathione peroxidase (GSH-Px, EC.1.11.1.6) activity was assessed spectrophotometrically using the method of Paglia and Valentine [34]. GSH-Px activity was assayed by measuring the conversion of NADPH to NADP$^+$. One unit of GSH-Px activity was defined as the amount of enzyme catalyzing the oxidation of 1 μmol NADPH/min at 25 °C and pH 7.4. Enzyme specific activity was expressed in units per mg of protein.

Glutathione reductase (GSSG-R, EC.1.6.4.2) activity was measured according to the method of Mize and Longdon [35] by monitoring the oxidation of NADPH at 340 nm. One unit of GSSG-R oxidized 1 mmol of NADPH/min at 25 °C and pH 7.4. Enzyme specific activity was expressed in milliunits per milligram of protein.

The thioredoxin reductase (TrxR, EC. 1.8.1.9) activity was measured using a commercially available kit (Sigma-Aldrich, St. Louis, MO, USA). The principle on which this kit is based is the NADPH-mediated reduction of 5,5′-dithiobis(2-nitrobenzoic) acid (DTNB) to 5-thio-2-nitrobenzoic acid (TNB), which produces a strong yellow color that is measured at 412 nm [36]. Enzyme activity was expressed in units per milligram of protein.

2.5. Determination of Non-Enzymatic Antioxidant Level

Glutathione was quantified using the capillary electrophoresis (CE) method of Maeso and co-authors [37]. Samples were sonicated in Eppendorf tubes with 2 mL of a mixture containing ACN/H_2O (62.5:37.5, *v*/*v*) and centrifuged at 29,620× *g* for 10 min. The supernatant was immediately subjected to CE. The separation was performed on a capillary with 47 cm total length (40 cm effective length) and 50 m i.d. and was operated at 27 kV with UV detection at 200 ± 10 nm. The GSH concentration was determined using a calibration curve range: 1–120 nmol/L (r^2 = 0.9985) and the level of GSH was expressed as nanomoles per milligram of protein.

Thioredoxin level was quantified using the ELISA method [38]. Prepared standards and cell lysates were loaded into ELISA plate wells (Nunc Immuno Maxisorp, Thermo Scientific, Waltham, MA, USA) and incubated at 4 °C overnight with anti-thioredoxin primary antibody (Abcam, Cambridge, MA, USA). After washing, the plates were incubated for 30 min with peroxidase blocking solution (3% H_2O_2, 3% fat free dry milk in PBS) at RT. The goat anti-rabbit secondary antibody solution (Dako, Carpinteria, CA, USA) was then added to each well, and the plates were incubated for 1 h at RT. Chromogen substrate solution (0.1 mg/mL TMB, 0.012% H_2O_2) was added to each well, and the plates were incubated for 40 min at RT. The reaction was stopped by addition of 2M sulfuric acid. Absorption was read at 450 nm with the reference filter set to 620 nm. Level of thioredoxin was expressed as the micrograms per milligram of protein.

2.6. Determination of Lipid Peroxidation Product

Lipid peroxidation was estimated by measuring the level of 4-hydroxynonenal (4-HNE). Aldehyde was measured by GC/MSMS (Agilent Technologies, Palo Alto, CA, USA), as the *O*-PFB-oxime-TMS derivative, using a modified method of Liu and co-authors [39]. Benzaldehyde-D_6 was added to the cell lysates as an internal standard, and aldehydes were derivatized by the addition of *O*-(2,3,4,5,6-pentafluoro-benzyl) hydroxyamine hydrochloride (0.05 M in PIPES buffer, 200 μL) and incubating for 60 min at RT. After incubation, samples were deproteinized by the addition of 1 mL of methanol and *O*-PFB-oxime aldehyde derivative was extracted by the addition of 2 mL of hexane. The top hexane layer was transferred into borosilicate tubes, and evaporated under a stream of argon gas, followed by the addition of *N*,*O*-bis(trimethylsilyl)trifluoroacetamide in 1% trimethylchlorosilane. A 1 μL aliquot was injected on the column. Derivatized aldehyde was analyzed using a 7890A GC–7000 quadruple MS/MS (Agilent Technologies, Palo Alto, CA) equipped with a HP-5ms capillary column (0.25 mm internal diameter, 0.25 μm film thickness, 30 m length). Derivatized aldehyde was detected in selected ion monitoring (SIM) mode. The ions used were: *m*/*z* 333.0 and 181.0 for 4-HNE-PFB-TMS and *m*/*z* 307.0 for IS derivative. Level of 4-HNE was expressed as the nanomoles per milligram of protein.

2.7. Determination Of Anti-Inflammatory Eicosanoid

15-deoxy-Δ12,14-prostaglandin J2 (15d-PGJ2) was estimated using ultra-performing liquid chromatography tandem mass spectrometry (LCMS 8060, Shimadzu, Kyoto, Japan) [40]. PGF2α-d4 was used as an internal standard for quantification. Lipid mediators were extracted using SPE and analyzed in negative-ion mode (MRM). The precursor to the product ion transition was *m*/*z* 315.2→271.2 for 15d-PGJ2. The level of 15d-PGJ2 was expressed as picomoles per milligram of protein.

2.8. Determination of Protein Expression

Western blot analysis of cell protein was performed according to Eissa and Seada [41]. Whole cell lysates or membrane fractions were mixed with sample loading buffer (Laemmli buffer containing 5% 2-mercaptoethanol), heated at 95 °C for 10 min, and separated using 10% Tris-Glycine SDS-PAGE. Separated proteins in the gels were electrophoretically transferred to nitrocellulose membrane. The membrane was blocked with 5% skim milk in TBS-T buffer (Tris buffered saline with 5% Tween-20) for 1 h. Primary monoclonal antibodies raised against phospho-Nrf2 (pSer40), HO-1, Keap1, WTX, DPP3, CBP, TNFα, NFκB (p52), IKKα, IKKβ, phosphor-IκB (pSer36), p38, phospho-MAPK (phospho-p44/42), phospho-ASK-1 (pThr845) PGAM5, NLRP3, β-actin (Sigma-Aldrich), and Bach1, KAP1, p21, p62, Ref-1, NFκB (p65) (Santa Cruz Biotechnology, Santa Cruz, CA, USA) were used at a concentration of 1:1000. Alkaline phosphatase-labeled secondary antibodies against mouse, rabbit or goat were used at a concentration of 1:1000 (Sigma-Aldrich, St. Louis, MO, USA). Protein bands were visualized using the BCIP/NBT Liquid substrate system (Sigma-Aldrich) and were quantitated using the Versa Doc System and Quantity One software (Bio-Rad Laboratories Inc., CA, USA). Level of proteins was expressed as the percentage of the value determined from the control cells

2.9. Determination of Keap1 Structure

Human Keap1 (Sino Biological, cat. 11981-H20B) was dissolved in 25 mM AMBIC and divided into portions containing 30 μg of protein. One sample was incubated for 1 h at 4 °C with 10 μM CBD, while an untreated sample was used as a control. Each sample was digested with trypsin (1:50) for 4 h at 37 °C. Digestion was stopped by addition of 10% formic acid (FA) to a final concentration 0.1%. Samples were dried, resuspended in 50 μl ACN + 0.1% FA, and separated using an Ultimate 3000 (Dionex, Idstein, Germany) onto a 150 mm × 75 mm PepMap RSLC capillary analytical C18 column with 2 μm particle size (Dionex, LC Packings, Idstein, Germany). The peptides eluted from the column were analyzed using a Q Exactive HF mass spectrometer with an electrospray ionization source (ESI) (Thermo Fisher Scientific, Bremen, Germany) as described previously [42]. Data were acquired with the Xcalibur software (Thermo Fisher Scientific, Bremen, Germany).

2.10. Determination of Protein Localization

Cells were seeded in BD Falcon™ 96-well clear bottom black tissue culture plates optimized for imaging applications at 10,000 cells per well. After treatment with UV radiation and/or a 24 h incubation with CBD, cells were fixed with 3.7% formaldehyde solution for 10 min and permeabilized with 0.1% Triton X-100 solution for 5 min. Next, the cells were washed twice with PBS, and non-specific binding was blocked by adding a 3% FBS solution for 30 min. After that time, the cells were incubated with an anti-Nrf2 rabbit polyclonal antibody or an anti-NFκB mouse polyclonal antibody (Sigma-Aldrich; 1:1000) for 1 h at RT, washed three times with PBS and incubated with fluorescent (FITC) anti-rabbit or anti-mouse secondary antibody (BD Pharmingen, San Diego, CA) for 60 min in the dark. After washing, nuclei were stained with Hoechst 33342 (2 μg/mL) and analyzed using microscopy imaging. Images of FITC-labeled cells were acquired using a 488/10 excitation laser and a 515LP emission laser.

2.11. Statistical Analysis

Data were analyzed using standard statistical analyses, including ANOVA, and the results are expressed as the mean ± standard deviation (SD) for $n = 5$. P-values less than 0.05 were considered significant.

3. Results

The assessment of the effect of different concentrations of CBD on the survival of keratinocytes by means of the MTT test indicated that concentrations of up to 1 μM CBD did not affect the survival of control keratinocytes (Figure 2). Higher concentrations caused a gradual decrease in cell survival, so that at a concentration of 10 μM it reached values of about 75% of control cells viability. In the case of cells irradiated with UVA radiation, both in the absence of CBD and in the presence of CBD up to 4 μM, the survival rate was maintained at 70%, and at higher concentrations was further reduced. In contrast, keratinocytes irradiated with UVB radiation were characterized by approximately 70% survival from 0 to 10 μM CBD and at higher CBD concentrations, the viability was somewhat reduced.

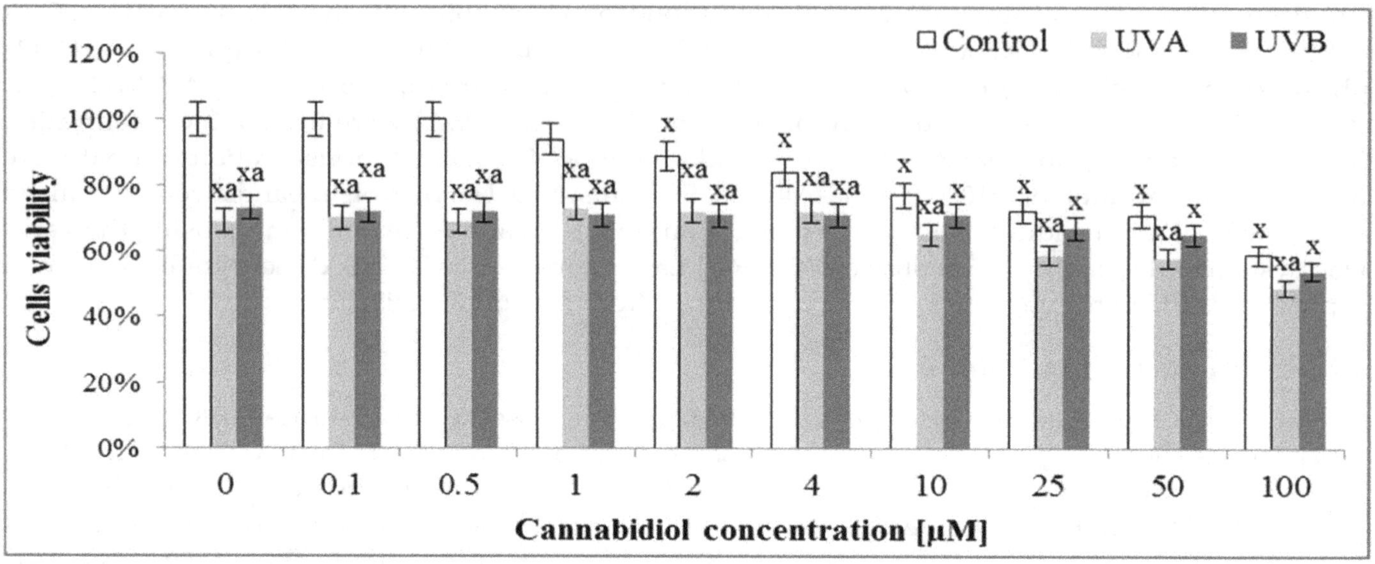

Figure 2. The influence of cannabidiol (CBD) on viability of keratinocytes exposed to UVA (30 J/cm^2) and UVB (60 mJ/cm^2) irradiation measured by MTT test. Mean values ± SD of five independent experiments are presented: ^{x} statistically significant differences vs. group without radiation or chemical treatment, $p < 0.05$; ^{a} statistically significant differences vs. UVA or UVB group, $p < 0.05$.

Results showed that keratinocytes irradiated with UVA or UVB were characterized by an altered redox balance associated with a more than 2-fold increase in superoxide anion production and a significant decrease in Cu, Zn-SOD and TrxR activity, especially after UVB irradiation (Table 1). Moreover, UV radiation enhanced the cytoprotective factors phospho-ASK-1 and Ref-1 expression by about 2- and 3-fold (Table 1, Figure 3), respectively. These changes were accompanied by a significant reduction in the level of non-enzymatic antioxidant parameters, such as GSH and Trx. As a consequence, despite the increase in the activity of enzymes involved in the GSH metabolism, oxidative metabolism of phospholipids increased 2-3 times after UVA or UVB exposure, which was manifested by an increased level of the lipid peroxidation product, 4-HNE. Treating keratinocytes with CBD significantly reduced the production of ROS, especially after UVA irradiation. In addition, CBD increased the antioxidant capacity manifested by significantly increased activity of Cu, Zn-SOD and TrxR and the level of Trx. On the other hand, the UV-increased Ref-1 and phospho-ASK-1 expression was partially restored following CBD treatment. However, after the use of CBD, the GSH-Px and GSSG-R activity was reduced, as well as GSH levels in control or irradiated cells. The consequence of the observed CBD-induced changes in redox parameters was the reduction in the intensity of lipid peroxidation, which was seen as a significant decrease in the level of 4-HNE, especially in keratinocytes treated with CBD after UVA irradiation.

Table 1. The influence of cannabidiol (CBD) on ROS generation (by ESR) and antioxidant capacity estimated as protein (Cu,Zn-SOD, GSH-Px, GSSG-R, TrxR by spectrometry) and non-protein antioxidants (GSH by CE, Trx by Elisa, Ref-1 and phospho-ASK-1 by Western blotting) as well as lipid peroxidation (4-HNE by GCMS) of keratinocytes exposed to UVA (30 J/cm^2) and UVB (60 mJ/cm^2) irradiation.

Determined Parameter	Groups					
	Control	CBD	UVA	UVA + CBD	UVB	UVB + CBD
Superoxide anion (nmol/min/mg protein)	90 ± 7	58 ± 5 ^{x}	240 ± 12 ^{x}	87 ± 9 ^{a}	216 ± 11 ^{x}	105 ± 9 ^{x,a}
Cu.Zn-SOD [mU/mg protein]	28.0 ± 1.5	48.4 ± 3.1 ^{x}	23.1 ± 1.4 ^{x}	51.4 ± 2.9 ^{x,a}	21.5 ± 1.3 ^{x}	53.6 ± 3.1 ^{x,a}
GSH-Px [U/mg protein]	14.6 ± 0.9	13.5 ± 1.2	34.4 ± 1.9 ^{x}	22.8 ± 1.7 ^{x,a}	37.5 ± 2.2 ^{x}	26.5 ± 1.5 ^{x,a}
GSSG-R [mU/mg protein]	14.3 ± 1.2	7.2 ± 0.4 ^{x}	18.6 ± 0.9 ^{x}	26.5 ± 1.8 ^{x,a}	20.2 ± 1.0	14.7 ± 0.9
GSH [nmol/mg protein]	17.1 ± 0.8	11.7 ± 0.7 ^{x}	11.5 ± 0.6 ^{x}	10.3 ± 0.6 ^{x,a}	10.9 ± 0.5 ^{x}	9.3 ± 0.5 ^{x,a}
TxrR [U/mg protein]	14.0 ± 0.7	17.2 ± 0.8 ^{x}	10.2 ± 0.6 ^{x}	16.1 ± 0.7 ^{x,a}	8.2 ± 0.3 ^{x}	15.4 ± 0.6 ^{x,a}
Trx [g/mg protein]	1.63 ± 0.09	2.74 ± 0.17 ^{x}	1.05 ± 0.05 ^{x}	1.79 ± 0.11 ^{x,a}	0.84 ± 0.04 ^{x}	1.16 ± 0.06 ^{x,a}
Ref-1 [vs. control]	1.00 ± 0.06	0.96 ± 0.04	3.10 ± 0.07 ^{x}	1.60 ± 0.06 ^{x,a}	3.10 ± 0.09 ^{x}	1.70 ± 0.06 ^{x,a}
phospho-ASK-1 [vs. control]	1.00 ± 0.04	1.10 ± 0.04	2.10 ± 0.07 ^{x}	1.70 ± 0.06 ^{x,a}	2.30 ± 0.09 ^{x}	1.60 ± 0.06 ^{x,a}
4-HNE [nmol/mg protein]	26.1 ± 1.3	23.5 ± 1.4	52.4 ± 2.7 ^{x}	35.6 ± 1.9 ^{x,a}	73.4 ± 3.7 ^{x}	64.7 ± 3.5 ^{x,a}

Mean values ± SD of five independent experiments are presented. ^{x} statistically significant differences vs. control group, $p < 0.05$; ^{a} statistically significant differences vs. UVA or UVB group, $p < 0.05$.

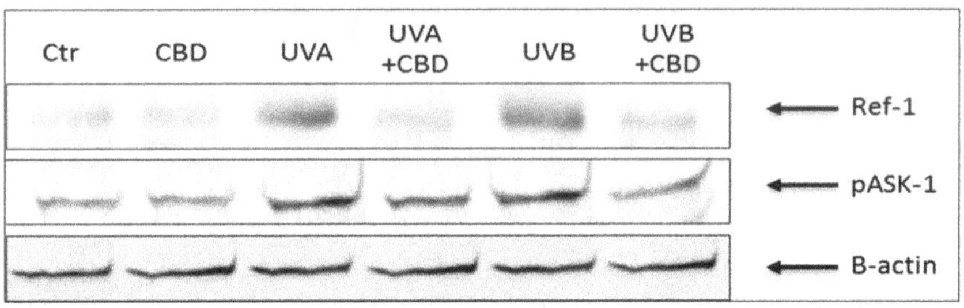

Figure 3. The influence of cannabidiol (CBD) on Ref-1 and phospho-ASK-1 (pThr845) levels in keratinocytes exposed to UVA (30 J/cm^2) and UVB (60 mJ/cm^2) irradiation.

Irradiation of keratinocytes with UV not only changed the activity and level of antioxidants, but also the expression and transcriptional activity of Nrf2 and NFκB, two transcription factors whose biological activity depends on each other and which participate in the modulation of redox imbalance and inflammation. UVA and UVB radiation increased the expression of NFκB (p52 and p65), while CBD treatment only partially, but significantly reduced this increase (Figure 4). However, keratinocytes treated with CBD alone were characterized by increased expression of NFκB. In addition, CBD used alone or after irradiation enhanced the expression of TNFα, an NFκB transcriptional activity product. Analyzing the NFκB activators and inhibitors, it can be concluded that the expression of the phosphorylated form of the basic NFκB inhibitor, IκB, was lowered by UV radiation and was significantly increased by CBD, particularly after UVB irradiation. In contrast, in the case of NFκB activators, CBD acted in a different way. It inhibited the expression of factors associated with the activation of inflammation, such as NLRP3 and PGAM5, and simultaneously activated IKK components, especially IKKα, whose expression was lowered by UV radiation as well as the protein p62 that is a direct linking element of the NFκB and Nrf2 pathways.

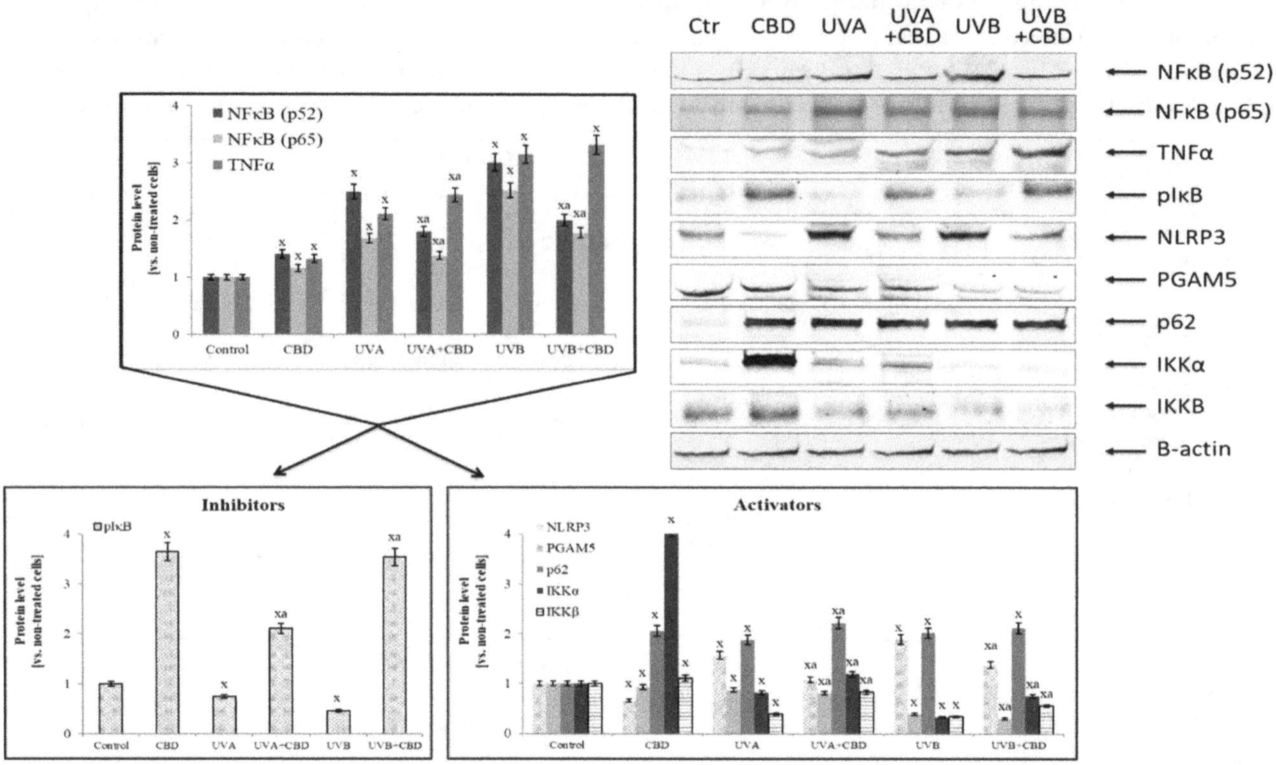

Figure 4. The influence of cannabidiol (CBD) on NFκB [p52] and NFκB [p65] transcriptional activity and its activators and inhibitors expression in keratinocytes exposed to UVA (30 J/cm^2) and UVB (60 mJ/cm^2) irradiation. Mean values ± SD of five independent experiments are presented. ^{x} statistically significant differences vs. control group, $p < 0.05$; ^{a} statistically significant differences vs. group without cannabidiol, $p < 0.05$.

Increased expression of NFκB (p52 and p65) by CBD promoted the increased expression of Nrf2 responsible for the transcription of cytoprotective proteins, which is confirmed by the increased biosynthesis of HO-1 and Trx and TrxR (Table 1, Figure 5). At the same time, the level of basic inhibitors Nrf2, cytosolic Keap1 and the nuclear Bach1 was decreased, as well as the level of PGAM5 that participate in inflammatory processes. Moreover, CBD stimulated the expression of Nrf2 activators (such as KAP1, p21, p62) in control keratinocytes two-fold. In contrast, in cells irradiated with UV, CBD only slightly increased the expression of p21 and p62 proteins.

Analyzing the decrease in the expression of the basic Nrf2 cytosolic inhibitor Keap1, it can be concluded that it was associated with a disruption of both inhibitors and activators of this factor (Table 2). In the case of the activators, the most unambiguous changes were observed in the DPP3 level, the expression of which was increased two-fold in all groups after the use of CBD. On the other hand, CBD significantly reduced the expression of another Keap1 activator, protein WTX. In the case of Keap1 inhibitors, CBD strongly increased 15d-PGJ$_2$ level in the control cells, but after exposure to UV radiation the CBD effect was observed as a reduction in the level of this eicosanoid.

In contrast, CBD in all cases reduced the level of inhibitors such as 4-HNE and PGAM5. Additionally, CBD influenced Keap1 activity by adduct formation on cysteine-288, which is responsible for Keap1 conformation and its interaction with Nrf2, as well as on cysteine-151 that is present in a Keap1 region responsible for Cul3 binding and indirectly contributes to Nrf2 ubiquitination (Figure 6).

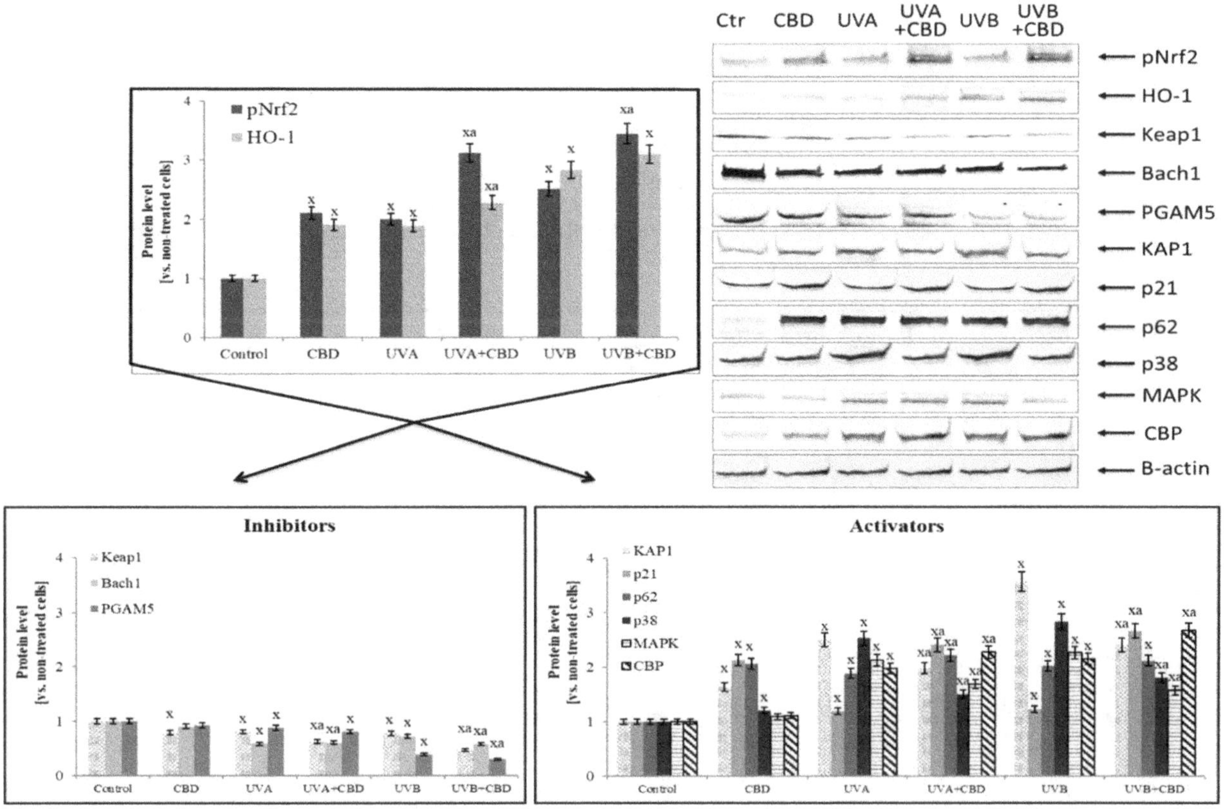

Figure 5. Effect of cannabidiol (CBD) on Nrf2 transcriptional activity and the level of its activators and inhibitors in keratinocytes exposed to UVA (30 J/cm^2) and UVB (60 mJ/cm^2) irradiation. Mean values ± SD of five independent experiments are presented: ^{x} statistically significant differences vs. control group, $p < 0.05$; ^{a} statistically significant differences vs. group without cannabidiol, $p < 0.05$.

Table 2. Effect of cannabidiol (CBD) on Keap1 expression and the level of its activators (determined by Western blotting) and inhibitors (determined by Western blotting–PGAM5 and by LCMS–4-HNE and 15dPGJ$_2$) in keratinocytes exposed to UVA (30 J/cm^2) and UVB (60 mJ/cm^2) irradiation.

Determined Parameter	Groups					
	Control	CBD	UVA	UVA + CBD	UVB	UVB + CBD
Keap1 [vs. control]	1.00 ± 0.05	0.79 ± 0.04 ^{x}	0.81 ± 0.04 ^{x}	0.63 ± 0.03 ^{x,a}	0.78 ± 0.04 ^{x}	0.47 ± 0.02 ^{x,a}
Inhibitors						
4-HNE [nmol/mg protein]	26.1 ± 1.3	23.5 ± 1.4	52.4 ± 2.7 ^{x}	35.6 ± 1.9 ^{x,a}	73.4 ± 3.7 ^{x}	64.7 ± 3.5 ^{x,a}
15d-PGJ$_2$ [pg/mg protein]	6.2 ± 0.36	11.8 ± 0.86 ^{x}	13.8 ± 0.81 ^{x}	10.1 ± 0.60 ^{x,a}	15.4 ± 0.91 ^{x}	11.2 ± 0.72 ^{x,a}
PGAM5 [vs. control]	1.00 ± 0.04	0.93 ± 0.05	0.88 ± 0.04 ^{x}	0.81 ± 0.04	0.39 ± 0.02 ^{x}	0.30 ± 0.02 ^{x,a}
Activators						
WTX [vs. control]	1.00 ± 0.04	0.92 ± 0.05	0.82 ± 0.04 ^{x}	0.75 ± 0.04 ^{x}	0.75 ± 0.04 ^{x}	0.45 ± 0.02 ^{x,a}
DPP3 [vs. control]	1.00 ± 0.05	1.81 ± 0.09 ^{x}	0.66 ± 0.03 ^{x}	1.47 ± 0.07 ^{x,a}	0.44 ± 0.02 ^{x}	1.83 ± 0.11 ^{x,a}

Mean values ± SD of five independent experiments are presented. ^{x} statistically significant differences vs. control group, $p < 0.05$; ^{a} statistically significant differences vs. group without cannabidiol, $p < 0.05$.

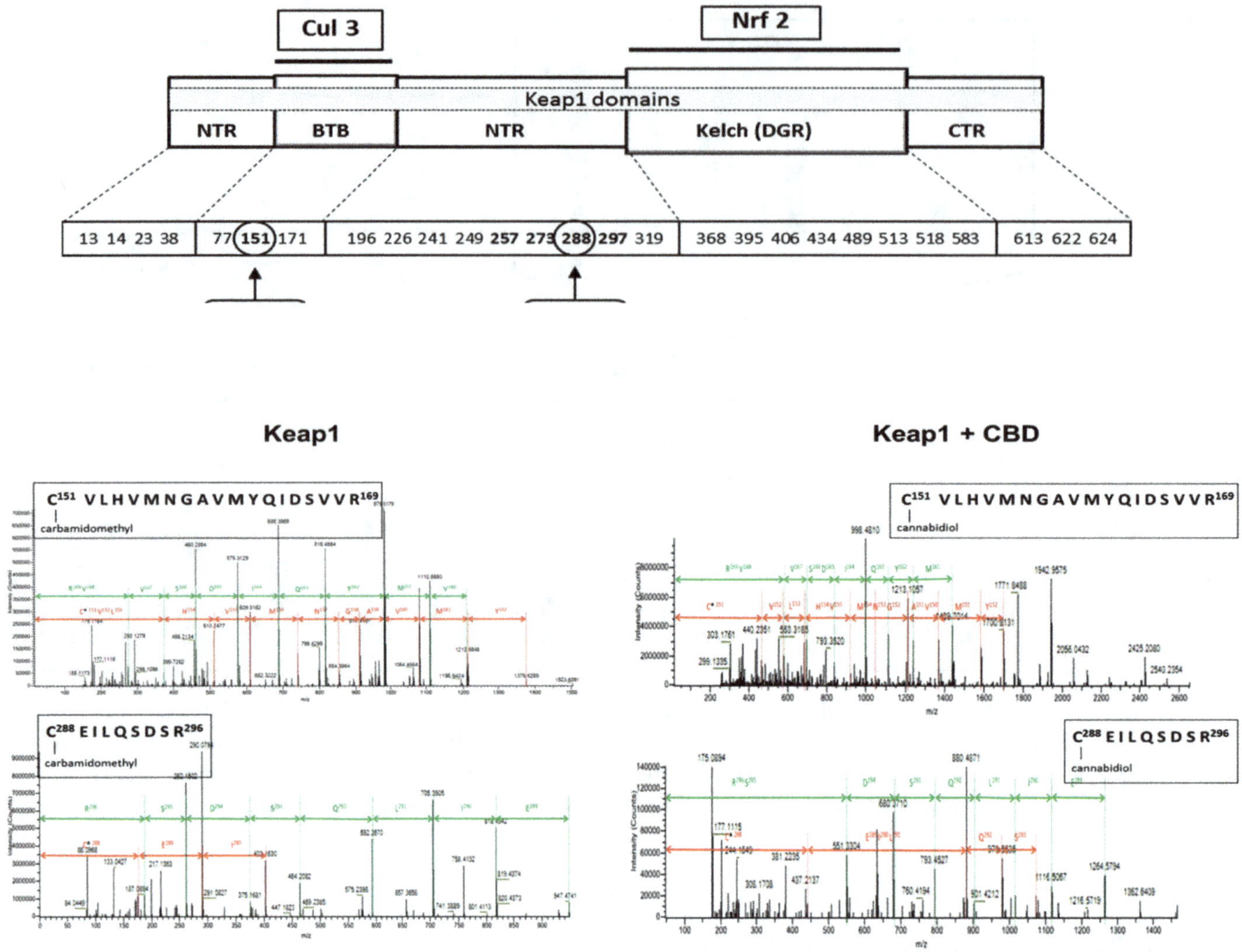

Figure 6. The localization of cysteine residues within Keap1 domains that form adducts with cannabidiol (CBD). Bolded cysteine residues are the most sensitive for oxidation.

Moreover, observed changes in proteins levels were accompanied with changes in translocation of transcriptional factors (Nrf2 and NFκB) to the nucleus. CBD led to the Nrf2 nuclear translocation in non-irradiated cells (Figure 7). However, the UVA induced Nrf2 nuclear translocation was slightly reduced by CBD, which was observed as increased Nrf2 level mainly in the cytosol. This effect was not observed in the case of cells treated with UVB and CBD. Different results were observed in the case of NFκB, where CBD treatment slightly enhanced NFκB levels in the cytoplasm but did not lead to its nuclear translocation. In the case of keratinocytes exposed to UV radiation the significant increase in NFκB level in the nucleus was observed, however, CBD treatment following UV radiation enhanced NFκB level in whole cells, but, especially in the case of UVA, prevented its accumulation in the nucleus.

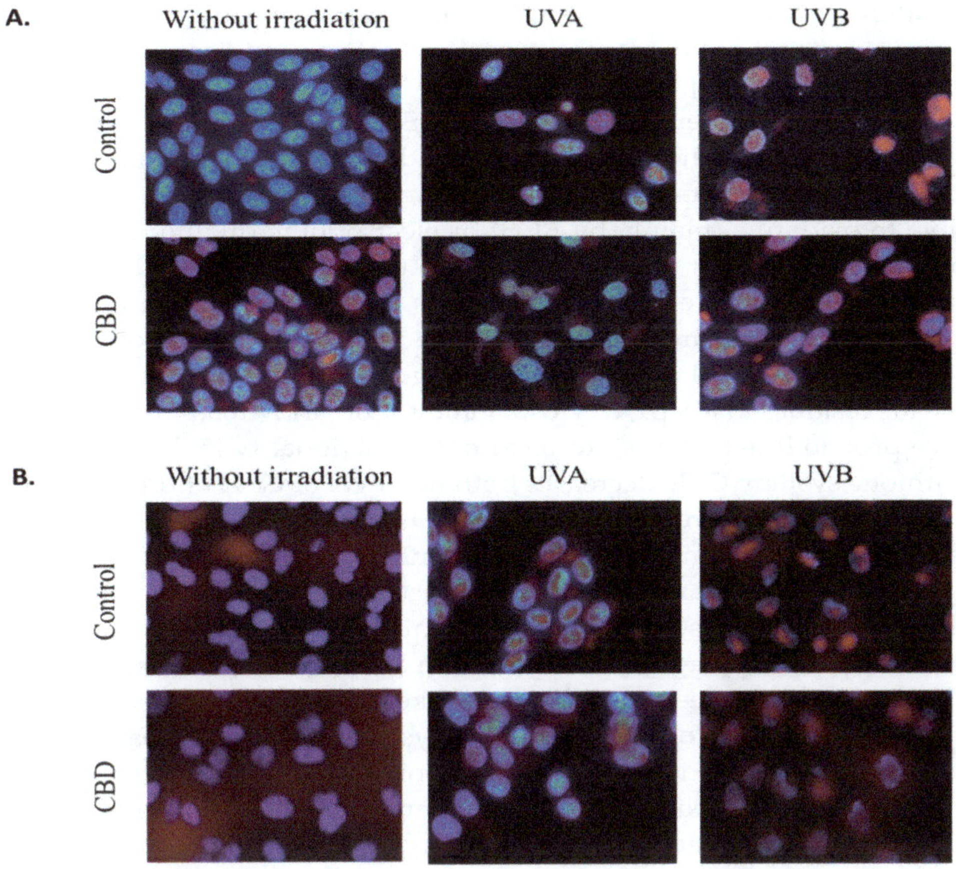

Figure 7. The influence of cannabidiol (CBD, 1 μM) on Nrf2 (**A**) and NFκB (p52) (**B**) nuclear localization in keratinocytes (control and exposed to UVA (30 J/cm^2) and UVB (60 mJ/cm^2) radiation). Nuclei were stained with Hoechst 33342 (blue) and antibodies against Nrf2 or NFκB (p52) were detected with fluorescent (FITC) secondary antibody (red dots).

4. Discussion

UVA and UVB radiation show different biological effects on skin cells, but the common characteristic of their action is to cause a redox imbalance, with a shift towards oxidation reactions [4]. This situation leads to the formation of oxidative stress as a result of which the metabolism of membrane phospholipids is increased and causes intensified generation of lipid mediators, which are products of reactions dependent on ROS or enzymes [43,44]. The main group of compounds generated during the enzymatic metabolism of phospholipids are endocannabinoids, which indirectly, by activating cannabinoid receptors (CB1/2), may modify the redox balance and inflammation in the cells, including UV irradiated cells [4,45,46]. Therefore, cannabinoids appear to be very promising therapeutic compounds.

One of the naturally occurring phytocannabinoids, and therefore a compound with potential protective effects, is CBD, which has antioxidant and anti-inflammatory properties [11,12]. It was demonstrated that this cannabinoid is a stronger antioxidant than α-tocopherol and ascorbic acid [47]. It has been also shown that in various diseases CBD reduces the intensity of ROS generation and oxidative stress formation, which decreases lipid mediators [48,49].

The results of this study indicate that CBD used under physiological conditions causes slight increases in the level of proinflammatory factors in the NFκB pathway and partially modifies the cytoprotective activity of Nrf2 in keratinocytes. This effect is not visible in cells with UV-induced metabolic stress, in which CBD partially normalizes redox balance and reduces inflammation in keratinocytes exposed to UVA and UVB radiation, which experience oxidative stress resulting from excessive production of ROS and consequently enhanced lipid peroxidation with 4-HNE as a main reactive electrophilic aldehyde as well as reduced antioxidant capacity, which reveals cytoprotective CBD (anti-inflammatory and antioxidant) properties that may indicate the need to adapt cells to pathophysiological conditions [50].

However, a completely different response of GSH- and thioredoxin-dependent antioxidant systems to CBD is observed for both unexposed and irradiated cells. The thioredoxin system consists of thioredoxin and thioredoxin reductase, which contains in its active site a highly reactive selenocysteine that is susceptible to oxidative (mainly by ROS) or electrophilic (mainly by 4-HNE) attack [51,52]. This makes it an important sensor of the redox cell state. Adducts created by 4-HNE can influence proliferation, differentiation and even cells apoptosis [53]. Therefore, in keratinocytes after CBD treatment, when ROS and lipid mediators levels are reduced, thioredoxin level and thioredoxin reductase activity are enhanced. Moreover, the data suggest that the thioredoxin system might reduce the oxidized cysteine residues in Keap1, a cytosol inhibitor of Nrf2 (the transcription factor responsible for cytoprotective protein biosynthesis), to restore its functionality [54]. On the other hand, in the case of the glutathione system, CBD decreases both the level of GSH and the activity of glutathione peroxidase. This may indicate a greater affinity of CBD for probably selenocysteine residues of these elements of the glutathione system. Moreover, CBD enhancing Nrf2 transcriptional activity leads to increase in the level of thioredoxin and thioredoxin reductase that levels are regulated by transcriptional action of Nrf2 [51]. Therefore, it is believed that the thioredoxin system is essential for the survival of keratinocytes, their protection against UV radiation and the improvement of wound healing [55]. However, this protection is not seen in CBD-treated keratinocytes because, as our results indicate, CBD modifies the structure of two of five essential cysteines in Keap1, leading to a decrease in its biological activity. Another explanation for this phenomenon may be the greater affinity of 4-HNE for glutathione than the thioredoxin system components and the interaction with the nucleophilic elements of this active/catalytic centers [56]. That may be confirmed by the decreased level of free 4-HNE in keratinocytes after CBD treatment.

These data show that CBD effectively supports the thioredoxin system, which is mainly responsible for maintaining the redox balance in the cytosol, while decreasing the efficiency of the GSH-dependent system, which is responsible for maintaining the redox balance in the mitochondrium and the endoplasmic reticulum [57]. Regardless of its direct antioxidant activity, cytoplasmic Trx regulates the activity of transcription factors, either directly (NFκB), or indirectly, i.e., AP-1, through interaction with the Ref-1 nuclear factor [58]. It demonstrates enhanced expression after UV radiation but reduced levels following CBD treatment. Under the influence of oxidative stress, both Trx and NFκB migrate from the cytoplasm to the nucleus, where Trx reduces the NFκB cysteine needed for its transcriptional activity [59]. Therefore, the CBD-induced increase in Trx level may protect against NFκB enhanced expression. Another equally important function of the thioredoxin-dependent system is participation in the regulation of gene expression by affecting apoptosis-regulating kinase (ASK-1) [60], whose expression is inhibited by a system of thioredoxins in UV-irradiated and CBD treated keratinocytes, thus protecting the cells from apoptosis, which may be the result of oxidative stress caused by radiation.

Increased expression of NFκB after UV irradiation and CBD treatment promotes enhanced expression of Nrf2, which induces the transcription of cytoprotective proteins required for resistance to stress [61]. However, Nrf2 activity depends on various activators and inhibitors [4]. Nrf2 is physiologically complexed with Keap1, causing this transcription factor to be the target for proteasomal degradation [62]. We showed that CBD modifies Keap1 Cys^{151} that participates in the Cul3-Keap1 complex formation that is needed for interaction with Nrf2. Because of decreased expression and structural modification of Keap1 in irradiated as well as CBD-treated keratinocytes, Nrf2 is transferred to the nucleus to a greater extent than in the control. Moreover, CBD increases the levels of Nrf2 activators, such as p21, KAP1 and p62 in control keratinocytes, while p21 is also increased in UV irradiated cells. In a non-canonical pathway, p62 activates Nrf2 in an autophagy-dependent manner, while Nrf2 positively regulates the expression of p62 genes and over expression of p62 is sufficient to inactivate Keap1 [63]. In addition, CBD tended to reduce the expression of PGAM5 phosphatase, which can interact with both Keap1 and Nrf2 [64]. At the same time, the expression of DPP3, a Keap1 activator, is increased. The combined effect of CBD on activators and inhibitors results in a significant reduction in Keap1 biological activity, which is seen as the enhanced expression of Nrf2.

As a consequence, increased Nrf2 target gene expression supports the stress resistance of irradiated cells and induces changes in metabolic pathways that participate in cellular defense by enhancing the removal of cytotoxic ROS and electrophiles including 4-HNE [65]. One of the classic activities of Nrf2 is the regulation of the HO-1 gene. It was earlier shown that HO-1 and its metabolites, such as CO and bilirubin, have significant anti-inflammatory activity mediated by Nrf2 [66]. However, HO-1 expression after UV radiation and CBD treatment is smaller than it can be expected after Nrf2 enhanced level. The results of this study confirm that the increased expression of Nrf2 and HO-1 after keratinocyte treatment with CBD results in a decrease in ROS generation, which in turn may inhibit the activation of the NLRP3 inflammasome. This mediates the secretion of the pro-inflammatory cytokine IL-1B, which induces cell death [67]. We have shown that CBD activates Nrf2 and partially inhibits UV-induced activation of NLRP3 and NFκB. However, the enhanced level of Trx as a result of CBD treatment could additional decrease the activation of the NLRP3 inflammasome. Therefore, the observed expression of the NLRP3 inflammasome is probably a result of CBD and Trx effects, which ultimately results in the inhibition of the expression of NLRP3. It has been previously shown that other natural and synthetic compounds (EGCG, biochanin) also demonstrate an inhibitory effect of Nrf2 on the activation of the NLRP3 inflammasome [68].

CBD under physiological conditions may stimulate the generation of many signaling factors-including the anti-inflammatory prostaglandin 15d-PGJ2, which can regulate COX activity. 15d-PGJ2 inhibits COX2 activity during inflammation but increases it under physiological conditions [69]. It was earlier shown that CBD reduces the activity of the NFκB pathway with a decrease in TNFα level [70]. Moreover, 15d-PGJ2 as an electrophilic molecule may form adducts with proteins, changing their structure and functions. This triggers important biological responses, including protection against inflammation as well as oxidative stress. $15dPGJ_2$ may exert anti-inflammatory activity by stimulating the Nrf2 pathway through the formation adducts with Keap1 and facilitate the dissociation of Nrf2 from the Keap1 adduct and consequent translocation of the transcription factor into the nucleus. This promotes up-regulation of cytoprotective genes, including HO-1 [71], which is also observed in this study after CBD treatment of UV irradiated keratinocytes. A similar effect may be demonstrated by interaction of electrophilic 4-HNE and PGJ_2 with Keap1 [43,72].

On the other hand, oxidative stress and inflammation are believed to cause Nrf2 overexpression to protect cells from the effects of ROS overproduction as well as proinflammatory cytokines [73]. Therefore, if CBD is used, the observed overexpression of Nrf2 may be the result of the overproduction of TNFα, while the expression of TNFα may be dependent on both NFκB and Nrf2 activity. Nrf2 and NFκB cellular signaling pathways interact to control the transcription and functions of different proteins and it have been shown that direct or indirect activation and inhibition occur between members of the Nrf2 and NFκB pathways [39]. At the same time, suggested the role of Keap1 not only in Nrf2 but also in NFκB inhibition by complex formation [74], as well as Keap1 structure modifications by CBD additionally explain the observed increase in the level of both Nrf2 and NFκB factors. These study indicate that in UV and CBD treated cells, Nrf2 activation partially leads to inhibition the NFκB signaling pathway.

The transcription factor NFκB is in equilibrium with its inhibitor IκB, which is degraded in response to stress stimuli, such as after UV irradiation, and its expression increases after treatment with CBD. One of the inhibitors of NFκB-IκB complex is 4-HNE, which level increases under stress conditions but partially is decreased by CBD [43]. In addition, IκB activated kinase (IKK) results in the phosphorylation of NFκB, which promotes its release and translocation into the nucleus and transcription of pro-inflammatory mediators such as IL-6, TNF-α, iNOS and IL-1 [75]. It is also known that the Nrf2 inhibitor complex, Keap1/Cul3, may target the IKKβ element of IKK for ubiquitination and degradation [76]. A modified Keap1 level by CBD treatment indicated by our study and structure by 4-HNE and 15d-PGJ_2 as was shown in the literature [43,77] due to UV and CBD treatment reduces the possibility of forming the Nrf2-Keap1/Cul3 complex and increases the transcriptional activity of both Nrf2 and NFκB. On the other hand, the increased expression of NFκB p65/RelA subunit after UV radiation and CBD treatment simultaneously imports Keap1 into the nucleus, where it terminates Nrf2 transcription of the genes by exporting this factor back to the cytoplasm [78]. However, when both transcription factors reach the nucleus, they need CREB binding protein (CBP) coactivators that are necessary for gene transcription [79], but the expression of CBP is already significantly increased by UV radiation and additionally by CBD treatment, so it probably does not require activators. Nevertheless, due to the action of UV and CBD, these two pathways can also be activated mutually, oxidative stress induced by NFκB activation facilitates Nrf2 translocation into the nucleus to protect cells against inflammation and oxidative damage.

5. Conclusions

These findings suggest that CBD activates Nrf2 and partially inhibits activation of NFκB. This activity is supported by cellular activators and inhibitors including lipid mediators such as 4-HNE and 15d-PGJ_2, whose elevated level is a response to metabolic stress accompanying both UV irradiation and CBD treatment and which may interact with both Nrf2 and NFκB pathways.

Author Contributions: Conceptualization, E.S.; formal analysis and investigation, A.J.; A.G.; validation and visualization, A.J.; A.G.; writing—review & editing, E.S.; project administration and funding acquisition, E.S.

References

1. Mori, S.; Shiraishi, A.; Epplen, K.; Butcher, D.; Murase, D.; Yasuda, Y.; Murase, T. Characterization of skin function associated with obesity and specific correlation to local/systemic parameters in American women. *Lipids Health Dis.* **2017**, *16*, 214–224. [CrossRef] [PubMed]
2. Chen, H.; Weng, Q.Y.; Fisher, D.E. UV signaling pathways within the skin. *J. Investig. Dermatol.* **2014**, *134*, 2080–2085. [CrossRef] [PubMed]
3. Brash, D.E. UV signature mutations. *Photochem. Photobiol.* **2015**, *91*, 15–26. [CrossRef] [PubMed]
4. Gęgotek, A.; Skrzydlewska, E. The role of transcription factor Nrf2 in skin cells metabolism. *Arch. Dermatol. Res.* **2015**, *307*, 385–396. [CrossRef] [PubMed]
5. Nita, M.; Grzybowski, A. The role of the reactive oxygen species and oxidative stress in the pathomechanism of the age-related ocular diseases and other pathologies of the anterior and posterior eye segments in adults. *Oxid. Med. Cell. Longev.* **2016**, *2016*. [CrossRef] [PubMed]
6. Rojo De La Vega, M.; Krajisnik, A.; Zhang, D.; Wondrak, G. Targeting NRF2 for improved skin barrier function and photoprotection: Focus on the achiote-derived apocarotenoid bixin. *Nutrients* **2017**, *9*, 1371–1386. [CrossRef] [PubMed]
7. He, X.; Ma, Q. NRF2 cysteine residues are critical for oxidant/electrophile-sensing, Kelch-like ECH-associated protein-1-dependent ubiquitination-proteasomal degradation, and transcription activation. *Mol. Pharmacol.* **2009**, *76*, 1265–1278. [CrossRef] [PubMed]
8. Huang, H.C.; Nguyen, T.; Pickett, C.B. Regulation of the antioxidant response element by protein kinase C-mediated phosphorylation of NF-E2-related factor 2. *Proc. Natl. Acad. Sci. USA* **2000**, *97*, 12475–12480. [CrossRef] [PubMed]
9. Huang, Y.; Li, W.; Su, Z.Y.; Kong, A.T. The complexity of the Nrf2 pathway: Beyond the antioxidant response. *J. Nutr. Biochem.* **2015**, *26*, 1401–1413. [CrossRef] [PubMed]
10. Sova, M.; Saso, L. Design and development of Nrf2 modulators for cancer chemoprevention and therapy: A review. *Drug Des. Devel. Ther.* **2018**, *12*, 3181–3197. [CrossRef] [PubMed]

11. Ligresti, A.; De Petrocellis, L.; Di Marzo, V. From phytocannabinoids to cannabinoid receptors and endocannabinoids: Pleiotropic physiological and pathological roles through complex pharmacology. *Physiol. Rev.* **2016**, *96*, 1593–1659. [CrossRef] [PubMed]
12. Di Marzo, V. New approaches and challenges to targeting the endocannabinoid system. *Nat. Rev. Drug Discov.* **2018**, *17*, 623–639. [CrossRef] [PubMed]
13. Crippa, J.A.; Guimarães, F.S.; Campos, A.C.; Zuardi, A.W. Translational investigation of the therapeutic potential of cannabidiol (CBD): Toward a new age. *Front. Immunol.* **2018**, *9*, 2009–2024. [CrossRef] [PubMed]
14. Fagherazzi, E.V.; Garcia, V.A.; Maurmann, N.; Bervanger, T.; Halmenschlager, L.H.; Busato, S.B.; Hallak, J.E.; Zuardi, A.W.; Crippa, J.A.; Schröder, N. Memory-rescuing effects of cannabidiol in an animal model of cognitive impairment relevant to neurodegenerative disorders. *Psychopharmacol. (Berl.)* **2012**, *219*, 1133–1140. [CrossRef] [PubMed]
15. Borges, R.S.; da Silva, A.B. Cannabidiol as an antioxidant. In *Handbook of Cannabis and Related Pathologies*, 1st ed.; Academic Press: San Diego, CA, USA, 2017; pp. 123–130.
16. Blessing, E.M.; Steenkamp, M.M.; Manzanares, J.; Marmar, C.R. Cannabidiol as a potential treatment for anxiety disorders. *Neurotherapeutics* **2015**, *12*, 825–836. [CrossRef] [PubMed]
17. Borges, R.S.; Batista, J., Jr.; Viana, R.B.; Baetas, A.C.; Orestes, E.; Andrade, M.A.; Honório, K.M.; da Silva, A.B. Understanding the molecular aspects of tetrahydrocannabinol and cannabidiol as antioxidants. *Molecules* **2013**, *18*, 12663–12674. [CrossRef] [PubMed]
18. Pan, H.; Mukhopadhyay, P.; Rajesh, M.; Patel, V.; Mukhopadhyay, B.; Gao, B.; Haskó, G.; Pacher, P. Cannabidiol attenuates cisplatin-induced nephrotoxicity by decreasing oxidative/nitrosative stress, inflammation, and cell death. *J. Pharmacol. Exp. Ther.* **2009**, *328*, 708–714. [CrossRef]
19. Peres, F.F.; Lima, A.C.; Hallak, J.E.; Crippa, J.A.; Silva, R.H.; Abílio, V.C. Cannabidiol as a promising strategy to treat and prevent movement disorders? *Front. Pharmacol.* **2018**, *9*. [CrossRef]
20. Liou, G.I.; Auchampach, J.A.; Hillard, C.J.; Zhu, G.; Yousufzai, B.; Mian, S.; Khan, S.; Khalifa, Y. Mediation of cannabidiol anti-inflammation in the retina by equilibrative nucleoside transporter and A2A adenosine receptor. *Invest. Ophthalmol. Vis. Sci* **2008**, *49*, 5526–5531. [CrossRef]
21. Leweke, F.M.; Piomelli, D.; Pahlisch, F.; Muhl, D.; Gerth, C.W.; Hoyer, C.; Klosterkötter, J.; Hellmich, M.; Koethe, D. Cannabidiol enhances anandamide signaling and alleviates psychotic symptoms of schizophrenia. *Transl. Psychiatry* **2012**, *2*, e94. [CrossRef]
22. Petrosino, S.; Verde, R.; Vaia, M.; Allarà, M.; Iuvone, T.; Marzo, V.D. Anti-inflammatory properties of cannabidiol, a non-psychotropic cannabinoid, in experimental allergic contact dermatitis. *J. Pharmacol. Exp. Ther.* **2018**, *365*, 652–663. [CrossRef] [PubMed]
23. Zou, S.; Kumar, U. Cannabinoid receptors and the endocannabinoid system: Signaling and function in the central nervous system. *Int. J. Mol. Sci.* **2018**, *19*, 833–855.
24. Laun, A.S.; Song, Z.H. GPR3 and GPR6, novel molecular targets for cannabidiol. *Biochem. Biophys. Res. Commun.* **2017**, *490*, 17–21. [CrossRef] [PubMed]
25. Mato, S.; Victoria Sánchez-Gómez, M.; Matute, C. Cannabidiol induces intracellular calcium elevation and cytotoxicity in oligodendrocytes. *Glia* **2010**, *58*, 1739–1747. [CrossRef] [PubMed]
26. Russo, C.; Ferk, F.; Mišík, M.; Ropek, N.; Nersesyan, A.; Mejri, D.; Holzmann, K.; Lavorgna, M.; Isidori, M.; Knasmüller, S. Low doses of widely consumed cannabinoids (cannabidiol and cannabidivarin) cause DNA damage and chromosomal aberrations in human-derived cells. *Arch. Toxicol.* **2019**, *93*, 179–188. [CrossRef]
27. Burstein, S. Cannabidiol (CBD) and its analogs: A review of their effects on inflammation. *Bioorg. Med. Chem.* **2015**, *23*, 1377–1385. [CrossRef] [PubMed]
28. Pucci, M.; Rapino, C.; Di Francesco, A.; Dainese, E.; D'Addario, C.; Maccarrone, M. Epigenetic control of skin differentiation genes by phytocannabinoids. *Br. J. Pharmacol.* **2013**, *170*, 581–591. [CrossRef]
29. Fotakis, G.; Timbrell, J.A. In vitro cytotoxicity assays: Comparison of LDH, neutral red, MTT and protein assay in hepatoma cell lines following exposure to cadmium chloride. *Toxicol. Lett.* **2006**, *160*, 171–177. [CrossRef]
30. Bradford, M.M. A rapid and sensitive method for the quantitation of microgram quantities of protein utilizing the principle of protein-dye binding. *Anal. Biochem.* **1976**, *72*, 248–254. [CrossRef]
31. Kuzkaya, N.; Weissmann, N.; Harrison, D.G.; Dikalov, S. Interactions of peroxynitrite, tetrahydrobiopterin, ascorbic acid, and thiols: Implications for uncoupling endothelial nitricoxide synthase. *J. Biol. Chem.* **2003**, *278*, 22546–22554. [CrossRef]

32. Misra, H.P.; Fridovich, I. The role of superoxide anion in the autoxidation of epinephrine and a simple assay for superoxide dismutase. *J. Biol. Chem.* **1972**, *247*, 3170–3175. [PubMed]
33. Sykes, J.A.; McCormac, F.X.; O'Breien, T.J. Preliminary study of the superoxide dismutase content of some human tumors. *Cancer Res.* **1978**, *38*, 2759–2762. [PubMed]
34. Paglia, D.E.; Valentine, W.N. Studies on the quantitative and qualitative characterization of erythrocyte glutathione peroxidase. *J. Lab. Clin. Med.* **1967**, *70*, 158–169. [PubMed]
35. Mize, C.E.; Longdon, R.G. Hepatic glutathione reductase. Purification and general kinetic properties. *J. Biol. Chem.* **1962**, *237*, 1589–1595. [PubMed]
36. Holmgren, A. Thioredoxin and thioredoxin reductase. *Methods Enzymol.* **1995**, *252*, 199–208. [PubMed]
37. Maeso, N.; Garcia-Martinez, D.; Ruperez, F.J.; Cifuentes, A.; Barbas, C. Capillary electrophoresis of glutathione to monitor oxidative stress and response to antioxidant treatments in an animal model. *J. Chromatogr. B.* **2005**, *822*, 61–69. [CrossRef] [PubMed]
38. Lovell, M.A.; Xie, C.; Gabbita, S.P.; Markesbery, W.R. Decreased thioredoxin and increased thioredoxin reductase levels in Alzheimer's disease brain. *Free Radic. Biol. Med.* **2000**, *28*, 418–427. [CrossRef]
39. Liu, G.H.; Qu, J.; Shen, X. NF-κB/p65 antagonizes Nrf2-ARE pathway by depriving CBP from Nrf2 and facilitating recruitment of HDAC3 to MafK. *Biochim. Biophys. Acta (BBA)-Mol. Cell Res.* **2008**, *1783*, 713–727. [CrossRef]
40. Morgenstern, J.; Fleming, T.; Kadiyska, I.; Brings, S.; Groener, J.B.; Nawroth, P.; Hecker, M.; Brune, M. Sensitive mass spectrometric assay for determination of 15-deoxy-Δ12,14-prostaglandin J2 and its application in human plasma samples of patients with diabetes. *Anal. Bioanal. Chem.* **2018**, *410*, 521–528. [CrossRef]
41. Eissa, S.; Seada, L.S. Quantitation of bcl-2 protein in bladder cancer tissue by enzyme immunoassay: Comparison with Western blot and immunohistochemistry. *Clin. Chem.* **1998**, *44*, 1423–1429.
42. Gęgotek, A.; Domingues, P.; Skrzydlewska, E. Proteins involved in the antioxidant and inflammatory response in rutin-treated human skin fibroblasts exposed to UVA or UVB irradiation. *J. Dermatol. Sci.* **2018**, *90*, 241–252. [CrossRef] [PubMed]
43. Łuczaj, W.; Gęgotek, A.; Skrzydlewska, E. Antioxidants and HNE in redox homeostasis. *Free Radic. Biol. Med.* **2017**, *111*, 87–101. [CrossRef] [PubMed]
44. Chanda, D.; Neumann, D.; Glatz, J.F.C. The endocannabinoid system: Overview of an emerging multi-faceted therapeutic target, *Prostaglandins Leukot. Essent. Fatty Acids* **2019**, *140*, 51–56. [CrossRef] [PubMed]
45. Lipina, C.; Harinder, S. Hundal Modulation of cellular redox homeostasis by the endocannabinoid system. *Open Biol.* **2016**, *6*, 150276. [CrossRef]
46. Ambrożewicz, E.; Wójcik, P.; Wroński, A.; Łuczaj, W.; Jastrząb, A.; Žarković, N.; Skrzydlewska, E. Pathophysiological alterations of redox signaling and endocannabinoid system in granulocytes and plasma of psoriatic patients. *Cells* **2018**, *7*, 159–176. [CrossRef]
47. Crouzin, N.; de Jesus Ferreira, M.C.; Cohen-Solal, C.; M'Kadmi, C.; Bernad, N.; Martinez, J.; Guiramand, J. α-Tocopherol and α-tocopheryl phosphate interact with the cannabinoid system in the rodent hippocampus. *Free Radic. Biol. Med.* **2011**, *51*, 1643–1655. [CrossRef]
48. Rajesh, M.; Mukhopadhyay, P.; Bátkai, S.; Patel, V.; Saito, K.; Matsumoto, S.; Haskó, G. Cannabidiol attenuates cardiac dysfunction, oxidative stress, fibrosis, and inflammatory and cell death signaling pathways in diabetic cardiomyopathy. *J. Am. Coll. Cardiol.* **2010**, *56*, 2115–2125. [CrossRef]
49. Fouad, A.A.; Albuali, W.H.; Al-Mulhim, A.S.; Jresat, I. Cardioprotective effect of cannabidiol in rats exposed to doxorubicin toxicity. *Environ. Toxicol. Pharmacol.* **2013**, *36*, 347–357. [CrossRef]
50. Romano, B.; Borrelli, F.; Pagano, E.; Cascio, M.G.; Pertwee, R.G.; Izzo, A.A. Inhibition of colon carcinogenesis by a standardized Cannabis sativa extract with high content of cannabidiol. *Phytomedicine* **2014**, *21*, 631–639. [CrossRef]
51. Cebula, M.; Schmidt, E.E.; Arner, E.S. TrxR1 as a potent regulator of the Nrf2-Keap1 response system. *Antioxid. Redox Signal.* **2015**, *23*, 823–853. [CrossRef]
52. Cassidy, P.B.; Edes, K.; Nelson, C.C.; Parsawar, K.; Fitzpatrick, F.A.; Moos, P.J. Thioredoxin reductase is required for the inactivation of tumor suppressor p53 and for apoptosis induced by endogenous electrophiles. *Carcinogenesis* **2006**, *27*, 2538–2549. [CrossRef] [PubMed]
53. Gasparovic, A.C.; Milkovic, L.; Sunjic, S.B.; Zarkovic, N. Cancer growth regulation by 4-hydroxynonenal. *Free Radic. Biol. Med.* **2017**, *111*, 226–234. [CrossRef] [PubMed]

54. Dinkova-Kostova, A.T.; Kostov, R.V.; Canning, P. Keap1, the cysteine-based mammalian intracellular sensor for electrophiles and oxidants. *Arch. Biochem. Biophys.* **2017**, *617*, 84–93. [CrossRef] [PubMed]
55. Telorack, M.; Meyer, M.; Ingold, I.; Conrad, M.; Bloch, W.; Werner, S.A. Glutathione-Nrf2-thioredoxin cross-talk ensures keratinocyte survival and efficient wound repair. *PLoS Genet.* **2016**, *12*, e1005800. [CrossRef] [PubMed]
56. Alviz-Amador, A.; Galindo-Murillo, R.; Pineda-Alemán, R.; Pérez-González, H.; Rodríguez-Cavallo, E.; Vivas-Reyes, R.; Méndez-Cuadro, D. 4-HNE carbonylation induces local conformational changes on bovine serum albumin and thioredoxin. A molecular dynamics study. *J. Mol. Graph. Model.* **2019**, *86*, 298–307. [CrossRef] [PubMed]
57. Poet, G.J.; Oka, O.B.; Van Lith, M.; Cao, Z.; Robinson, P.J.; Pringle, M.A.; Bulleid, N.J. Cytosolic thioredoxin reductase 1 is required for correct disulfide formation in the ER. *EMBO J.* **2017**, *36*, 693–702. [CrossRef] [PubMed]
58. Matsuzawa, A. Thioredoxin and redox signaling: Roles of the thioredoxin system in control of cell fate. *Arch. Biochem. Biophys.* **2017**, *617*, 101–105. [CrossRef] [PubMed]
59. Branco, V.; Carvalho, C. The thioredoxin system as a target for mercury compounds. *Biochim. Biophys. Acta (BBA)-Gen. Subj.* **2018**. S0304-4165. [CrossRef] [PubMed]
60. Holmgren, A.; Lu, J. Thioredoxin and thioredoxin reductase: Current research with special reference to human disease. *Biochem. Biophys. Res. Commun.* **2010**, *396*, 120–124. [CrossRef] [PubMed]
61. Loboda, A.; Damulewicz, M.; Pyza, E.; Jozkowicz, A.; Dulak, J. Role of Nrf2/HO-1 system in development, oxidative stress response and diseases: An evolutionarily conserved mechanism. *Cell Mol. Life Sci.* **2016**, *73*, 3221–3247. [CrossRef]
62. Bellezza, I.; Giambanco, I.; Minelli, A.; Donato, R. Nrf2-Keap1 signaling in oxidative and reductive stress. *Biochim. Biophys. Acta (BBA)-Mol. Cell Res.* **2018**, *1865*, 721–733. [CrossRef] [PubMed]
63. Helwa, I.; Choudhary, V.; Chen, X.; Kaddour-Djebbar, I.; Bollag, W.B. Anti-psoriatic drug monomethylfumarate increases nuclear factor erythroid 2-related factor 2 levels and induces aquaporin-3 mRNA and protein expression. *J. Pharmacol. Exp. Ther.* **2017**, *362*, 243–253. [CrossRef] [PubMed]
64. O'Mealey, G.B.; Plafker, K.S.; Berry, W.L.; Janknecht, R.; Chan, J.Y.; Plafker, S.M. A PGAM5–KEAP1–Nrf2 complex is required for stress-induced mitochondrial retrograde trafficking. *J. Cell Sci.* **2017**, *130*, 3467–3480. [CrossRef] [PubMed]
65. Kimura, S.; Warabi, E.; Yanagawa, T.; Ma, D.; Itoh, K.; Ishii, Y.; Ishii, T. Essential role of Nrf2 in keratinocyte protection from UVA by quercetin. *Biochem. Biophys. Res. Commun.* **2009**, *387*, 109–114. [CrossRef] [PubMed]
66. Pae, H.O.; Lee, Y.C.; Chung, H.T. Heme oxygenase-1 and carbon monoxide: Emerging therapeutic targets in inflammation and allergy. *Recent Pat. Inflamm. Allergy Drug Discov.* **2008**, *2*, 159–165. [CrossRef] [PubMed]
67. Jourdan, T.; Godlewski, G.; Cinar, R.; Bertola, A.; Szanda, G.; Liu, J.; Ju, C. Activation of the Nlrp3 inflammasome in infiltrating macrophages by endocannabinoids mediates beta cell loss in type 2 diabetes. *Nat. Med.* **2013**, *19*, 1132–1143. [CrossRef] [PubMed]
68. Lim, H.; Min, D.S.; Park, H.; Kim, H.P. Flavonoids interfere with NLRP3 inflammasome activation. *Toxicol. Appl. Pharmacol.* **2018**, *355*, 93–102. [CrossRef] [PubMed]
69. Gilroy, D.W. Eicosanoids and the endogenous control of acute inflammatory resolution. *Int. J. Biochem. Cell Biol.* **2010**, *42*, 524–528. [CrossRef] [PubMed]
70. Kozela, E.; Pietr, M.; Juknat, A.; Rimmerman, N.; Levy, R.; Vogel, Z. Cannabinoids Δ9-tetrahydrocannabinol and cannabidiol differentially inhibit the lipopolysaccharide-activated NF-κB and interferon-β/STAT proinflammatory pathways in BV-2 microglial cells. *J. Biol. Chem.* **2010**, *285*, 1616–1626. [CrossRef]
71. Oh, J.Y.; Giles, N.; Landar, A.; Darley-Usmar, V. Accumulation of 15-deoxy-D12,14- prostaglandin J2 adductformation with Keap1 over time: Effects on potency for intracellular antioxidant defence induction. *Biochem. J.* **2008**, *411*, 297–306. [CrossRef]
72. Pajaud, J.; Kumar, S.; Rauch, C.; Morel, F.; Aninat, C. Regulation of signal transduction by glutathione transferases. *Int. J. Hepatol.* **2012**, *2012*. [CrossRef] [PubMed]
73. Kobayashi, E.H.; Suzuki, T.; Funayama, R.; Nagashima, T.; Hayashi, M.; Sekine, H.; Yamamoto, M. Nrf2 suppresses macrophage inflammatory response by blocking proinflammatory cytokine transcription. *Nat. Communi.* **2016**, *7*, 11624–11637. [CrossRef] [PubMed]
74. Stefanson, A.; Bakovic, M. Dietary regulation of Keap1/Nrf2/ARE pathway: Focus on plant-derived compounds and trace minerals. *Nutrients* **2014**, *6*, 3777–3801. [CrossRef] [PubMed]

75. Perkins, N.D. Integrating cell-signalling pathways with NF-κB and IKK function. *Nat. Rev. Mol. Cell Biol.* **2007**, *8*, 49–62. [CrossRef] [PubMed]
76. Kim, J.E.; You, D.J.; Lee, C.; Ahn, C.; Seong, J.Y.; Hwang, J.I. Suppression of NF-κB signaling by KEAP1 regulation of IKKβ activity through autophagic degradation and inhibition of phosphorylation. *Cell. Signal.* **2010**, *22*, 1645–1654. [CrossRef]
77. Chen, K.; Li, J.; Li, S.; Feng, J.; Wu, L.; Liu, T.; Zhou, S. 15d-PGJ2 alleviates ConA-induced acute liver injury in mice by up-regulating HO-1 and reducing hepatic cell autophagy. *Biomed. Pharmacother.* **2016**, *80*, 183–192. [CrossRef] [PubMed]
78. Itoh, K.; Wakabayashi, N.; Katoh, Y.; Ishii, T.; Igarashi, K.; Engel, J.D.; Yamamoto, M. Keap1 represses nuclear activation of antioxidant responsive elements by Nrf2 through binding to the amino-terminal Neh2 domain. *Genes Dev.* **1999**, *13*, 76–86. [CrossRef]
79. Ambrozova, N.; Ulrichova, J.; Galandakova, A. Models for the study of skin wound healing. The role of Nrf2 and NF-κB. *Biomed. Pap. Med. Fac. Palacky. Olomouc. Czech. Repub.* **2017**, *161*, 1–13. [CrossRef]

4

Differential Sensitivity of Two Endothelial Cell Lines to Hydrogen Peroxide Toxicity: Relevance for In Vitro Studies of the Blood–Brain Barrier

Olufemi Alamu [1,2], Mariam Rado [1], Okobi Ekpo [1] and David Fisher [1,*]

1 Department of Medical Bioscience, University of the Western Cape, Bellville, Cape Town 7530, South Africa; olufemialamu@gmail.com (O.A.); maryamadem99@gmail.com (M.R.); oekpo@uwc.ac.za (O.E.)
2 Anatomy Department, Ladoke Akintola University of Technology, Ogbomoso 210241, Nigeria
* Correspondence: dfisher@uwc.ac.za

Abstract: Oxidative stress (OS) has been linked to blood–brain barrier (BBB) dysfunction which in turn has been implicated in the initiation and propagation of some neurological diseases. In this study, we profiled, for the first time, two endothelioma cell lines of mouse brain origin, commonly used as in vitro models of the blood–brain barrier, for their resistance against oxidative stress using viability measures and glutathione contents as markers. OS was induced by exposing cultured cells to varying concentrations of hydrogen peroxide and fluorescence microscopy/spectrometry was used to detect and estimate cellular glutathione contents. A colorimetric viability assay was used to determine changes in the viability of OS-exposed cells. Both the b.End5 and bEnd.3 cell lines investigated showed demonstrable content of glutathione with a statistically insignificant difference in glutathione quantity per unit cell, but with a statistically significant higher capacity for the b.End5 cell line for de novo glutathione synthesis. Furthermore, the b.End5 cells demonstrated greater oxidant buffering capacity to higher concentrations of hydrogen peroxide than the bEnd.3 cells. We concluded that mouse brain endothelial cells, derived from different types of cell lines, differ enormously in their antioxidant characteristics. We hereby recommend caution in making comparisons across BBB models utilizing distinctly different cell lines and require further prerequisites to ensure that in vitro BBB models involving these cell lines are reliable and reproducible.

Keywords: oxidative stress; blood–brain barrier; bEnd5; bEnd.3; glutathione; viability

---

1. Introduction

The blood–brain barrier (BBB) is a functional and morphological interface between the systemic blood circulation and the CNS. The central regulatory component is the brain capillary endothelial cell (BEC), which is assisted by a number of cellular entities, viz. pericytes and astrocytes, together forming an interface, described as the neuro-vasculo-glial unit (NVU). The NVU has the dynamic ability to respond to the homeostatic changes in the brain interstitium ensuring a stable environment for neuronal functionality [1]. From a research point of view, it is therefore of interest to understand the capability of the BEC to withstand oxidative stress (OS), and to scrutinize the BBB in vitro models used to study these physiological processes. However, from a therapeutic point of view the strict regulatory mechanisms for molecules to cross the barrier provide a serious clinical challenge to molecules of desired therapeutic interest to reach their neural target sites [2–5].

Endothelial dysfunction-mediated vascular diseases such as BBB dysfunction have been well linked to excess production of reactive oxygen species (ROS) [6,7]. A major source of ROS in these conditions is the upregulation of nicotinamide adenine dinucleotide phosphate oxidase (Nox, especially Nox2) activity [8,9]. Excess superoxide (O_2^-) produced primarily from increased Nox activity undergoes both spontaneous and enzymatic dismutation by superoxide dismutases (SOD), resulting in increase and/or accumulation of its dismutation product, hydrogen peroxide (H_2O_2) [10]. Keeping vascular endothelial cells in redox balance involves the activities of several endogenous antioxidants whose activities may be enzymatic or non-enzymatic [10]. Superoxide dismutase (SOD) is an enzymatic, first-line, intracellular antioxidant that catalyzes the conversion of superoxide to hydrogen peroxide [11]. Other important enzymatic antioxidants such as glutathione peroxidase (GPx), and catalase (CAT), help to neutralize H_2O_2 to water and oxygen [11]

Endothelial nitric oxide synthase (eNOS) is another enzyme which normally plays a protective role in the vascular endothelial cell through its production of nitric oxide (NO) while oxidizing its substrate L-arginine to L-citrulline using molecular oxygen (O_2) [12,13]. However, parallel upregulation of eNOS and Nox as occurs in many vascular diseases results in eNOS uncoupling with subsequent conversion of eNOS to an O_2^--producing enzyme that contributes to further vascular oxidative stress [14–16]. Furthermore, free radicals such as superoxide can attack polyunsaturated fatty acids (PUFAs) such as arachidonic acid in the cell membranes of vascular endothelial cells to yield advanced lipid peroxidation end products (ALEs) [17]. One of ALEs of relevance is 4-hydroxynonenal (4-HNE), an electrophile that is highly reactive towards nucleophilic thiols and amino groups [17,18]. It readily reacts with proteins (4-HNE-protein adduction), lipids, and nucleic acids of DNA. Its reaction with proteins modifies their activity thus acting as a second messenger in various biologic activities including modification of enzymatic actions or transcription factor modulations that determine cell survival or death [18]. Oxidative stress (OS) occurs in the cells when an unbalanced accumulation of ROS exists within the cell and/or in its immediate environment [19]. BBB dysfunction in relation to OS has been implicated in the initiation and propagation of several neurological conditions such as epilepsy, stroke, and degenerative neuropathies such as Alzheimer's disease, Parkinson's disease, and multiple sclerosis [20,21]. Thus, OS has been scientifically documented as a common factor to both the etiologies of these neurological disorders as well as abnormal BBB function. Cell modeling has, to date, provided a robust tool in the in vitro study of the BBB [22,23]. The brain microvascular endothelial cells are the principal cells of the BBB which have had several cell models characterized for use as in vitro models in the study of the BBB [24]. This study focused on two mouse-derived cell lines, b.End5 and bEnd.3, established for use as in vitro models of the BBB [25,26]. It is of interest to understand how the endothelial cells of the BBB respond to oxidative stress as well as the endothelial cell-specific events that underlie the abnormalities of BBB permeability. It is, however, disconcerting that most of the in vitro cell models in use for BBB studies have not been characterized for their oxidative/antioxidant features, which are necessary for the definition of OS, as well as for features that characterize changes in oxidative stress responsible for the observation of abnormal permeability under OS conditions. In this study, we profiled for the first time, b.End5 and bEnd.3 cells, both mouse-derived cell lines obtained after immortalizing primary mouse brain endothelial cells by infection with middle T antigen-expressing polyoma virus, for their resistance against a suitably-selected ROS, hydrogen peroxide (H_2O_2), which can permeate all intracellular membranes and thus exert its effects on organelles [27]. Cellular glutathione content (reduced/oxidized form [GSH/GSSG]), a well reported marker of cellular oxidative status, is a tripeptide, L-γ-glutamine-L-cysteinyl-glycine, that acts as an endogenous cellular antioxidant either by direct neutralization of ROS or as cofactor for the antioxidant enzyme, glutathione peroxidase (GPx) [28]. [Hereafter, we refer to the full component of reduced and oxidized glutathione as 'total glutathione while GSH and GSSG refer to the 'reduced glutathione' and 'glutathione disulfide/oxidized glutathione' fractions respectively]. This functional capability of GSH is conferred by its active thiol group residing in its cysteine residue [29]. Glutathione exists within cells in either the reduced (GSH) form or in oxidized form as glutathione disulfide (GSSG). It is usually synthesized as GSH but oxidized to GSSG upon participating in a redox reaction. Changes in cellular reduced glutathione content as well as changes in cell viabilities were used to assess cellular capacities to respond to varying levels of ROS. This protocol is robustly useful whenever it is desired to use these cells to study oxidant effects on the BBB.

2. Materials and Methods

2.1. Bio-Reagents

Analytical grade reagents were used for all experiments. These included monochlorobimane (Molecular Probe M1381MP, Eugene, OR 97402, USA), trypan blue, (Gibco 1520-061, Gaithersburg MD 20877, USA), tris (2-carboxyethyl) phosphine hydrochloride, TCEP (Sigma C4706, Laramie, WY 82070, USA), GSH-Glo^{TM} Glutathione Assay Kit (Promega V6911/2, Madison, WI 53711, USA), Trolox ((±)-6-Hydroxy-2,5,7,8-tetramethylchromane-2-carboxylic acid, Sigma 238813, Laramie, WY 82070, USA), Cell Proliferation Kit II (XTT) (Roche, Sigma 11465015001, Laramie, WY 82070, USA), and hydrogen peroxide (30%, Merck Millipore 107209, Feldbergstraße 80, 64293 Darmstadt, Germany).

2.2. Cell Cultures

Two mouse brain endothelioma cell lines (b.End5 and bEnd.3) were used in this study. The b.End5 (ECACC 96091930, Salisbury, Wiltshire SP4 0JG, UK) cells were cultured in Dulbecco's modified Eagle's medium (DMEM Lonza BE 12-719F, Salisbury, MD 21801, USA.) supplemented with 10% fetal bovine serum (FBS Biowest12010S181G, Rue du Vieux Bourg, 49340 Nuaillé, France), 100 U/mL penicillin/streptomycin (Lonza DE17-602E), 1 mM sodium pyruvate solution (Lonza BE13-115E), and 1% non-essential amino acids (NEAA Lonza BE13-114E) at 37 °C and 5% CO_2 in a humidified incubator. For all experiments, b.End5 cells at passages 5–20 were used and culture medium was changed every 2–3 days. Prior to experimentation, cells were rinsed in 1× phosphate buffer solution (PBS). The adherent cells were detached by the addition of 0.25% trypsin-EDTA after which equal volume of fresh media was added and the cell suspension aspirated into 15 mL conical tubes. The cell suspensions were then centrifuged for 5 min at rpm of 2500 to obtain a cell pellet. The supernatant was then removed by gentle aspiration and thereafter, 5 mL of fresh media added to bring the cell pellets back in suspension. The bEnd.3 (ATCC® CRL-2299, Gaithersburg, MD 20877, United States) cells were cultured in Dubelcco's Eagle's medium (Gibco 11320074) supplemented with 10% fetal bovine serum (FBS Biowest S12010S181G), 100 U/mL penicillin/streptomycin (Lonza DE17-602E), and L-glutamine to a final concentration of 4.1 mM (Invitrogen 25030081, Camarillo, CA 93012, United States) and at 37 °C and 5% CO_2 in a humidified incubator. The bEnd.3 cells used for all experiments were at passages 5–20 and the medium was changed every 2–3 days. Prior to cell seeding for experiments, adherent bEnd.3 cells were similarly brought to suspension as described for the b.End5 cells.

2.3. Hydrogen Peroxide Treatments

A stock solution of H_2O_2 (9.8 M, Merck Millipore, Feldbergstraße 80, 64293 Darmstadt, Germany) was diluted to the required concentrations in complete media. Cultured b.End5 and bEnd.3 cells were divided into labelled treatment wells exposed to H_2O_2 concentrations ranging between 10 μM and 2 mM and cells that were unexposed to H_2O_2 served as control. To determine the antioxidant capacity of cultured cells both b.End5 and bEnd.3 cells were seeded in separate transparent 96-well plates at 1×10^4 cells per well in 200 μL of normal media and allowed to attach overnight. Media were then aspirated and replaced with either 100 μL of fresh media to serve as control or 100 μL of media dosed with the appropriate concentrations of hydrogen peroxide for another 24 h. Experiments were repeated thrice and average values of parameters were recorded. Viability of the cells exposed to different concentrations of hydrogen peroxide were then analysed and compared against viability measures for the control cells under same duration in culture.

2.4. Viability Assay

Cells were cultured in a transparent 96-well plate as described for hydrogen peroxide treatments. Equal numbers of b.End5 cells were seeded into eighteen treatment (H_2O_2) wells (n = 3) starting from control (unexposed) and treatment with [H_2O_2] in multiples of 50 μM up to a maximum of 850 μM. For cultured bEnd.3 cells, equal numbers of cells were seeded into sixteen sets of 3 wells (n = 3) and treated as control (unexposed), then [H_2O_2] in multiples of 10 μM up to 100 μM and then in multiples of 100 μM up to a maximum of 500 μM. A blank column of three wells was also included in both treatment plates to facilitate the determination of relative absorbance units. The XTT [30] viability assay kit (Roche) was used to quantify cell viability after treatment for 24 h. The XTT reagent was reconstituted by mixing 100 μL of electron-coupling reagent (0.383 mg/mL) with 5 mL of XTT labelling reagent (1 mg/mL) to activate it as per manufacturer's recommendation. Reconstituted XTT, 50 μL, was then added to each well containing 100 μL of cell culture and incubated for 4 h at 37 °C in a CO_2 incubator. Absorbance was then read for each well at 450 nm and blank-corrected values obtained using a GloMax–Multi Detection System (Promega, Madison, WI 53711, USA). The absorbance measures directly correlated with the viability of the cells in each well.

2.5. *Fluorescent Detection of Glutathione in Cultured Cells*

Equal numbers of b.End5 and bEnd.3 cells were cultured under standard conditions on microscopic glass slides in separate Petri dishes. The cells were then allowed to attach overnight in all Petri dishes and cells on each slide were used to demonstrate glutathione. Briefly, the medium was removed from the attached cells and were rinsed twice with PBS solution, pH, 7.4, and then incubated with monochlorobimane solution (mBCl, Molecular Probe™ M1381MP) 60 μM in complete DMEM for 30 min [31]. Following mBCl loading, slides were fixed using a mixture of 4% paraformaldehyde (PFA) and 0.2% glutaraldehyde (GA) in PBS solution at pH 7.4 for 10 min and following fixation, cells were nuclear-counterstained by incubating slides with 20 μg/mL propidium iodide (PI) solution for 15 min. DABCO (1,4-diazobicyclo-[2,2,2]-octane) mountant, 20 μL, was added to each slide mounted with cover slips. Cells on each slide were then viewed and imaged under a Nikon Eclipse 50i fluorescent microscope at $\lambda_{ex}/\lambda_{em}$ of 365/490 nm and 439/636 nm for mBCl and PI, respectively.

2.6. *Quantification of Total Cellular Glutathione in bEnd5 Cells*

To accurately quantify the total amount of glutathione in a single b.End5/bEnd.3 cell, we used a GSH-Glo™ Glutathione Assay Kit which works by a luminescence assay to detect and quantify glutathione [32]. The assay is based on the conversion of a luciferin derivative into luciferin in the presence of glutathione, catalyzed by glutathione-S-transferase (GST). The reaction is further coupled with a firefly luciferase which leads to the generation of luminescence signal proportional to the amount of glutathione in the sample. To estimate glutathione fairly accurately in 1×10^4 cells, according to manufacturer's recommendation and to control for cell proliferation occurring alongside cell attachment, cells were plated in white 96-well plates and incubated at 37 °C and 5% CO_2 at a density of 4×10^3 cells per well for the b.End5 cells and 4.5×10^3 cells per well for the bEnd.3 cells, based on an optimized number of the respective cells that gave the target density at 24 h in culture (based on our data from proliferation study, not shown). Cells were plated in columns of four wells ($n = 4$) in a 96-well white bottom plate and 100 μL of prepared 1X GSH-Glo reagent was transferred to each well. In order to measure the total glutathione (GSH + GSSG), 100 μL of 1 mM tris (2-carboxyethyl) phosphine (TCEP) was added to a group of four wells in addition to the GSH-Glo reagent according to the GSSG recycling method [33]. The contents of the wells were agitated briefly on an orbital shaker before incubation at room temperature for 30 min. Then, 100 μL of reconstituted luciferin detection reagent was transferred to each well, and the plate was mixed briefly on an orbital shaker before incubation at room temperature for 15 min. Luminescence values were then read using a GloMax–Multi Detection System (Promega, Madison, WI 53711, USA). Luminescence readings were converted to GSH concentration using a standard curve generated from a 5 mM GSH standard supplied by the manufacturer.

2.7. *Statistical Analysis*

Statistical analysis was done using Graph Pad Prism.5 statistical analysis software. Data were expressed as mean ± SEM and significant differences in data were accepted at $p < 0.05$

3. Results

In this study differences in antioxidant capacities of two brain endothelial cell lines from the same animal species were observed. We firstly established baseline data for both cell lines in terms of their endogenous glutathione concentrations, and secondly, we profiled the response of these cell lines to an exogenous ROS stress, H_2O_2, with respect to the content of the endogenous antioxidant, glutathione, using fluorescent imaging, and fluoro-spectrometric quantification.

3.1. Both b.End5 and bEnd.3 Cells Demonstrated Glutathione Presence on Fluorescent Microscopy

We first investigated presence and distribution of glutathione in each of the selected cell lines. Monochlorobimane with propidium iodide nuclear counterstaining revealed blue fluorescence due to the presence of glutathione while the nuclei fluoresced red (Figure 1). Both cells appeared intensely fluorescent for glutathione, however, bEnd.3 showed less cytoplasmic glutathione possibly due to a higher nucleo-cytoplasmic ratio. Variable segments and rings of blue fluorescence were observed around the periphery of the nuclei in both cells (Figure 1, Plates A2 and B2) which are evidence of glutathione presence within the nuclear structure. This was more prominently observed at the nucleo-cytoplasmic interface in both cell types and especially in the bEnd.3 cell (Figure 1, Plate B2).

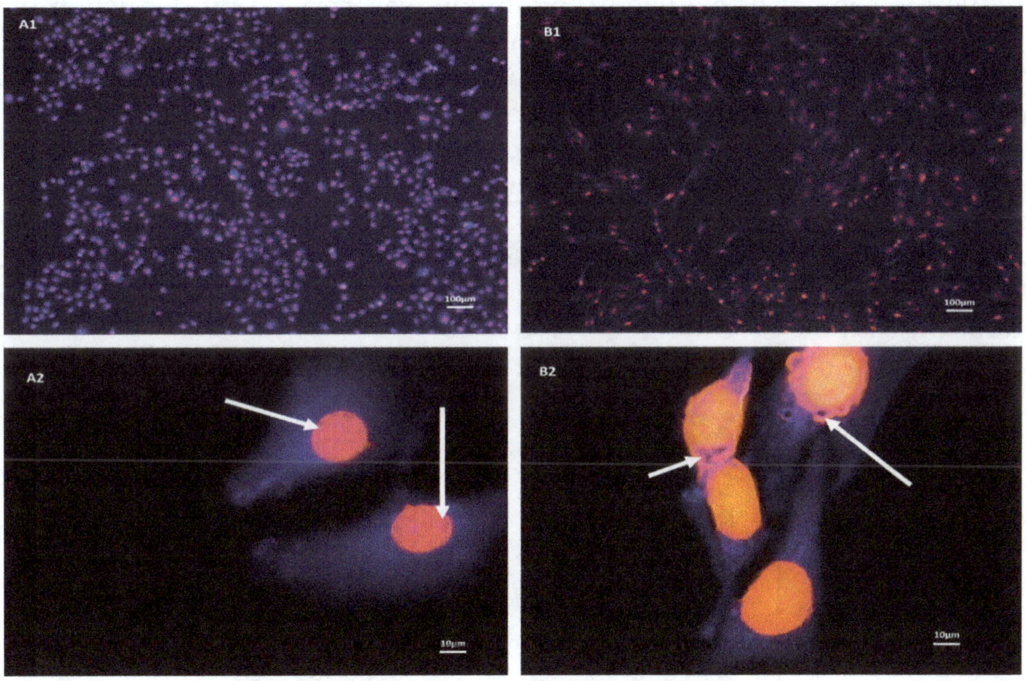

Figure 1. Micrographs show fluorescent images of b.End5 (**A1**) and (**A2**) and bEnd.3 cells (**B1**) and (**B2**) in normal culture. Both cells showed blue monochlorobimane solution (mBCl) fluorescence (due to binding with reduced glutathione (GSH)) in their cytoplasm, though at the higher magnification b.End5 appeared more deeply stained. Furthermore, plate A2 revealed a lower nucleo-cytoplasmic ratio in b.End5 cells suggesting more cytoplasmic GSH content than in bEnd.3 cells. Furthermore, multiple segments and rings of blue fluorescence (white arrows) were indicative of glutathione observed within the nuclei and nucleo-cytoplasmic interface in the cells of both cell types (Plates A2 and B2).

3.2. Glutathione Contents of Both b.End5 and bEnd.3 Cells Are Comparable in Normal Culture

The difference in blue fluorescence between the two cell types provided micrographical evidence for the presence and distribution of glutathione. The objective quantification of the levels of glutathione in each cell type was required to compare the cells' ability to respond to OS (Figure 2).

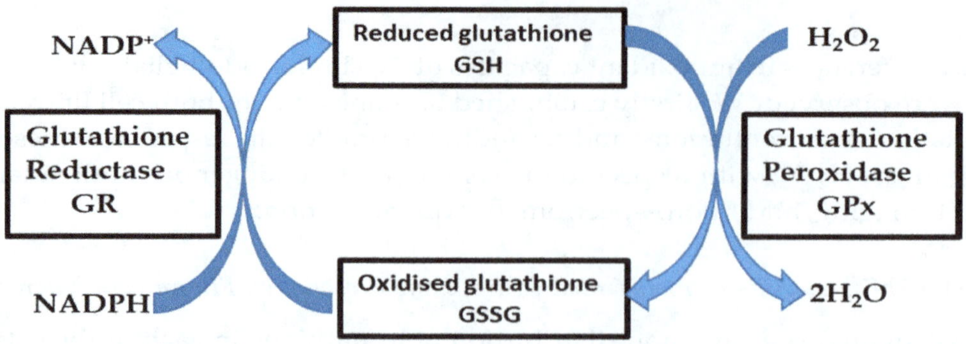

Figure 2. The above diagram illustrates the redox buffering reaction of the glutathione system. Glutathione peroxidase (GPx) enzymatically converts H_2O_2 to $2H_2O$ using reduced glutathione (GSH) as substrate which is then converted to its oxidized form, glutathione disulfide (GSSG) in the process. The GSSG in a second reaction involving glutathione reductase enzyme is converted back to GSH and thus GSH is recycled. The glutathione reductase reaction contributes significantly to the cellular maintenance of pooled reduced glutathione for redox defense.

Total and reduced fraction of the glutathione content in both cell types were experimentally determined while the content of oxidized glutathione was derived by deduction of the reduced glutathione values from the respective total glutathione values. The value of the total glutathione was obtained by reducing the GSSG fraction in each sample using TCEP which also has the ability to recover protein-bound GSH. However, because GSH exists within cells either freely or bound to proteins the recovered GSH still constituted a fraction of the total glutathione pool. In b.End5 cells the amount of glutathione per unit cell was higher than that of bEnd.3, however, the difference was not statistically significant (p = 0.1325) (Figure 3). Estimated values were 2.769 ± 0.113 fM/cell for b.End5 and 2.305 ± 0.219 fMol/cell for bEnd.3. Corresponding GSSG values for both b.End5 and bEnd.3 cells were respectively 0.139 ± 0.006 fM/cell and 0.115 ± 0.011 fM/cell. GSH/GSSG ratio in both cells approximated to 95%/5%.

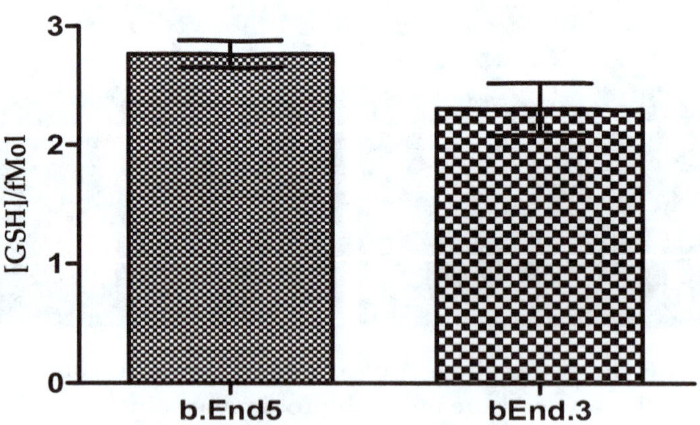

Figure 3. A comparison of GSH concentration in the two types of cells revealed no statistical difference between the two means (Student's t test: p = 0.1325).

3.3. Antioxidant Capacity Is Higher in b.End5 Cells

Although both cell lines tested showed that glutathione was abundantly present in both the cytoplasm and nucleoplasm, the response of these cell lines to ROS stress remains to be investigated. Both bEnd.3 and b.End5 cells treated with increasing concentrations of hydrogen peroxide showed changes in viability which correlated with the relative absorbance units obtained from XTT proliferation assay. The assay is based on the cleavage of the yellow tetrazolium salt, XTT, to form an orange formazan dye by mitochondrial dehydrogenases in metabolically active cells. Because an increase in the number of cells results in an increase in the overall activity of mitochondrial dehydrogenases in the samples, these changes correlated to the amount of orange formazan formed, a parameter that was monitored by the relative changes in the absorbance. The viability changes in both cells were normalized as percentages of the control, unexposed, cells and plotted against the logarithmic values of the various concentrations and the half-maximal inhibitory concentrations (IC_{50}) were determined and compared for the two cells (Figure 4A,B). Experiments were repeated three times and average values were analyzed. Results showed that the IC_{50} for the b.End5 cell was significantly higher than for the bEnd.3 cell (Figure 4C).

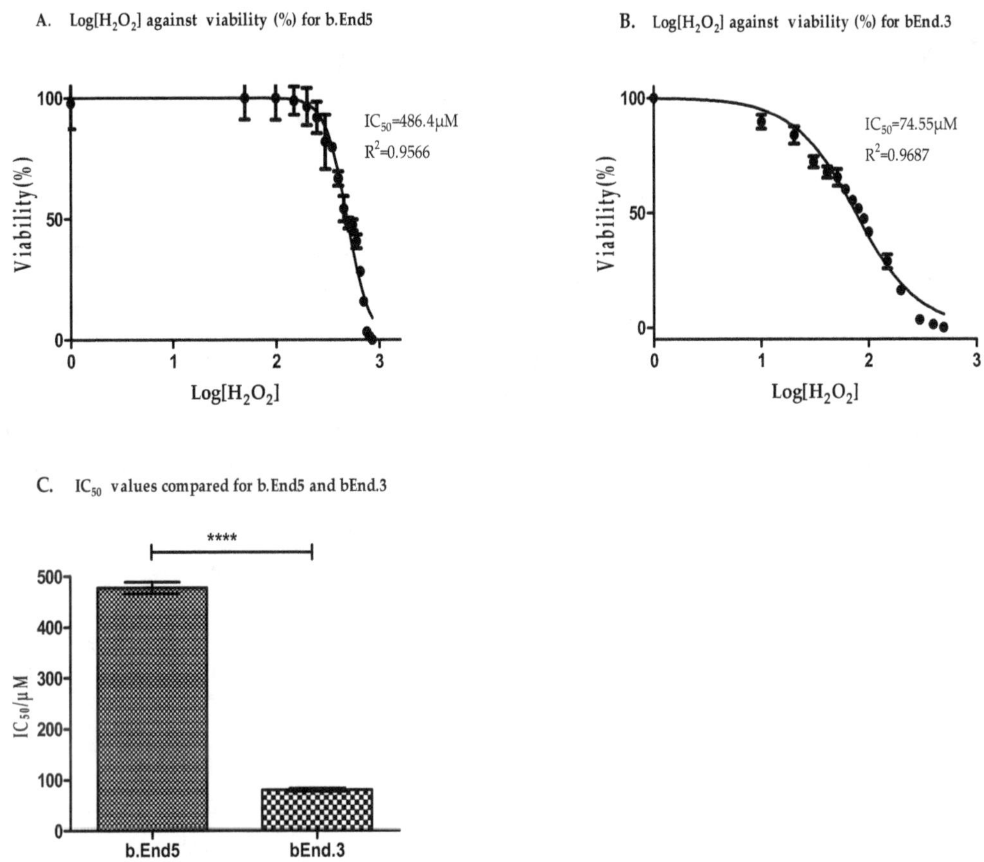

Figure 4. A non-linear regression analysis of logarithmic values of $[H_2O_2]$ against normalized viability was used to determine the hydrogen peroxide concentration that caused 50% inhibition of cell viability in b.End5 cells. Cells were exposed to $[H_2O_2]$ that ranged between 0 (control) and 850 μM for 24 h in flat-bottom transparent 96-well plates. Viability changes correlated to the absorbance measured at 450 nm from each well following incubation with XTT reagent for 4 h. (**A**) Results showed IC_{50} for b.End5 cell as equivalent of 486.4 μM at $r^2 = 0.9566$. (**B**) Data for bEnd.3 cells was obtained by exposing cultured bEnd.3 cells to $[H_2O_2]$ ranged from 0 (control) to 500 μM and non-linear regression analysis done as described above. Results showed IC_{50} for bEnd.3 cell to be 74.55 μM at $r^2 = 0.9687$ μM. (**C**) Graph of the IC_{50} values for the b.End5 compared with the same values for the bEnd.3 cells (annotation * denotes statistically significant difference between the values shown). IC_{50} values for b.End5 and bEnd.3 cells were statistically compared using the Student's *t* test. The analyzed data showed that the IC_{50} value was significantly higher for b.End5 cell than for the bEnd.3 cell ($p < 0.0001$).

3.4. Glutathione Was More Resistant to Oxidant Depletion in b.End5

Changes in the glutathione content of equal number of cells were plotted against the hydrogen peroxide concentrations. This allowed for the analysis of the physiological response (endogenous antioxidant response) of the two cell lines to increased H_2O_2 concentrations. The profile in glutathione depletion against increasing concentrations of hydrogen peroxide for the two cell lines was examined (Figure 5A,B). The data showed a steady decline in the glutathione content of the bEnd.3 cells (Figure 5B) while the b.End5 cells, in contrast, showed an initial significant increase in glutathione content and then a decline (Figure 5A). In b.End5 cells, glutathione increase was sustained either higher or at par with the control cells until about a concentration of 500 μM which was close to its IC_{50} value (Figure 5A). Also, the glutathione content of b.End5 cells were observed to decline to a steady lowest value at about 1 mM hydrogen peroxide concentration and thereafter remained constant.

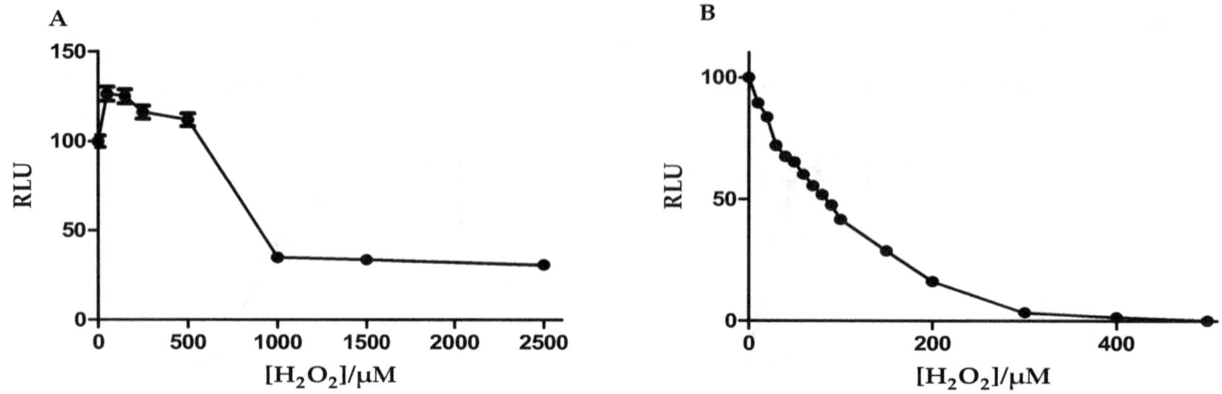

Figure 5. (**A**) and (**B**) show trends in the glutathione contents of b.End5 and bEnd3 cells with exposure to increasing $[H_2O_2]$. Cells were exposed to increasing concentrations of $[H_2O_2]$ for 24 h and average cellular glutathione contents estimated using the GSH-Glo kit. Glutathione contents of the cells correlated directly to the relative luminescent values (RLU) obtained following incubation with the optimized reagents of the GSH-Glo assay kit described. Data in Figure 4A represents b.End5 cells and shows an upward trend in the glutathione content of the b.End5 cells upon exposure to $[H_2O_2]$ of 0–250 μM. Above this concentration was observed a downward trend though values remained higher or at par with the starting point until $[H_2O_2]$ higher than 500 μM. From this point a steady decline occurred until about 1 mM $[H_2O_2]$ followed by a plateau but not a complete depletion. Data in Figure 4B represent the trend in bEnd.3 glutathione changes with increasing $[H_2O_2]$. A steady decline was observed until complete depletion at about 400 μM $[H_2O_2]$.

4. Discussion

Although several previous reports have reported the existence of strain-specific genes in contributing to some differences in the physiological characteristics of different organs and/or systems in animals of the same species [34–36], researchers have largely ignored this evidence in choosing cell lines to model a physiological system. This is particularly evident in the use of cell lines modelling the BBB, where a variety of primary and immortalized cell lines have been used to study its physiological functions. This phenomenon creates un-necessary murkiness in comparing data between different studies. In this study, we were able to illustrate convincingly that differences in the antioxidant characteristics of the two cell lines (b.End5 and bEnd.3) (Figure 4A–C) of the same species, have distinctly different response profiles to an escalating ROS stressor. These two cells were derived from primary brain endothelial cells of two different mouse strains, BALB/c and SV129 respectively, by infection with a retrovirus coding for the Polyoma middle T-antigen [37,38]. Both have been used extensively to study BBB function [39–41], all with the underlying assumption, that there is physiological parity between these cell lines.

We investigated the cellular content of the glutathione antioxidant in the two cell types under normal culture conditions (as per the instructions provided by the suppliers), first by fluorescent microscopy, and although we easily established the cellular presence for glutathione, we were unable to distinguish clear quantitative differences between the two cell lines. Close examination of the fluorescence micrographs clearly illustrated the even distribution of glutathione throughout the cytoplasm of both cell types, however, the higher ring of fluorescence just inside the nuclear membrane seem to suggest that glutathione plays an important role in neutralizing ROS entering into the nucleus. Given that DNA fragmentation is prone under conditions of OS [42,43], this provides solid circumstantial evidence as to the role of glutathione in protecting the nuclear material from ROS compromise. Although fluorescence demonstrated the presence of glutathione in both types of cells, the presence of blue fluorescence provided subjective evidence of increased cytoplasmic glutathione within b.End5 cells as evidenced by a lower nucleo-cytoplasmic ratio in the b.End5 cells with consequently more cytoplasmic content of glutathione.

Our quantitative data of the glutathione content of the b.End5 and bEnd.3 cell types substantiated our visual observations and, although b.End5 cells had a slightly higher mean glutathione content, statistically no significant difference in glutathione quantity was found between the respective cell lines (Figure 3). Calculations from data indicated that the GSH concentration per unit cell for the b.End5 and bEnd.3 cell types were within the range of 2.769 ± 0.113 fMol and 2.305 ± 0.219 fMol respectively. Both cell types also have GSH/GSSG ratio of approximately 95%/5% which indicated that both cells were redox-stable in normal culture. Comparisons of the average cellular glutathione per single cell for both the b.End5 and bEnd.3 cell lines predict that these cells are well suited for OS resistance in that their GSH contents are in the range of cells specialized for detoxification, such as the liver cancer cell line, HepG2 with GSH content of 2.9 fMol/cell [44]. These data, taken independently endorses the use of these cell lines for BBB modeling. It is presumptuous to assume that simply on the basis that the cell lines have similar basal levels of endogenous antioxidants that these cells may indeed respond to incremental ROS stress in a similar manner. This is all the more complicated given the arbitrary concentrations of ROS stressors used in experiments to demonstrate OS on the BBB [45,46].

To evaluate the response of the endogenous antioxidant glutathione to incremental concentrations of H_2O_2 in each cell-line, we then determined the relative levels of OS that causes decompensation of the GSH/GSSG antioxidant system in the cell after 24 h exposure (Figure 4A,B). In this study we profiled the cell lines using the glutathione system which has been reported as a reliable and sensitive marker of oxidative cellular status [47,48]. Furthermore, because the physiological variable of cell viability was used as a general indicator of the response of the cells to ROS (H_2O_2), it is scientifically plausible that viability changes will be a reliable indicator of the total cellular antioxidant capacity, and therefore, a singular marker (glutathione) would provide enough insight on the differences in the total antioxidant capacities of the two cell lines. Using the IC_{50} [49]values for comparison, we documented for the first time evidence of a significantly greater capability of the b.End5 cells to neutralize higher ROS concentrations than the bEnd.3 cells ([$p < 0.0001$] Figure 4C). When we studied the profile in the GSH depletion within the cells exposed to increasing ROS load the data showed that the bEnd3 cell type has a very limited ability for de novo upregulation of glutathione synthesis under condition of increased ROS accumulation (Figure 5B), whereas the b.End5 cell type showed ability to sustain adequate levels of glutathione, to elevate levels of GSH in response to initial low concentration of ROS, and to maintain this sustained endogenous GSH levels despite several subsequent increases in the concentration of ROS (Figure 5A). The mechanism for this divergent reaction of the glutathione system in the two cell types against increasing concentration of ROS is not clear, and requires further study. However, this observed difference is suggestive of inter-strain differences in the system that regulates ROS accumulation within these cells, as has been reported for several diverse physiological characteristics in different strains of cells from the same species of animal [35,49,50]. Differentially expressed genes in specific domains of the system that regulate ROS in intracellular milieu or perhaps strain-specific genetic alteration following viral oncogenic transformation of the primary cells could be responsible [51]. Mouse strain-specific differences in neuro-behavior, neuronal excitability, susceptibility to fibrosis, anti-inflammatory response, and bone density have been previously reported [50]. These previous reports have strengthened our position that strain-specific genes are very important in shaping the phenotypic differences in the two cell types investigated with respect to their glutathione system plasticity to OS. This finding has an important bearing on experimental designs aimed at studying effects of OS on the BBB. However, whether these two mouse cells respond differently to OS in the in vivo situation is not known. We propose here that a genome-wide genetic screen of the cells in use for evaluating the physiology of the BBB will identify important differences in the genes that control several key functions of the BBB [52]. Also, further research is required to determine these properties in primary cells of the mouse brain endothelial cells with the potential for the unraveling of the true behavior of the mouse brain endothelial cells during OS conditions with respect to capacity for ROS neutralization and the glutathione system response.

Given that under control conditions we are confident that the BECs would be in redox balance, it is, therefore, unimportant to localize the source of endogenous ROS production in response to exogenous ROS exposure, neither does it make scientific sense to measure ROS production from the cell organelles when the source of ROS was clearly defined as the experimental treatment (exogenous H_2O_2). Exogenous ROS treatment would gauge the physiological capacity of the cells to respond to ROS exposure from an exogenous source. Exogenous exposure would be additional to the normal ROS load generated by normal physiological cellular processes. Thus, the measuring of ROS levels while exposing cells to exogenous H_2O_2 in this study would be superfluous. Nevertheless, it might be assumed that the study largely ignored the spatiotemporal localization and quantification of endogenous ROS production, but we are aware of the lack of the use of these techniques, which is intended to be the focus of our future study, in which we propose to measure ROS produced endogenously by cellular organelles using a combination of immuno-electron microscopy as qualitative and ELISA as semi-quantitative methods [53–55].

5. Conclusions

The glutathione system responses and ROS buffering capacities are clearly linked and they determine the magnitude of ROS that induces OS in the b.End5 and bEnd.3 mouse brain endothelial cell line models of the BBB. Strain-specific differences in the different cells will result in different definitions of OS in different models of different animal strains within the same species. Thus it is important to establish experimental parameters that best define OS for each endothelial cell model of the BBB for reproducibility. Such information will avail researchers of opportunity to verify and select appropriate models of BBB endothelial cells for specific redox investigations and enable them to draw comparable and reproducible conclusions.

Author Contributions: O.A.: conceptualization, investigation, methodology, formal analysis, and writing—original draft. M.R.: investigation assistance. O.E.: resources, writing—review, and editing. D.F.: resources, supervision, writing—review, and editing. All authors have read and agreed to the published version of the manuscript.

Acknowledgments: The authors wish to acknowledge the technical support of Shireen Mentor.

References

1. Abbott, N.J.; Patabendige, A.A.; Dolman, D.E.; Yusof, S.R.; Begley, D.J. Structure and function of the blood-brain barrier. *Neurobiol. Dis.* **2010**, *37*, 13–25. [CrossRef] [PubMed]
2. Ramirez, S.H.; Potula, R.; Fan, S.; Eidem, T.; Papugani, A.; Reichenbach, N.; Dykstra, H.; Weksler, B.B.; Romero, I.A.; Couraud, P.O.; et al. Methamphetamine disrupts blood brain barrier function by induction of oxidative stress in brain endothelial cells. *J. Cereb. Blood Flow Metab.* **2009**, *29*, 1933–1945. [CrossRef] [PubMed]
3. Betzer, O.; Shilo, M.; Motiei, M.; Popovtzer, R. Insulin-coated gold nanoparticles as an effective approach for bypassing the blood-brain barrier. *Nanoscale Imaging Sens. Actuation Biomed. Appl. XVI* **2019**, *10891*, 108911H.
4. Zhao, Z.; Nelson, A.R.; Betsholtz, C.; Zlokovic, B.V. Establishment and dysfunction of the blood-brain barrier. *Cell* **2015**, *163*, 1064–1078. [CrossRef]
5. Zhao, Z.; Sagare, A.P.; Ma, Q.; Halliday, M.R.; Kong, P.; Kisler, K.; Winkler, E.A.; Ramanathan, A.; Kanekiyo, T.; Bu, G. Central role for PICALM in amyloid-β blood-brain barrier transcytosis and clearance. *Nat. Neurosci.* **2015**, *18*, 978–987. [CrossRef]
6. Liebner, S.; Dijkhuizen, R.M.; Reiss, Y.; Plate, K.H.; Agalliu, D.; Constantin, G. Functional morphology of the blood–brain barrier in health and disease. *Acta Neuropathol.* **2018**, *135*, 311–336. [CrossRef]

7. Nation, D.A.; Sweeney, M.D.; Montagne, A.; Sagare, A.P.; D'Orazio, L.M.; Pachicano, M.; Sepehrband, F.; Nelson, A.R.; Buennagel, D.P.; Harrington, M.G. Blood–brain barrier breakdown is an early biomarker of human cognitive dysfunction. *Nat. Med.* **2019**, *25*, 270–276. [CrossRef]
8. Fan, L.M.; Cahill-Smith, S.; Geng, L.; Du, J.; Brooks, G.; Li, J.-M. Aging-associated metabolic disorder induces Nox2 activation and oxidative damage of endothelial function. *Free Radic. Biol. Med.* **2017**, *108*, 940–951. [CrossRef] [PubMed]
9. Breitenbach, M.; Rinnerthaler, M.; Weber, M.; Breitenbach-Koller, H.; Karl, T.; Cullen, P.; Basu, S.; Haskova, D.; Hasek, J. The defense and signaling role of NADPH oxidases in eukaryotic cells. *Wien. Med. Wochenschr.* **2018**, *168*, 286–299. [CrossRef]
10. Sies, H.; Berndt, C.; Jones, D.P. Oxidative Stress. *Annu. Rev. Biochem.* **2017**, *86*, 715–748. [CrossRef]
11. Mirończuk-Chodakowska, I.; Witkowska, A.M.; Zujko, M.E. Endogenous non-enzymatic antioxidants in the human body. *Adv. Med. Sci.* **2018**, *63*, 68–78. [CrossRef] [PubMed]
12. Forstermann, U.; Munzel, T. Endothelial nitric oxide synthase in vascular disease: From marvel to menace. *Circulation* **2006**, *113*, 1708–1714. [CrossRef] [PubMed]
13. Koch, S.R.; Choi, H.; Mace, E.H.; Stark, R.J. Toll-like receptor 3-mediated inflammation by p38 is enhanced by endothelial nitric oxide synthase knockdown. *Cell Commun. Signal.* **2019**, *17*, 33. [CrossRef] [PubMed]
14. Santhanam, A.V.R.; d'Uscio, L.V.; He, T.; Das, P.; Younkin, S.G.; Katusic, Z.S. Uncoupling of endothelial nitric oxide synthase in cerebral vasculature of Tg2576 mice. *J. Neurochem.* **2015**, *134*, 1129–1138. [CrossRef]
15. Drummond, G.R.; Cai, H.; Davis, M.E.; Ramasamy, S.; Harrison, D.G. Transcriptional and posttranscriptional regulation of endothelial nitric oxide synthase expression by hydrogen peroxide. *Circ. Res.* **2000**, *86*, 347–354. [CrossRef]
16. Toda, N.; Okamura, T. Cigarette smoking impairs nitric oxide-mediated cerebral blood flow increase: Implications for Alzheimer's disease. *J. Pharmacol. Sci.* **2016**, *131*, 223–232. [CrossRef]
17. Enciu, A.-M.; Gherghiceanu, M.; Popescu, B.O. Triggers and Effectors of Oxidative Stress at Blood-Brain Barrier Level: Relevance for Brain Ageing and Neurodegeneration. *Oxidative Med. Cell. Longev.* **2013**, *2013*, 1–12. [CrossRef]
18. Dalleau, S.; Baradat, M.; Gueraud, F.; Huc, L. Cell death and diseases related to oxidative stress: 4-hydroxynonenal (HNE) in the balance. *Cell Death Differ.* **2013**, *20*, 1615–1630. [CrossRef]
19. Sies, H. On the history of oxidative stress: Concept and some aspects of current development. *Curr. Opin. Toxicol.* **2018**, *7*, 122–126. [CrossRef]
20. Dohgu, S.; Takata, F.; Matsumoto, J.; Kimura, I.; Yamauchi, A.; Kataoka, Y. Monomeric α-synuclein induces blood–brain barrier dysfunction through activated brain pericytes releasing inflammatory mediators in vitro. *Microvasc. Res.* **2019**, *124*, 61–66. [CrossRef]
21. Solé, M.; Esteban-Lopez, M.; Taltavull, B.; Fábregas, C.; Fadó, R.; Casals, N.; Rodríguez-Álvarez, J.; Miñano-Molina, A.J.; Unzeta, M. Blood-brain barrier dysfunction underlying Alzheimer's disease is induced by an SSAO/VAP-1-dependent cerebrovascular activation with enhanced Aβ deposition. *Biochim. Et Biophys. Acta (BBA)-Mol. Basis Dis.* **2019**, *1865*, 2189–2202.
22. Gastfriend, B.D.; Palecek, S.P.; Shusta, E.V. Modeling the blood–brain barrier: Beyond the endothelial cells. *Curr. Opin. Biomed. Eng.* **2018**, *5*, 6–12. [CrossRef] [PubMed]
23. Campisi, M.; Shin, Y.; Osaki, T.; Hajal, C.; Chiono, V.; Kamm, R.D. 3D self-organized microvascular model of the human blood-brain barrier with endothelial cells, pericytes and astrocytes. *Biomaterials* **2018**, *180*, 117–129. [CrossRef] [PubMed]
24. Linville, R.M.; DeStefano, J.G.; Sklar, M.B.; Xu, Z.; Farrell, A.M.; Bogorad, M.I.; Chu, C.; Walczak, P.; Cheng, L.; Mahairaki, V. Human iPSC-derived blood-brain barrier microvessels: Validation of barrier function and endothelial cell behavior. *Biomaterials* **2019**, *190*, 24–37. [CrossRef]
25. Steiner, O.; Coisne, C.; Engelhardt, B.; Lyck, R. Comparison of immortalized bEnd5 and primary mouse brain microvascular endothelial cells as in vitro blood–brain barrier models for the study of T cell extravasation. *J. Cereb. Blood Flow Metab.* **2011**, *31*, 315–327. [CrossRef]
26. He, F.; Yin, F.; Peng, J.; Li, K.Z.; Wu, L.W.; Deng, X.L. Immortalized mouse brain endothelial cell line Bend.3 displays the comparative barrier characteristics as the primary brain microvascular endothelial cells. *Zhongguo Dang Dai Er Ke Za Zhi = Chin. J. Contemp. Pediatrics* **2010**, *12*, 474–478.
27. Glasauer, A.; Chandel, N.S. Ros. *Curr. Biol.* **2013**, *23*, R100–R102. [CrossRef]

28. Lushchak, V.I. Glutathione Homeostasis and Functions: Potential Targets for Medical Interventions. *J. Amino Acids* **2012**, *2012*, 1–26. [CrossRef]
29. Jones, D.P.; Park, Y.; Gletsu-Miller, N.; Liang, Y.; Yu, T.; Accardi, C.J.; Ziegler, T.R. Dietary sulfur amino acid effects on fasting plasma cysteine/cystine redox potential in humans. *Nutrition* **2011**, *27*, 199–205. [CrossRef]
30. Kazaks, A.; Collier, M.; Conley, M. Cytotoxicity of Caffeine on MCF-7 Cells Measured by XTT Cell Proliferation Assay (P06-038-19). *Curr. Dev. Nutr.* **2019**, *3*. [CrossRef]
31. Chatterjee, S.; Noack, H.; Possel, H.; Keilhoff, G.; Wolf, G. Glutathione levels in primary glial cultures: Monochlorobimane provides evidence of cell type-specific distribution. *Glia* **1999**, *27*, 152–161. [CrossRef]
32. Scherer, C.; Cristofanon, S.; Dicato, M.; Diederich, M. Homogeneous luminescence-based assay for quantifying the glutathione content in mammalian cells. *Cells Nots* **2008**, *22*, 7–9.
33. Tietze, F. Enzymic method for quantitative determination of nanogram amounts of total and oxidized glutathione: Applications to mammalian blood and other tissues. *Anal. Biochem.* **1969**, *27*, 502–522. [CrossRef]
34. Zhai, R.; Xue, X.; Zhang, L.; Yang, X.; Zhao, L.; Zhang, C. Strain-Specific Anti-inflammatory Properties of Two Akkermansia muciniphila Strains on Chronic Colitis in Mice. *Front. Cell. Infect. Microbiol.* **2019**, *9*, 239. [CrossRef]
35. Sinclair, J.L.; Barnes-Davies, M.; Kopp-Scheinpflug, C.; Forsythe, I.D. Strain-specific differences in the development of neuronal excitability in the mouse ventral nucleus of the trapezoid body. *Hear. Res.* **2017**, *354*, 28–37. [CrossRef]
36. Gooch, J.L.; Yee, D. Strain-specific differences in formation of apoptotic DNA ladders in MCF-7 breast cancer cells. *Cancer Lett.* **1999**, *144*, 31–37. [CrossRef]
37. Montesano, R.; Pepper, M.; Möhle-Steinlein, U.; Risau, W.; Wagner, E.; Orci, L. Increased proteolytic activity is responsible for the aberrant morphogenetic behavior of endothelial cells expressing the middle T oncogene. *Cell* **1990**, *62*, 435–445. [CrossRef]
38. Williams, R.L.; Risau, W.; Zerwes, H.-G.; Drexler, H.; Aguzzi, A.; Wagner, E.F. Endothelioma cells expressing the polyoma middle T oncogene induce hemangiomas by host cell recruitment. *Cell* **1989**, *57*, 1053–1063. [CrossRef]
39. Helms, H.C.; Abbott, N.J.; Burek, M.; Cecchelli, R.; Couraud, P.-O.; Deli, M.A.; Förster, C.; Galla, H.J.; Romero, I.A.; Shusta, E.V. In vitro models of the blood–brain barrier: An overview of commonly used brain endothelial cell culture models and guidelines for their use. *J. Cereb. Blood Flow Metab.* **2016**, *36*, 862–890. [CrossRef]
40. Yang, S.; Mei, S.; Jin, H.; Zhu, B.; Tian, Y.; Huo, J.; Cui, X.; Guo, A.; Zhao, Z. Identification of two immortalized cell lines, ECV304 and bEnd3, for in vitro permeability studies of blood-brain barrier. *PLoS ONE* **2017**, *12*, e0187017. [CrossRef]
41. Yang, T.; Roder, K.E.; Abbruscato, T.J. Evaluation of bEnd5 cell line as an in vitro model for the blood–brain barrier under normal and hypoxic/aglycemic conditions. *J. Pharm. Sci.* **2007**, *96*, 3196–3213. [CrossRef] [PubMed]
42. Agarwal, A.; Cho, C.-L.; Esteves, S.C.; Majzoub, A. Reactive oxygen species and sperm DNA fragmentation. *Transl. Androl. Urol.* **2017**, *6*, S695. [CrossRef] [PubMed]
43. Homa, S.T.; Vassiliou, A.M.; Stone, J.; Killeen, A.P.; Dawkins, A.; Xie, J.; Gould, F.; Ramsay, J.W. A comparison between two assays for measuring seminal oxidative stress and their relationship with sperm DNA fragmentation and semen parameters. *Genes* **2019**, *10*, 236. [CrossRef]
44. Yuan, Y.; Zhang, J.; Wang, M.; Mei, B.; Guan, Y.; Liang, G. Detection of glutathione in vitro and in cells by the controlled self-assembly of nanorings. *Anal. Chem.* **2013**, *85*, 1280–1284. [CrossRef] [PubMed]
45. Cao, C.; Dai, L.; Mu, J.; Wang, X.; Hong, Y.; Zhu, C.; Jin, L.; Li, S. S1PR2 antagonist alleviates oxidative stress-enhanced brain endothelial permeability by attenuating p38 and Erk1/2-dependent cPLA2 phosphorylation. *Cell. Signal.* **2019**, *53*, 151–161. [CrossRef]
46. Song, J.; Kang, S.M.; Lee, W.T.; Park, K.A.; Lee, K.M.; Lee, J.E. Glutathione protects brain endothelial cells from hydrogen peroxide-induced oxidative stress by increasing nrf2 expression. *Exp. Neurobiol.* **2014**, *23*, 93–103. [CrossRef]
47. Bains, V.K.; Bains, R. The antioxidant master glutathione and periodontal health. *Dent. Res. J.* **2015**, *12*, 389–405. [CrossRef]

48. Kranner, I.; Birtić, S.; Anderson, K.M.; Pritchard, H.W. Glutathione half-cell reduction potential: A universal stress marker and modulator of programmed cell death? *Free Radic. Biol. Med.* **2006**, *40*, 2155–2165. [CrossRef]
49. Doroshow, J.H.; Juhasz, A. Modulation of selenium-dependent glutathione peroxidase activity enhances doxorubicin-induced apoptosis, tumour cell killing and hydroxyl radical production in human NCI/ADR-RES cancer cells despite high-level P-glycoprotein expression. *Free Radic. Res.* **2019**, *53*, 882–891. [CrossRef]
50. Shidara, K.; Mohan, G.; Lay, Y.-A.E.; Jepsen, K.J.; Yao, W.; Lane, N.E. Strain-specific differences in the development of bone loss and incidence of osteonecrosis following glucocorticoid treatment in two different mouse strains. *J. Orthop. Transl.* **2019**, *16*, 91–101. [CrossRef]
51. Sandberg, R.; Yasuda, R.; Pankratz, D.G.; Carter, T.A.; Del Rio, J.A.; Wodicka, L.; Mayford, M.; Lockhart, D.J.; Barlow, C. Regional and strain-specific gene expression mapping in the adult mouse brain. *Proc. Natl. Acad. Sci. USA* **2000**, *97*, 11038–11043. [CrossRef] [PubMed]
52. Cao, J.Y.; Poddar, A.; Magtanong, L.; Lumb, J.H.; Mileur, T.R.; Reid, M.A.; Dovey, C.M.; Wang, J.; Locasale, J.W.; Stone, E. A genome-wide haploid genetic screen identifies regulators of glutathione abundance and ferroptosis sensitivity. *Cell Rep.* **2019**, *26*, 1544–1556. [CrossRef] [PubMed]
53. Tanikawa, K.; Torimura, T. Studies on oxidative stress in liver diseases: Important future trends in liver research. *Med. Mol. Morphol.* **2006**, *39*, 22–27. [CrossRef] [PubMed]
54. Grasso, G.; Komatsu, H.; Axelsen, P. Covalent modifications of the amyloid beta peptide by hydroxynonenal: Effects on metal ion binding by monomers and insights into the fibril topology. *J. Inorg. Biochem.* **2017**, *174*, 130–136. [CrossRef]
55. Jaganjac, M.; Milkovic, L.; Gegotek, A.; Cindric, M.; Zarkovic, K.; Skrzydlewska, E.; Zarkovic, N. The relevance of pathophysiological alterations in redox signaling of 4-hydroxynonenal for pharmacological therapies of major stress-associated diseases. *Free Radic. Biol. Med.* **2019**. [CrossRef]

5

Intermittent Hypoxia Prevents Myocardial Mitochondrial Ca^{2+} Overload and Cell Death during Ischemia/Reperfusion: The Role of Reactive Oxygen Species

Jui-Chih Chang [1,2], Chih-Feng Lien [3], Wen-Sen Lee [4], Huai-Ren Chang [2,5], Yu-Cheng Hsu [6], Yu-Po Luo [1], Jing-Ren Jeng [2,5], Jen-Che Hsieh [2,5] and Kun-Ta Yang [7,*]

1 Department of Surgery, Buddhist Tzu Chi General Hospital, Hualien 97002, Taiwan; medraytw@hotmail.com (J.-C.C.); 103327102@gms.tcu.edu.tw (Y.-P.L.)
2 School of Medicine, Tzu Chi University, Hualien 97004, Taiwan; huairenchang@mail.tcu.edu.tw (H.-R.C.); jrj4511@gmail.com (J.-R.J.); jenchehsieh@gmail.com (J.-C.H.)
3 Institute of Medical Sciences, Tzu Chi University, Hualien 97004, Taiwan; 98327101@gms.tcu.edu.tw
4 Graduate Institute of Medical Sciences, College of Medicine, Taipei Medical University, Taipei 11031, Taiwan; wslee@tmu.edu.tw
5 Division of Cardiology, Department of Internal Medicine, Buddhist Tzu Chi General Hospital, Hualien 97002, Taiwan
6 Master Program in Medical Physiology, School of Medicine, Tzu Chi University, Hualien 97004, Taiwan; washwolf007@gmail.com
7 Department of Physiology, School of Medicine, Tzu Chi University, Hualien 97004, Taiwan
* Correspondence: ktyang@mail.tcu.edu.tw

Abstract: It has been documented that reactive oxygen species (ROS) contribute to oxidative stress, leading to diseases such as ischemic heart disease. Recently, increasing evidence has indicated that short-term intermittent hypoxia (IH), similar to ischemia preconditioning, could yield cardioprotection. However, the underlying mechanism for the IH-induced cardioprotective effect remains unclear. The aim of this study was to determine whether IH exposure can enhance antioxidant capacity, which contributes to cardioprotection against oxidative stress and ischemia/reperfusion (I/R) injury in cardiomyocytes. Primary rat neonatal cardiomyocytes were cultured in IH condition with an oscillating O_2 concentration between 20% and 5% every 30 min. An MTT assay was conducted to examine the cell viability. Annexin V-FITC and SYTOX green fluorescent intensity and caspase 3 activity were detected to analyze the cell death. Fluorescent images for DCFDA, Fura-2, Rhod-2, and TMRM were acquired to analyze the ROS, cytosol Ca^{2+}, mitochondrial Ca^{2+}, and mitochondrial membrane potential, respectively. RT-PCR, immunocytofluorescence staining, and antioxidant activity assay were conducted to detect the expression of antioxidant enzymes. Our results show that IH induced slight increases of $O_2^{-\cdot}$ and protected cardiomyocytes against H_2O_2- and I/R-induced cell death. Moreover, H_2O_2-induced Ca^{2+} imbalance and mitochondrial membrane depolarization were attenuated by IH, which also reduced the I/R-induced Ca^{2+} overload. Furthermore, treatment with IH increased the expression of Cu/Zn SOD and Mn SOD, the total antioxidant capacity, and the activity of catalase. Blockade of the IH-increased ROS production abolished the protective effects of IH on the Ca^{2+} homeostasis and antioxidant defense capacity. Taken together, our findings suggest that IH protected the cardiomyocytes against H_2O_2- and I/R-induced oxidative stress and cell death through maintaining Ca^{2+} homeostasis as well as the mitochondrial membrane potential, and upregulation of antioxidant enzymes.

Keywords: intermittent hypoxia; mitochondria; Ca^{2+}, ROS; antioxidant

1. Introduction

Obstructive sleep apnea (OSA), also known as intermittent hypoxia (IH), is characterized by repetitive episodic obstructions of airflow during sleep [1]. It has been indicated that IH is related to cardiovascular diseases, including hypertension, stroke, and myocardial infarction [2,3]. On the other hand, some studies have demonstrated that IH can protect against the cell death induced by ischemia or ischemia reperfusion (I/R) injury [4–6]. Several mechanisms have been proposed to be involved in the IH-induced protective effects, including activation of hypoxia-responsive genes, amelioration of coronary circulation, activation of protein kinase C, balance of Ca^{2+} handling activity, and inhibition of mitochondrial permeability transition pores (mPTP) opening [6–10]. However, the complete effects and underlying mechanisms of IH on cardiac function are still unclear.

Ca^{2+} homeostasis, which is dependent on a complex network of ion transporters, channels, and regulatory proteins, is important for maintaining cardiac functions. It has been demonstrated that the activities of L-type Ca^{2+} channel (LTCC), ryanodine receptors (RyR), sarcoplasmic/endoplasmic reticulum Ca^{2+}-ATPase (SERCA) and phospholamban were regulated by Ca^{2+}/calmodulin-dependent kinase II (CaMKII) and cAMP-dependent protein kinase A (PKA), and these proteins have been demonstrated to be involved in regulating intracellular Ca^{2+} homeostasis under physiological and pathophysiological conditions [11,12]. In recent years, post-translational oxidative modification of LTCC, RyR and SERCA caused by ROS also has been reported [13]. ROS has been considered an important signaling molecule in regulating cardioprotection. It has been shown that dimethylthiourea (DMTU), a potent hydroxyl radical scavenger, inhibits the ischemic preconditioning-mediated tyrosine kinase phosphorylation and cardioprotection [14]. Furthermore, the cardioprotective effects induced by diazoxide, a selective opener of the mitochondrial ATP-sensitive potassium channel, are abolished by *N*-acetyl cysteine, a ROS scavenger [15]. ROS has also been demonstrated to regulate other signaling molecules or channels, including protein tyrosine kinase, MAP kinase, phospholipase c (PLC), NFκB, IP3 receptor, ryanodine receptor, and Na^{+}/Ca^{2+} exchanger (NCX) [16]. These findings suggest that ROS play a crucial role in the preconditioning-induced cardioprotection.

Intracellular Ca^{2+} plays an important role in regulating cardiomyocyte behavior under physiological and pathophysiological conditions, such as myocyte excitation-contraction coupling, cell proliferation and differentiation, and cell death. Disruptions in Ca^{2+} handling can contribute to the pathogenesis of many diseases, such as Alzheimer's disease, Huntington's disease, and congestive heart failure [17]. Previous studies reported that IH can prevent the I/R-induced cell death via ameliorating Ca^{2+} homeostasis by increasing the activities of RyR, SERCA, and NCX during I/R [6,9]. I/R injury can induce burst ROS production to trigger cytosolic Ca^{2+} overload, leading to an excessive increase in mitochondrial Ca^{2+}, which in turn induces mPTP to open and depolarizes the mitochondrial membrane potential. The opening of mPTP leads to a loss of ATP, mitochondrial swelling, and release of cytochrome c, resulting in apoptosis [18]. Zhu et al. reported that IH improves mitochondrial tolerance to Ca^{2+} overload and delays oxidative stress-induced mPTP opening [7]. Additionally, it has been indicated that IH can increase the expression of superoxide dismutase (SOD) and glutathione peroxidase (GPx) [19]. These findings reveal that oxidative stress might play a key role in inducing intracellular Ca^{2+} overload, and IH probably enhances the antioxidant capacity to prevent oxidative stress-induced intracellular Ca^{2+} overload. However, the relationship between IH-increased antioxidant enzyme expression and the balance of Ca^{2+} handling activity is still unclear. In this study, we aimed to determine whether IH induced cardioprotective effects through improving the intracellular Ca^{2+} balance via ROS-increased antioxidant capacity.

2. Materials and Methods

2.1. Chemicals

Fluorescent indicators were purchased from Molecular Probes (Eugene, OR, USA). Fetal bovine serum (FBS), penicillin, and trypsin were from Gibco/Life Technologies (Rockville, MD, USA). All other chemicals were purchased from Sigma (St. Louis, MO, USA).

2.2. *Preparation of Neonatal Rat Cardiomyocytes*

Neonatal rat cardiomyocytes were prepared and cultured as described previously [20]. Briefly, 1–2-day-old Sprague-Dawley rats (both sexes) were sacrificed by decapitation. The ventricles were pooled from several hearts and minced into small pieces. Cardiac tissues were digested using 0.051% pancreatin and 0.01% collagenase in Hank's solution, and then incubated with F-12 medium containing 80% F-12 nutrient mixture, 10% horse serum, 10% FBS, and 1% penicillin (Gibco/Life Technologies) to inactivate enzymatic digestion. Cells were pre-plated on a 10-cm dish for 1 h at 37 °C in a 5% CO_2 incubator to remove fibroblasts. Subsequently, cardiomyocytes were seeded on 24-mm collagen-coated cover slips in F-12 medium, and then added 10 μM cytosine arabinoside to inhibit fibroblast proliferation. The medium was replaced daily during the experiments. All animal studies were performed following the recommended procedures approved by the Institutional Animal Care and Use Committee of Tzu Chi University (PPL number: 104107) and conform to the guidelines from Directive 2010/63/EU of the European Parliament on the protection of animals used for scientific purposes or the NIH guidelines.

2.3. *IH Exposures*

Neonatal rat cardiomyocytes were placed in Plexiglas box chambers (length 25 cm, width 30 cm, and height 15 cm). The cardiomyocytes were exposed to room air (RA)/normoxia (20% O_2, 5% CO_2, and 75% N_2) or IH (5% O_2, 5% CO_2, and 90% N_2 for 30 min alternating with 30 min RA) for 1–4 days using a timer solenoid valve control [20]. Oxygen fractions in the chambers were continuously monitored by an oxygen detector. A micro dissolved oxygen electrode from Lazar Research Laboratories (DO-166MT-1, Los Angeles, CA, USA) was used to detect fluctuations in oxygen concentrations in the medium and the chamber.

2.4. *Ischemia and Reperfusion (I/R) Injury*

Simulated I/R in cultured cardiomyocytes was performed using a modified protocol described previously [21]. Briefly, cardiomyocytes were stabilized at 37 °C in Normal Tyrode (NT) buffer (140 mM NaCl, 4.5 mM KCl, 2.0 mM $CaCl_2$, 1.2 mM $MgCl_2$, 11 mM glucose, and 10 mM HEPES, with pH adjusted to 7.4 using NaOH) for 10 min, transferred to 100% N_2-saturated ischemia buffer (123 mM NaCl, 8 mM KCl, 2.5 mM $CaCl_2$, 0.9 mM NaH_2PO_4, 0.5 mM $MgSO_4$, 20 mM Na-lactate, and 20 mM HEPES with pH adjusted to 6.0 using NaOH) for 6 h, and then reperfused with a culture medium for 12 h in a 5% CO_2 incubator.

2.5. *Analysis of Cell Death by Flow Cytometry*

Apoptosis/necrosis was determined using Annexin V-FITC Apoptosis Detection Kit (BioVision, Inc., Mountain View, CA, USA) according to the manufacturer's recommendations. Cardiomyocytes were washed with NT, dissociated by trypsin gently, harvested by centrifugation, and stained with Annexin V-FITC and SYTOX green (BioVision, Inc.) in binding buffer for 10 min at RT. Fluorescence was detected on a Gallios Flow Cytometer (Beckman Coulter, Indianapolis, IN, USA). The excitation/emission wavelengths for Annexin V-FITC and SYTOX were 488/520 nm. Based on the staining intensity, the cell population was divided into three groups: live cells (M1), apoptotic cells (M2), and necrotic cells (M3).

2.6. *Cell Viability Assay*

Cell viability was measured by the MTT assay preformed in triplicate. Briefly, cardiomyocytes were washed with NT, and incubated in culture medium containing 50 μg/mL MTT for 1 h at 37 °C in a 5% CO_2 incubator. After removal of the MTT medium, 1 mL of DMSO was added to each dish to dissolve the precipitate for 10 min, and then detected by a Multiskan EX ELISA Reader (Thermo Scientific, Rockford, IL, USA). The absorbance at 570 nm of the control group cells was considered to be 100%.

2.7. Imaging of Intracellular Reactive Oxygen Species (ROS)

Intracellular levels of ROS were detected using 5-(and-6)-chloromethyl-2′,7′-dichlorodihydrofluorescein diacetate acetyl ester (CM-H_2DCFDA; DCFDA). DCFDA fluorescence is triggered by oxidation via hydroxides (OH^-), hydrogen peroxides (H_2O_2) or hydroxyl radicals ($OH^{\cdot}$). Cardiomyocytes were loaded with 5 μM DCFDA for 30 min at RT, and washed twice with NT buffer. Using a real-time fluorescence imaging system including Sutter LAMBDA-LS/30 (Leica, Microsystems, Wetzlar, Germany), 1.10 set Filter wheel with Smart Shutter (Leica), and inverted fluorescent microscope (DMI3000 B; Leica) to detect fluorescence signal. Cells perfused continuously with NT buffer (37 °C) were used for fluorescent imaging. The wavelength of 480 nm was used for DCFDA excitation and 535 nm for emission. Signal increases are presented as the peak/basal fluorescence ratio (F/F_0).

2.8. Flow Cytometric Analysis for Oxidative Stress Detection

Intracellular levels of ROS were detected by DCFDA. After IH or H_2O_2 treatment, cardiomyocytes were washed with NT solution, loaded with 5 μM DCFDA for 30 min at RT. Subsequently, cells were trypsinized, resuspended, and placed on ice. Fluorescence was measured on a Gallios Flow Cytometer (Beckman Coulter). The excitation/emission wavelengths for DCFDA was 488/540 nm (FL1).

2.9. Quantitative Real-Time PCR

Total RNA was isolated from neonatal rat cardiomyocytes using TRIzol reagent (Life Technologies) according to the manufacturer's protocol. cDNA was synthesized using a Verso™ cDNA Kit (Thermo). Total RNA (3 μg) was used for quantitative real-time PCR, which was performed using Maxima SYBR Green qPCR Master Mix (2×) (Thermo) with a LightCycler 480 (Roche). The primer sequences were as follows: *CuZnSOD*: forward 5′-TGG GAG AGC TTG TCA GGT G-3′, reverse 5′-CAC CAG TAG CAG GTT GCA GA-3′; *MnSOD*: forward 5′-GCC TCC CTG ACC TGC CTT AC-3′, reverse 5′-GCA TGA TCT GCG CGT TAA TG-3′; *Catalase*: forward 5′-CCC AGA AGC CTA AGA ATG CAA-3′, reverse 5′-GCT TTT CCC TTG GCA GCT ATG-3′; *GPx*: forward 5′-GTG TTC CAG TGC GCA GAT ACA-3′, reverse 5′-GGG CTT CTA TAT CGG GTT CGA-3′; and *GAPDH*: forward 5′-ATG TTC CAG TAT GAC TCC ACT CAC G-3′, reverse 5′-GAA GAC ACC AGT AGA CTC CAC GAC A-3′. All gene expression was analyzed using the comparative Ct method ($2^{-\Delta\Delta Ct}$); $\Delta\Delta Ct = \Delta Ct$ (sample) − ΔCt (reference) relative to β-actin levels.

2.10. Measurement of Total Antioxidant Capacity and Catalase and Glutathione Peroxidase Activity

Cardiomyocytes were scraped on ice and centrifuged at 1400× *g* at 4 °C for 10 min. The cell pellet was sonicated in assay buffer on ice, and centrifuged at 10,600× *g* at 4 °C for 15 min. Protein concentration in supernatants was quantified using the Protein Assay kit (Biorad, Hercules, CA, USA). Equal amounts of protein lysates were used for determination of total antioxidant capacity (TAC), catalase and glutathione peroxidase (GPx) activity. TAC was measured using the Antioxidant Assay Kit (Cayman Chemical, Ann Arbor, MI, USA). Catalase activity was measured using the Catalase Assay Kit (Cayman Chemical). GPx activity was measured using the Glutathione Peroxidase Assay Kit (Cayman Chemical). These assay kits were used according to the manufacturer's instructions. 96-Well assay plates were read using an ELISA plate reader (Thermo Scientific) at 405 nm for TAC, at 540 nm for CAT, and at 340 nm for GPx.

2.11. Immunocytofluorescence Staining

Cardiomyocytes were fixed with methanol at 4 °C for 10 min or 10% formalin in NT at RT for 1 h, incubated in 5% non-fat milk for 60 min to block the non-specific IgG followed by primary antibody for 60 min at 37 °C, and then incubated with secondary antibody (FITC-conjugated goat anti-rabbit IgG or anti-mouse IgG) for 60 min at 37 °C. Images were obtained by confocal microscopy with resolution of 512 × 512 pixels. Green fluorescence represented FITC with excitation/emission wavelengths at 488/530 nm. MnSOD (1:200) and CuZnSOD (1:200) antibodies were purchased from Upstate Biotechnology (Lake Placid, NY, USA).

2.12. Imaging of Intracellular Ca^{2+} Concentrations ($[Ca^{2+}]_i$)

Cardiomyocytes were loaded with 5 μM Fura-2 AM for 60 min at RT, and then washed twice with NT buffer. A real-time fluorescence imaging system, including Sutter LAMBDA-LS/30 (Leica), 1.10 set Filter wheel with Smart Shutter (Leica), and inverted fluorescent microscope (DMI3000 B; Leica), was used to detect fluorescence signal. Cells perfused continuously with NT buffer (37 °C) were used for fluorescent imaging. Fura-2 was excited alternately at 340/380 nm, and emission was monitored at 520 nm. $[Ca^{2+}]_i$ were expressed as a ratio of F340/F380. Simulated I/R in cultured cardiomyocytes were placed on inverted fluorescent microscope. Cardiomyocytes were placed in a plexiglass box chamber containing 100% N_2, and perfused with 100% N_2-saturated ischemia buffer for 30 min followed by reperfusion for 30 min with 100% O_2-saturated NT buffer.

2.13. Imaging of Mitochondrial Ca^{2+} Concentration ($[Ca^{2+}]_m$)

Changes in mitochondrial Ca^{2+} concentration were recorded using Rhod-2 AM. Cells were loaded with 5 μM Rhod-2 AM and 0.02% Pluronic F127 for 1 h at RT. The cells were also loaded with 300 nM MitoTracker Green FM (MTG) to detect the localization of mitochondria. Using time-lapse confocal laser scanning microscopy (TCS-SPII; Leica) to detect fluorescence signal. Cells perfused continuously with NT buffer (37 °C) were used for fluorescent imaging. Rhod-2 was excited at 514 nm, and emission was monitored at >530 nm. $[Ca^{2+}]_m$ were presented as the peak/basal fluorescence ratio (F/F_0). MTG was excited at 488 nm, and emission was monitored at >520 nm.

2.14. Measurement of Mitochondrial Membrane Potential Using Flow Cytometry

Changes in mitochondrial membrane potential (ΔΨm) were recorded with tetramethylrhodamine methyl ester perchlorate (TMRM). Cells were loaded with 500 nM TMRM and Pluronic F127 for 30 min at RT. After washing with NT, cardiomyocytes were harvested, and fluorescence was measured on a FACSCalibur Flow Cytometer (Becton Dickinson Biosciences, San Jose, CA, USA). The excitation/emission wavelengths for TMRM was 488/580 nm (FL2)

2.15. In Situ Labeling of Activated Caspase-3

In situ labeling of activated caspase-3 was carried out using a CaspGLOW Fluorescein Active Caspase-3 Staining Kit (BioVision, Inc.). After treatment, cells were incubated with FITC–DEVD–FMK for 1 h at RT. Cardiomyocytes were fixed with 4% formaldehyde for 1 h, and then incubated with Hoechst 33342 for 15 min. Images were obtained using confocal microscopy. FITC was represented by green fluorescence with excitation/emission wavelengths at 488/>530 nm. Hoechst 33342 fluorochrome was represented by blue fluorescence with excitation/emission wavelengths at 364/400–470 nm.

2.16. Statistics

All results are expressed as means and standard errors of the means (mean ± SEM). Statistical differences were compared using the unpaired t-test, taking p-values < 0.05 as significant. Data obtained from three or more groups were compared by one-way ANOVA followed by LSD tests. p-values < 0.05 were considered significant.

3. Results

3.1. Effects of IH on Cell Death

It has been demonstrated that pathogenic or beneficial effects of IH depend on hypoxic oxygen concentration, hypoxia duration, number of cycles, and IH pattern [22]. Initially, we examined whether IH can prevent cell death. Cardiomyocytes were treated with IH for 1–4 days, followed by 30 μM

H_2O_2 treatment for 40 min, and then the cell viability was measured using the MTT assay. As shown in Figure 1A, the viability of cells treated with four days IH (IH4) was not significantly different from the viability of those treated with RA. However, treatment with IH for 1–4 days time-dependently attenuated the H_2O_2-induced cell death. Furthermore, cytometric analysis for Annexin V-FITC and SYTOX green revealed that the percentage of apoptotic and necrotic cells was not significantly different between the RA-treated group and the IH4-treated group. In this study, the H_2O_2-treated group was used to serve as a positive control for the occurrence of apoptosis and necrosis (Figure 1B,C).

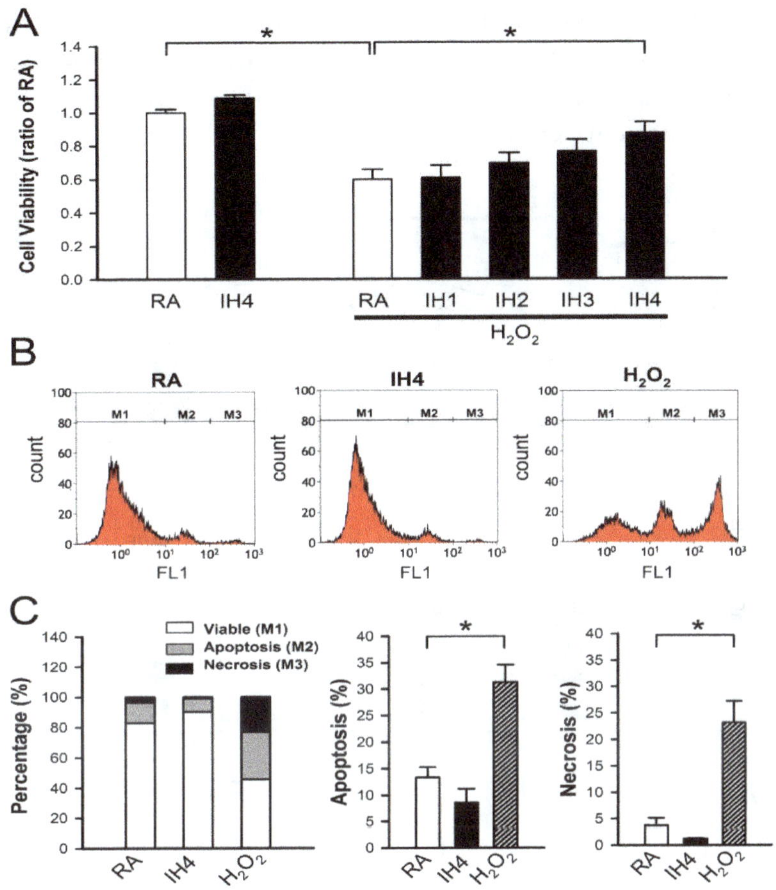

Figure 1. Effects of IH cell death. (**A**) Primary cultured neonatal cardiomyocytes were treated with RA or IH for 1–4 days followed by 30 μM H_2O_2 treatment for 40 min. Cell viability was determined using MTT assay. Treatment with IH for four days did not significantly affect the cell death. In contrast, treatment with H_2O_2 significantly reduced the cell viability However, treatment with IH for four days significantly reduced the H_2O_2-induced cell death. Data represent the means ± SEM. RA: $n = 10$, IH4: $n = 3$, RA + H_2O_2: $n = 10$, IH1 + H_2O_2: $n = 4$, IH2 + H_2O_2: $n = 4$, IH3 + H_2O_2: $n = 4$, IH4 + H_2O_2: $n = 10$. * $p < 0.05$. (**B**) The representative plot shows flow cytometry analysis of Annexin V-FITC and SYTOX green florescence intensity in the cells treated with RA, IH4 or H_2O_2. (**C**) The left panel shows the percentage of viable cells (M1), apoptotic cells (M2), and necrotic cells (M3) in the RA-, IH4-, and H_2O_2-treated groups. The middle panel shows that the percentage of apoptotic cells in the IH4-treated group was not significantly different from the RA-treated group. The right panel shows that the percentage of necrotic cells in the IH4-treated group was not significantly different from the RA-treated group. H_2O_2 treatment was serviced as a positive control for the induction of cell death. Data represent the means ± SEM. RA: $n = 6$, IH4: $n = 3$, H_2O_2: $n = 6$. * $p < 0.05$. RA, room air; IH4, intermittent hypoxia for four days.

3.2. Effects of IH on the Endogenous Antioxidant Defense in Cardiomyocytes

Previously, we demonstrated that IH can protect cardiomyocytes against the H_2O_2-induced cell death [21]. Here, we further investigated whether IH can enhance the antioxidant defense capacity to against oxidative stress. To address this issue, the cells were treated with IH for four days, followed by perfusion with 30 μM H_2O_2 to increase the intracellular ROS level, which was detected by fluorescence probe, DCFDA, and the signals were detected by a real-time fluorescence imaging system (Figure 2A). In the RA-treated group, H_2O_2 treatment increased the level of intracellular ROS approximately 5-fold compared with the baseline. IH4 significantly attenuated the H_2O_2-induced increases of the intracellular ROS levels (Figure 2B). Next, we examined the mRNA expression levels of antioxidant enzymes by Q-PCR. As shown in Figure 2C, IH4 significantly increased the mRNA levels of Cu/Zn SOD and MnSOD. However, the mRNA levels of catalase and GPx did not change significantly. We further measured the activity of catalase, GPx and TAC using assay kit. The activities of catalase (Figure 2D), GPx (Figure 2E), and TAC (Figure 2F) were significantly higher in the IH4-treated group than the RA-treated group. These data suggest that treatment with IH for four days increased the endogenous antioxidant defense capacity.

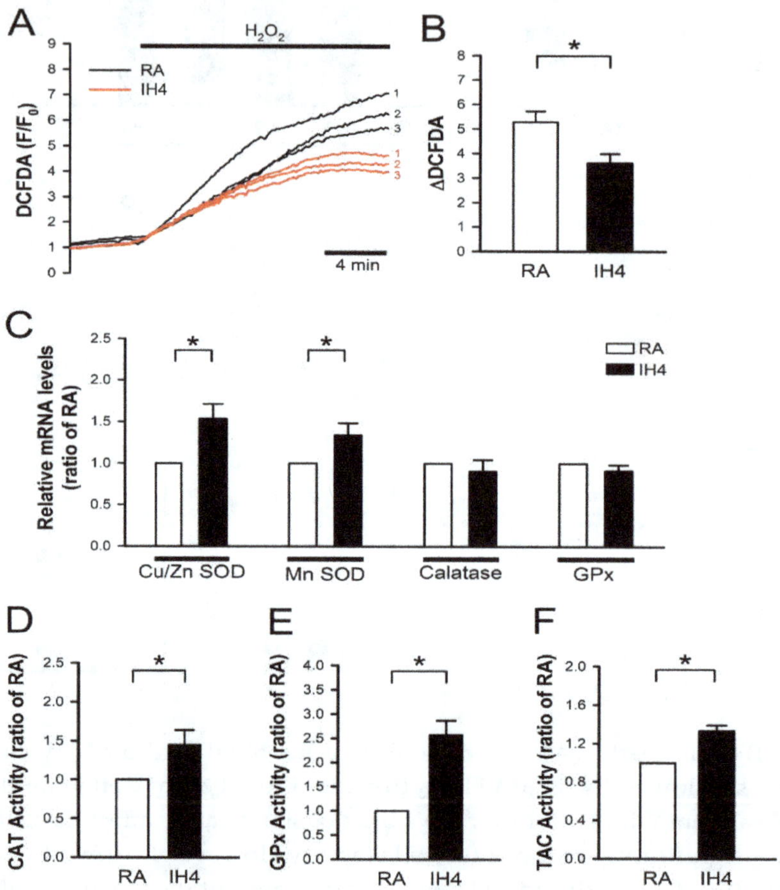

Figure 2. Effects of IH on endogenous antioxidant defense in cardiomyocytes. (**A**) The time course of intracellular ROS changes in cardiomyocytes perfused with H_2O_2 (30 μM) was recorded by a real-time fluorescence imaging system. Treatment of cardiomyocytes to IH for four days attenuated the H_2O_2-induced increases of intracellular ROS level. (**B**) The difference between baseline and plateau of DCFDA (F/F_0) was defined as ΔDCFDA. The quantitative results show that the H_2O_2-induced intracellular ROS increase was significantly lower in cardiomyocytes treated with IH4 than with RA. Data represent the means ± SEM. $n = 8$ for each group, * $p < 0.05$. (**C**) The levels of Cu/Zn SOD and MnSOD mRNA were significantly higher in the IH4-treated group than the RA-treated group. Data represent the means ± SEM. $n = 7$ for each group. * $p < 0.05$. (**D**) The activity levels of catalase (D), GPx (**E**), and TAC (**F**) were significantly higher in the IH4-treated group than in the RA-treated group. (The activities of catalase, GPx and TAC were determined by ELISA. Data represent the means ± SEM. RA: $n = 5$, IH4: $n = 5$. * $p < 0.05$.

3.3. *Effects of IH on Oxidative Stress-Induced Intracellular Ca^{2+} Disturbance and Mitochondrial Membrane Depolarization*

Previous studies have suggested that oxidative stress induces intracellular Ca^{2+} overload, which in turn causes mitochondrial membrane depolarization, hence leading cell death [23,24]. To test whether IH can prevent oxidative stress-mediated accumulation of intracellular Ca^{2+}, the cardiomyocytes were treated with IH for four days followed by 30 μM H_2O_2 treatment to increase the intracellular ROS level, and then subjected to the intracellular Ca^{2+} concentration detection in the IH- and RA-treated cells by fluorescence probe, Fura-2. Our data show that IH4 significantly attenuated the H_2O_2-induced increases of the intracellular Ca^{2+} levels (Figure 3A,B). Using fluorescence probe, Rhod-2, we further demonstrated that IH4 significantly attenuated the H_2O_2-induced increases of the mitochondrial Ca^{2+} levels (Figure 3C,D). It has been suggested that excessive ROS can open mitochondrial permeability transition pore, hence leading mitochondrial membrane depolarization [24]. We further examined whether IH can prevent oxidative stress-mediated mitochondrial membrane depolarization. The change of mitochondrial membrane potential was evaluated by flow cytometry analysis using a fluorescence probe, TMRM, which is a cell-permeant dye that accumulates in healthy mitochondria and can be detected a strong signal. When the mitochondrial membrane depolarizes, TMRM was released from the mitochondria and weak signal was detected. We measured the average of TMRM fluorescence intensity of 10,000 cells per well. The average of TMRM in the RA group was as 100% (Figure 3E). As illustrated in Figure 3F, the TMRM signals in the IH-treated group were significantly lower than in the RA group. Under H_2O_2 treatment conditions, however, the IH4-treated group showed significantly higher TMRM signals than the RA group. These data demonstrate that treatment of cardiomyocytes with IH for four days significantly attenuated the H_2O_2-induced intracellular Ca^{2+} overload and mitochondrial membrane depolarization.

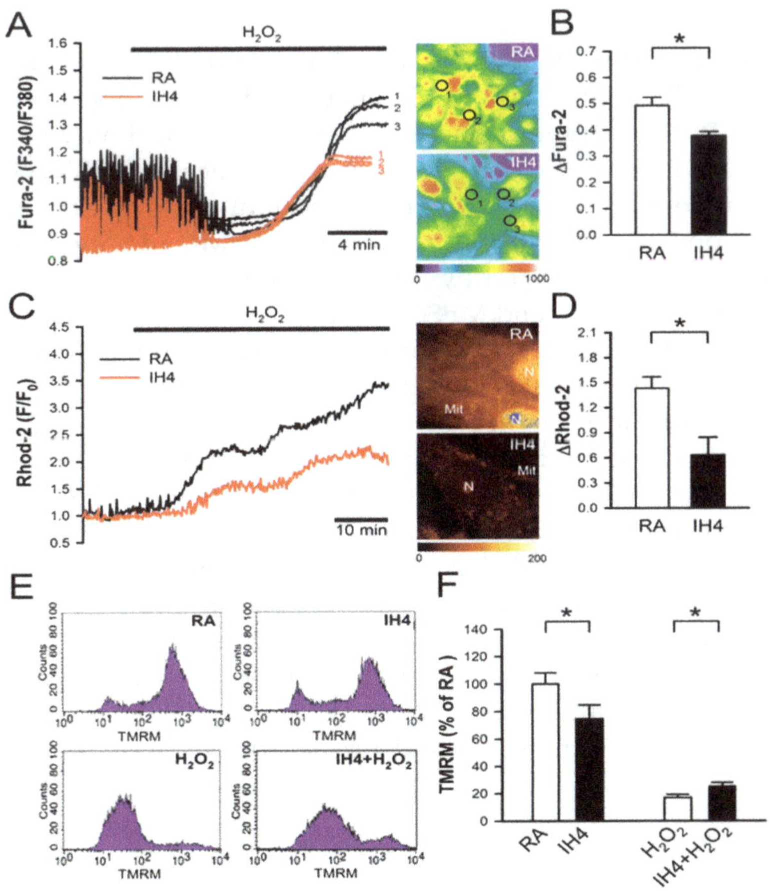

Figure 3. Effects of IH on H_2O_2-induced intracellular Ca^{2+} increase and mitochondrial membrane depolarization. (**A**) The left panel shows the time course of cytosolic Ca^{2+} changes in cardiomyocytes

perfused with H_2O_2 (30 μM), and recorded by a real-time fluorescence imaging system. Treatment of cardiomyocytes to IH for four days attenuated H_2O_2-induced increases of the cytosolic Ca^{2+} level. The right panel shows representative images of cytosolic Ca^{2+} stained by Fura-2 (pseudo color). The open circle indicated the area that was examined for the intensity of Fura-2 staining. (**B**) The difference between the baseline and plateau of Fura-2 (F340/F380) was defined as ΔFura-2. The quantitative results show that the H_2O_2-increased cytosolic Ca^{2+} was significantly lower in the cardiomyocytes treated with IH4 than with RA. Data represent the means ± SEM. $n = 6$ for each group. * $p < 0.05$. (**C**) The left panel shows the time course of mitochondrial Ca^{2+} changes in cardiomyocytes perfused with H_2O_2 (30 μM), and recorded by confocal microscope. Treatment of cardiomyocytes to IH for four days attenuated the H_2O_2-induced increases of mitochondrial Ca^{2+} level. The right panel shows representative images of mitochondrial Ca^{2+} stained by Rhod-2 (pseudo color). (**D**) The difference between baseline and plateau of Rhod-2 (F/F_0) was defined as ΔRhod-2. The quantitative results show that the H_2O_2-increased mitochondrial Ca^{2+} was significantly lower in the cardiomyocytes treated with IH4 than with RA. Data represent the means ± SEM. $n = 3$ for each group. * $p < 0.05$. (**E**) The representative plots show the mitochondrial membrane potential changes measured by flow cytometry analysis using mitochondrial membrane potential probe, TMRM. (**F**) The quantitative results are measured and normalized to RA. Mitochondrial membrane potential was significantly lower in the cardiomyocytes treated with IH4 than with RA. However, treatment of cardiomyocytes with IH for four days significantly attenuated the H_2O_2-induced depolarization of mitochondrial membrane potential. Data represent the means ± SEM. RA: $n = 10$, IH4: $n = 10$, H_2O_2: $n = 4$, IH4 + H_2O_2: $n = 4$. * $p < 0.05$. Mit, mitochondria; N, Nucleus.

3.4. Involvement of ROS Generation in the IH-Increased Antioxidant Defense and Cardioprotective Effects

It has been suggested that ROS can serve as a mediator to trigger signaling pathways involved in regulating cardioprotective effects [20,21]. Flow cytometry analysis demonstrated that IH4 significantly increased the intracellular ROS generation. However, treatment of cardiomyocytes with an antioxidant, such as Apo, Phe, or MnTBAP, abolished the IH4-increased intracellular ROS generation (Figure 4A,B). We further investigated whether IH can increase antioxidant enzyme expression. Cardiomyocytes were treated with IH for four days with or without Phe (100 nM), an antioxidant, and the expression of Cu/Zn SOD and MnSOD were detected by immunocytofluorescence staining. IH4 significantly increased the expression of Cu/Zn SOD and MnSOD. However, Phe abolished the IH4-increased Cu/Zn SOD and MnSOD expression (Figure 4C,D). We also measured the total antioxidant capacity (TAC) to clarify whether IH increased endogenous antioxidant defense system via intracellular ROS generation. The levels of TAC were significantly higher in the IH4-treated group than in the RA-treated group. However, treatment of cardiomyocytes with IH for four days with Phe (100 nM) or SOD (5 U) abolished the IH4-increased TAC (Figure 4E). To investigate whether ROS was a key mediator contributing to the IH-induced cardioprotective effects, cardiomyocytes were pre-treated for four days with IH and Apo (100 μM), Phe (100 μM), or MnTBAP (50 μM) together to prevent ROS generation, followed by 30 μM H_2O_2 treatment for 40 min, and then subjected to measure the cell viability by MTT assay. As shown in Figure 4F, treatment with IH for four days significantly attenuated the H_2O_2-decreased cell viability. However, treatment with IH together with an antioxidant, such as Apo, Phe, or MnTBAP, abolished the IH-protected cell death. These data suggest that IH increased ROS generation, which upregulated antioxidant enzyme expression and increased the total antioxidant capacity, thereby preventing cell death.

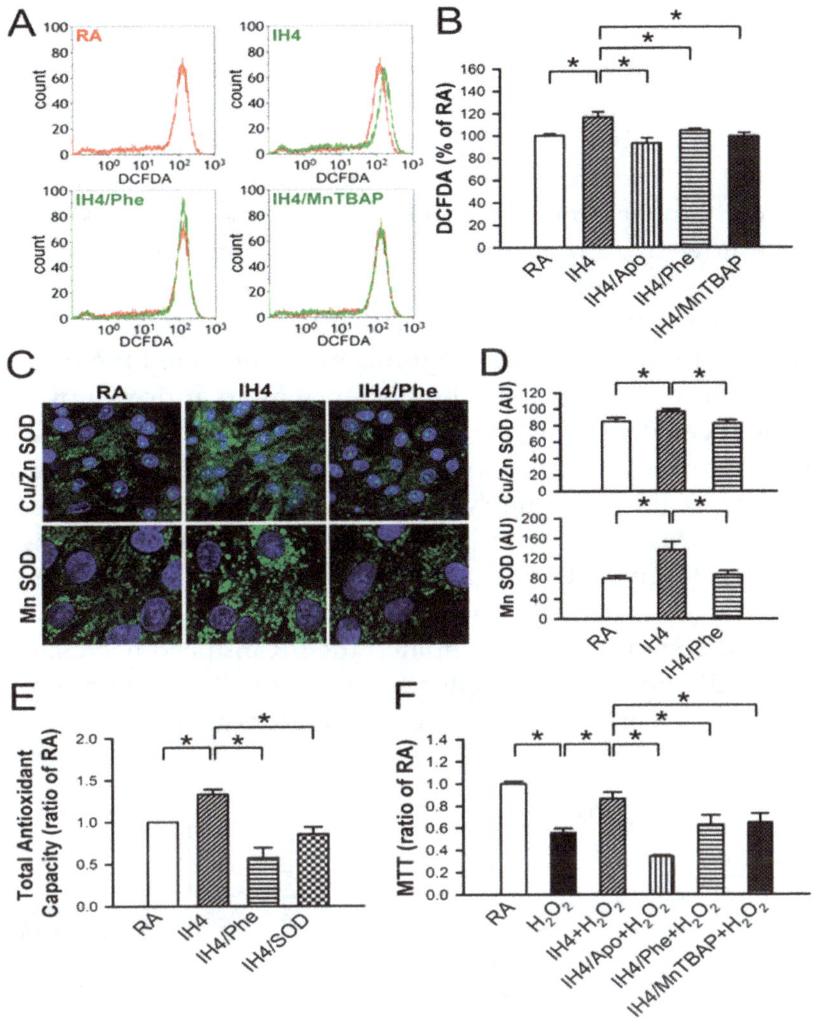

Figure 4. Involvement of ROS generation in the IH-increased antioxidant expression and cardioprotective effect. (**A**) The representative plots show the intracellular ROS change in various treatment measured by flow cytometry analysis using intracellular ROS probe, DCFDA. The result of cells treated with H_2O_2 (30 μM) in the RA was used as the positive control and shown with the red line, whereas treatment with IH4, IH4/Phe, or IH4/MnTBAP was shown with the green line. (**B**) The quantitative results show that the level of intracellular ROS was significantly higher in the cardiomyocytes treated with IH4 than with RA. However, co-treatment with an antioxidant, such as Apo, Phe, or MnTBAP, abolished the IH4-increased intracellular ROS level. Data represent the means ± SEM. RA: $n = 10$, IH4: $n = 6$, IH4/Apo: $n = 6$, IH4/Phe: $n = 6$, IH4/MnTBAP: $n = 6$. * $p < 0.05$. (**C**) The representative images show the expression of Cu/Zn SOD and MnSOD (green fluorescence) detected by immunocytofluorescence staining. The nucleus was stained by Hoechst 33342 (blue fluorescence). (**D**) The quantitative results show that the levels of Cu/Zn SOD and MnSOD were significantly higher in the cardiomyocytes treated with IH4 than with RA. However, co-treatment with Phe, an antioxidant, abolished the IH4-increased Cu/Zn SOD and MnSOD expression. Data represent the means ± SEM. ($n = 12$). * $p < 0.05$. (**E**) The levels of TAC were significantly higher in the IH4-treated group than the RA-treated group. However, co-treatment with an antioxidant, Phe or SOD, abolished the IH4-increased TAC. Total antioxidant capacity as determined by assay kit. Data represent the means ± SEM. RA: $n = 3$, IH4: = 3, IH4/Phe: $n = 3$, IH4/SOD: $n = 3$. * $p < 0.05$ (**F**) Neonatal cardiomyocytes were treated with IH for four days, and then co-treated with IH and an antioxidant, Apo, Phe or MnTBAP, followed by 30 μM H_2O_2 treatment for 40 min. Treatment with IH4 prevented the H_2O_2-induced cell death. However, co-treatment with Apo, Phe or MnTBAP abolished the IH-prevented cell death. Data represent the means ± SEM. RA: $n = 10$, RA + H_2O_2: $n = 10$, IH4 + H_2O_2: $n = 10$, IH4/Apo + H_2O_2: $n = 4$, IH4/Phe + H_2O_2: $n = 6$, IH4/MnTBAP + H_2O_2: $n = 6$. * $p < 0.05$.

3.5. Effects of IH on the I/R-Induced Cell Death in Cardiomyocytes

We previously demonstrated that Ca^{2+} overload plays an important role in the I/R-induced cell death [25]. In Figure 3, we demonstrated that IH4 significantly attenuated the H_2O_2-induced cytosolic Ca^{2+} overload. We further investigated whether IH could also prevent the I/R-induced cytosolic Ca^{2+} overload and cell death. After treatment with IH for four days, cardiomyocytes were placed in a plexiglass box chamber containing 100% N_2, perfused with 100% N_2-saturated ischemia buffer for 30 min, and followed by reperfusion for 30 min with 100% O_2-saturated NT solution. Fluorescence probe, Fura-2, was used to detect the changes of Ca^{2+} during I/R. Our results show that IH4 significantly attenuated the I/R-induced increases of the cytosolic Ca^{2+} levels (Figure 5A,B). Moreover, treatment with IH for four days also significantly attenuated the I/R-decreased cell viability. However, treatment with IH for four days in the presence of an antioxidant, Apo, Phe or MnTBAP, abolished the IH-prevented cell death (Figure 5C). We also detected the level of cleaved caspase-3 to analyze apoptosis status in cardiomyocytes. As shown in Figure 5D, IH4 significantly decreased the I/R-induced apoptosis. Finally, flow cytometry analysis was conducted to measure the percentage of live cells (M1), apoptotic cells (M2) and necrotic cells (M3) in cardiomyocytes (Figure 6A). I/R increased the occurrence of apoptosis (Figure 6B) and necrosis (Figure 6C), and this effect was significantly attenuated by IH4 treatment. However, the IH-prevented apoptosis were abolished by co-treatment with Phe or MnTBAP. IH4 also significantly attenuated the I/R-induced necrosis. Noticeably, treatment with Phe or MnTBAP alone did not induce apoptosis or necrosis. These data suggest that IH prevented the I/R-induced cytosolic Ca^{2+} overload and cell death via ROS generation.

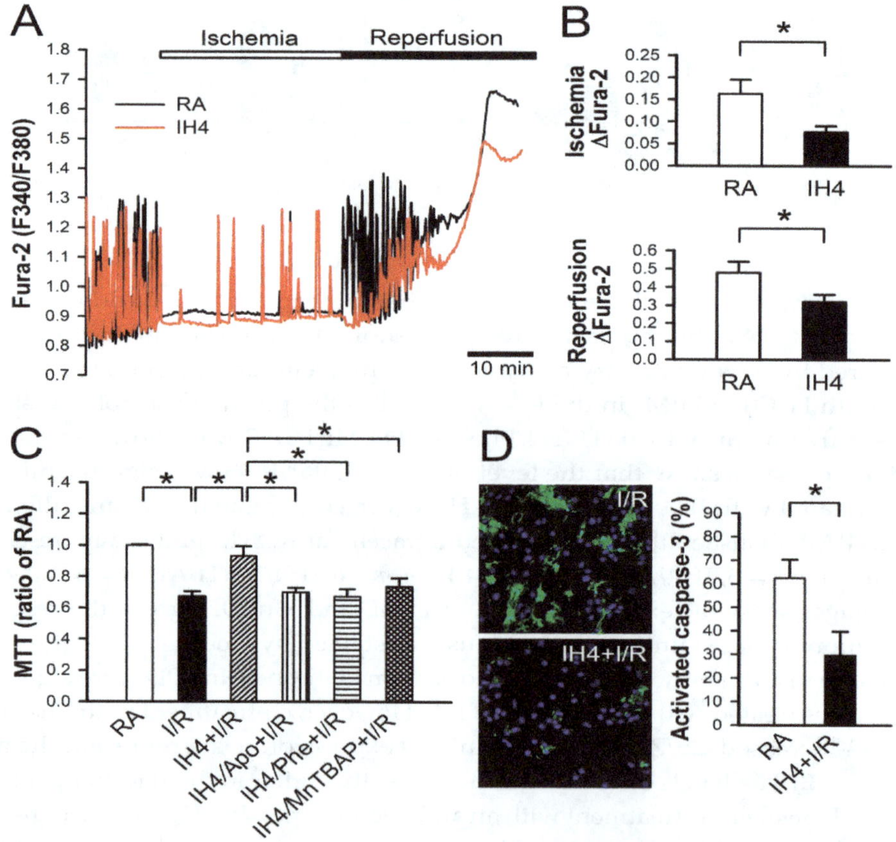

Figure 5. Effects of IH on the I/R-induced cytosolic Ca^{2+} increases and cell death. (**A**) The time course of cytosolic Ca^{2+} changes in cardiomyocytes perfused with ischemic and reperfusion buffer to mimic I/R injury, and recorded by a real-time fluorescence imaging system. Treatment of cardiomyocytes to IH for four days attenuated the ischemia and reperfusion-induced increases of cytosolic Ca^{2+} level. (**B**) The difference between the baseline and plateau of Fura-2 (F340/F380) was defined as ΔFura-2.

The quantitative results show that the ischemia and reperfusion-increased cytosolic Ca^{2+} was significantly lower in cardiomyocytes treated with IH4 than with RA. Data represent the means ± SEM. $n = 4$ for each group. * $p < 0.05$. (**C**) Neonatal cardiomyocytes were treated with IH for four days, and then co-treated with IH and antioxidant, Apo, Phe or MnTBAP, followed by I/R treatment. Treatment with IH4 prevented the I/R-induced cell death. However, co-treatment with Apo, Phe or MnTBAP abolished the IH-prevented cell death. Data represent the means ± SEM. $n = 5$ for each group. * $p < 0.05$. (**D**) The left panel shows the representative images of activated caspase-3 expression detected by CaspGLOW fluorescein active caspase-3 staining kit after I/R treatment. The right panel gives the quantitative results, showing that the expression of active caspase-3 was significantly lower in cardiomyocytes treated with I/R + IH4 than with I/R + RA. Data represent the means ± SEM. ($n = 14$). * $p < 0.05$. I/R, ischemia/reperfusion.

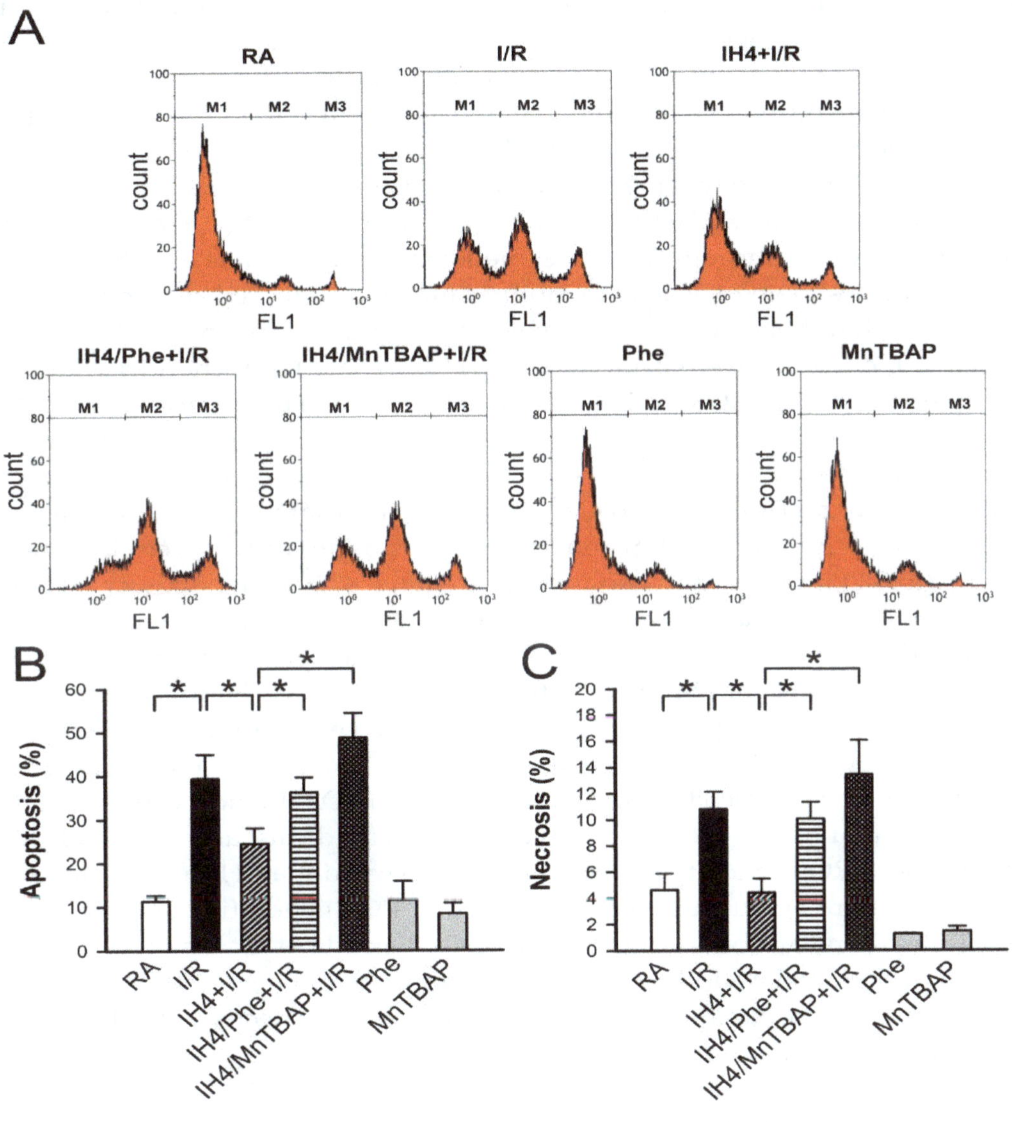

Figure 6. Effects of IH on the I/R-induced apoptosis and necrosis. (**A**) The representative plots show the flow cytometry analysis of Annexin V-FITC and SYTOX green florescence intensity in the cardiomyocytes treated with RA, IH4 and IH4 with an antioxidant, Phe or MnTBAP, followed by ischemia for 6 h and reperfusion for 12 h. M1, M2, and M3 represent the population of viable cells, apoptotic cells and necrotic cells, respectively. IH4 attenuated the I/R-induced apoptosis (**B**) and necrosis (**C**). IH4 with Phe or MnTBAP abolished the IH4-attenuated the I/R-induced apoptosis and necrosis.. Data represent the means ± SEM. RA: $n = 10$, RA + I/R: $n = 6$, IH4 + I/R: $n = 5$, IH4/Phe + I/R: $n = 6$, IH4/MnTBAP + I/R: $n = 6$, Phe: $n = 4$, MnTBAP: $n = 4$. * $p < 0.05$. I/R, ischemia/reperfusion; IH, intermittent hypoxia; Phe, Phenanthroline.

4. Discussion

In this study, we demonstrated that IH4 attenuated the H_2O_2-induced cytosolic and mitochondrial Ca^{2+} overload through ROS production, thereby preventing mitochondrial membrane depolarization. It has been reported that IH could prevent cytosolic Ca^{2+} overload through various pathways. To our knowledge, this is first demonstration that IH can protect cardiomyocytes against the H_2O_2- and I/R-induced oxidative stress and cell death through maintaining the Ca^{2+} homeostasis.

Gu et al. reported that IH could reduce myocardial Ca^{2+} overload and hypoxia/reoxygenation injury through upregulating PGC-1α and regulating the energy metabolism of glucose and lipid mediated by activating the HIF-1α-AMPK-PGC-1α signaling pathway [26]. HIF-1 plays a central role in cellular signaling in hypoxia adaption. It has been demonstrated that cyclic hypoxia-reoxygenation activates HIF-1 expression, which triggers expression of vascular endothelial growth factor and erythropoietin to promote angiogenesis and anti-infarct effects [10]. It has also been reported that HIF-1 could improve intracellular Ca^{2+} handling to prevent cell death. HIF-1stabilization increases protein expression of SERCA2, resulting in cytosolic Ca^{2+} reuptake into ER stores faster in thymocytes [27]. Furthermore, ischemic preconditioning activated HIF-1/Sp1 complex to increase NCX1 gene and protein expression to attenuate oxygen and glucose deprivation-induced neuronal cell death [28]. On the other hand, IH regulated the activities of Na^+/K^+ pump and Na^+/Ca^{2+} exchanger to prevent Ca^{2+} overload and I/R injury through the PKC-mediated pathway has also been reported [29–32]. Additionally, IH has been shown to be able to inhibit mitochondrial Ca^{2+} overload and mitochondrial membrane depolarization, hence preventing against reperfusion injury [7,33]. Reduction of Ca^{2+} overload has been demonstrated to contribute to the IH-prevented mitochondria-mediated cell death.

During I/R injury, a burst of ROS leads to cell death, mediated by intracellular Ca^{2+} overload [34]. Excessive ROS can directly modify many Ca^{2+}-handling proteins, such as RyR, SERCA, and NCX, to influence cardiomyocyte viability and heart functions [35]. We observed that treatment with IH for four days prevented the I/R- and H_2O_2-induced Ca^{2+} overload associated with an increase of antioxidant defense capacity through non-lethal ROS generation. IH can not only upregulate the expression of Cu/Zn SOD and MnSOD proteins but also increases catalase and GPx activity. Aguilar et al. also found that IH can increase SOD and GPx expression in cardiac tissue associated with a higher ejection and shortening fraction of the left ventricle function [19]. Increases of antioxidant capacity prevent intracellular Ca^{2+} overload through improving the intracellular Ca^{2+} handling capacity.

ROS can be both harmful and protective [36]. It has been shown that several intracellular sources including mitochondria, NADPH oxidase, xanthine oxidase, and uncoupled nitric oxide synthase can generate ROS [35]. Our data revealed that apocynin (Apo), a NADPH-oxidase inhibitor, Phenanthroline (Phe), an iron chelator that blocks the formation of hydroxyl radicals, and MnTBAP, a SOD mimetic, inhibited the IH-induced ROS generation. Our data also revealed that IH-induced ROS generation did not cause cell death (Figure 1). Furthermore, reduction of ROS generation abolished the IH-induced prevention of cell death in cardiomyocytes (Figure 5C). Estrada et al. also found that treatment of mongrel dogs with IH for 20 days induces robust cardioprotection against I/R, manifested by a 90% reduction in left ventricular infarct size. However, the cardioprotective effects induced by IH are abolished by pre-treatment with an antioxidant, *N*-acetylcysteine (NAC). They demonstrated that ROS are obligatory participants in regulating cardioprotection induced by IH [37]. It seems that non-lethal ROS may play an important role in the IH-induced cardioprotection.

In cells, large amount of ROS induces severe oxidative stress, which causes damages of DNA, lipids and proteins, thereby contributing to many different disease developments, including cardiovascular diseases [38]. However, it has also been demonstrated that ROS can serve as a redox signaling molecule in physiological conditions. Kelch-like ECH-associated protein-1 (Keap1) is a redox sensor associated with NFE2-like 2 (Nrf2). ROS oxidation of cysteines in Keap1 leads to Nrf2 release and translocation into the nucleus. Subsequently, the Nrf2-small Maf heterodimer binds to the antioxidant-responsive element (ARE), resulting in upregulation of antioxidant genes, including SOD and GPx [39–41]. In addition, PPARγ has been reported to be able to directly modulate the expression of several antioxidants, such as MnSOD and GPx3 [42]. Our data revealed that IH-increased non-lethal ROS generation upregulated the expression of Cu/Zn SOD and MnSOD proteins and increased catalase and GPx activity. However, the mechanisms underlying ROS-increased antioxidant enzyme expression in IH still require future investigation.

We also observed that the mRNA levels of CAT and GPx did not change (Figure 2D). However, the enzyme activity of CAT and GPx increased (Figure 2C). CAT and GPx mRNA expression was probably time-dependent [43]. It has been reported that mRNA half-lives were about several minutes to several hours. Therefore, mRNA was degraded, but the protein expression or activity still increased [44].

IH has been demonstrated to exert beneficial effects on myocardial infarction, cardiac function, arrhythmias, and coronary flow [10]. IH could cause greater left ventricle ejection fraction and fractional shortening [19]. IH also protects the myocardium from I/R injury and maintains contractility by upregulating ATF6 through the Akt-mediated signaling pathway [45]. Wang et al. reported that IH can improve post-ischemic recovery of myocardial contractile function through the Akt and the PKC-ε pathway via enhancing the production of ROS during early reperfusion [46]. Although there are many different protocols for IH treatment, the findings from the present study and other labs strongly suggest that IH can trigger multiple pathways to protect cardiomyocytes and the heart against I/R injury. In conclusion, the findings of the present study demonstrate that IH could protect cardiomyocytes against H_2O_2- and the I/R-induced oxidative stress and cell death through the maintenance of Ca^{2+} homeostasis, mitochondrial membrane potential, and upregulation of antioxidant enzymes via non-lethal ROS production.

Author Contributions: Conceptualization, K.-T.Y.; Methodology, C.-F.L., Y.-C.H. and Y.-P.L.; Software, C.-F.L.; Validation, J.-C.C., W.-S.L.; Formal Analysis, C.-F.L., Y.-C.H. and Y.-P.L.; Resources, H.-R.C., J.-R.J., J.-C.H and K.-T.Y.; Data Curation, J.-C.C., C.-F.L., Y.-C.H. and Y.-P.L.; Writing-Original Draft Preparation, J.-C.C. and C.-F.L.; Writing-Review & Editing, W.-S.L., H.-R.C. and K.-T.Y.; Visualization, C.-F.L., Y.-C.H. and Y.-P.L.; Supervision, H.-R.C., J.-R.J. and J.-C.H..; Project Administration, K.-T.Y.

References

1. Neubauer, J.A. Invited review: Physiological and pathophysiological responses to intermittent hypoxia. *J. Appl. Physiol.* **2001**, *90*, 1593–1599. [CrossRef] [PubMed]
2. Marin, J.M.; Carrizo, S.J.; Vicente, E.; Agusti, A.G. Long-term cardiovascular outcomes in men with obstructive sleep apnoea-hypopnoea with or without treatment with continuous positive airway pressure: An observational study. *Lancet* **2005**, *365*, 1046–1053. [CrossRef]
3. Shahar, E.; Whitney, C.W.; Redline, S.; Lee, E.T.; Newman, A.B.; Javier Nieto, F.; O'Connor, G.T.; Boland, L.L.; Schwartz, J.E.; Samet, J.M. Sleep-disordered breathing and cardiovascular disease: Cross-sectional results of the Sleep Heart Health Study. *Am. J. Respir. Crit. Care Med.* **2001**, *163*, 19–25. [CrossRef] [PubMed]
4. Cai, Z.; Manalo, D.J.; Wei, G.; Rodriguez, E.R.; Fox-Talbot, K.; Lu, H.; Zweier, J.L.; Semenza, G.L. Hearts from rodents exposed to intermittent hypoxia or erythropoietin are protected against ischemia-reperfusion injury. *Circulation* **2003**, *108*, 79–85. [CrossRef] [PubMed]
5. Beguin, P.C.; Joyeux-Faure, M.; Godin-Ribuot, D.; Levy, P.; Ribuot, C. Acute intermittent hypoxia improves rat myocardium tolerance to ischemia. *J. Appl. Physiol.* **2005**, *99*, 1064–1069. [CrossRef]
6. Chen, L.; Lu, X.Y.; Li, J.; Fu, J.D.; Zhou, Z.N.; Yang, H.T. Intermittent hypoxia protects cardiomyocytes against ischemia-reperfusion injury-induced alterations in Ca^{2+} homeostasis and contraction via the sarcoplasmic reticulum and Na^+/Ca^{2+} exchange mechanisms. *Am. J. Physiol. Cell Physiol.* **2006**, *290*, C1221–1229. [CrossRef] [PubMed]
7. Zhu, W.Z.; Xie, Y.; Chen, L.; Yang, H.T.; Zhou, Z.N. Intermittent high altitude hypoxia inhibits opening of mitochondrial permeability transition pores against reperfusion injury. *J. Mol. Cell Cardiol.* **2006**, *40*, 96–106. [CrossRef]

8. Guo, H.C.; Zhang, Z.; Zhang, L.N.; Xiong, C.; Feng, C.; Liu, Q.; Liu, X.; Shi, X.L.; Wang, Y.L. Chronic intermittent hypobaric hypoxia protects the heart against ischemia/reperfusion injury through upregulation of antioxidant enzymes in adult guinea pigs. *Acta Pharmacol. Sin.* **2009**, *30*, 947–955. [CrossRef]
9. Yeung, H.M.; Kravtsov, G.M.; Ng, K.M.; Wong, T.M.; Fung, M.L. Chronic intermittent hypoxia alters Ca^{2+} handling in rat cardiomyocytes by augmented Na^+/Ca^{2+} exchange and ryanodine receptor activities in ischemia-reperfusion. *Am. J. Physiol. Cell Physiol.* **2007**, *292*, C2046–2056. [CrossRef]
10. Mallet, R.T.; Manukhina, E.B.; Ruelas, S.S.; Caffrey, J.L.; Downey, H.F. Cardioprotection by intermittent hypoxia conditioning: Evidence, mechanisms, and therapeutic potential. *Am. J. Physiol. Heart Circ. Physiol.* **2018**, *315*, H216–H232. [CrossRef]
11. Bers, D.M. Cardiac excitation-contraction coupling. *Nature* **2002**, *415*, 198–205. [CrossRef] [PubMed]
12. Grueter, C.E.; Colbran, R.J.; Anderson, M.E. CaMKII, an emerging molecular driver for calcium homeostasis, arrhythmias, and cardiac dysfunction. *J. Mol. Med.* **2007**, *85*, 5–14. [CrossRef] [PubMed]
13. Gorlach, A.; Bertram, K.; Hudecova, S.; Krizanova, O. Calcium and ROS: A mutual interplay. *Redox Biol.* **2015**, *6*, 260–271. [CrossRef] [PubMed]
14. Das, D.K.; Maulik, N.; Sato, M.; Ray, P.S. Reactive oxygen species function as second messenger during ischemic preconditioning of heart. *Mol. Cell Biochem.* **1999**, *196*, 59–67. [CrossRef] [PubMed]
15. Forbes, R.A.; Steenbergen, C.; Murphy, E. Diazoxide-induced cardioprotection requires signaling through a redox-sensitive mechanism. *Circ. Res.* **2001**, *88*, 802–809. [CrossRef] [PubMed]
16. Kamata, H.; Hirata, H. Redox regulation of cellular signalling. *Cell Signal.* **1999**, *11*, 1–14. [CrossRef]
17. Karlstad, J.; Sun, Y.; Singh, B.B. Ca^{2+} signaling: An outlook on the characterization of Ca^{2+} channels and their importance in cellular functions. *Adv. Exp. Med. Biol.* **2012**, *740*, 143–157. [CrossRef] [PubMed]
18. Murphy, E.; Steenbergen, C. Mechanisms underlying acute protection from cardiac ischemia-reperfusion injury. *Physiol. Rev.* **2008**, *88*, 581–609. [CrossRef] [PubMed]
19. Aguilar, M.; Gonzalez-Candia, A.; Rodriguez, J.; Carrasco-Pozo, C.; Canas, D.; Garcia-Herrera, C.; Herrera, E.A.; Castillo, R.L. Mechanisms of Cardiovascular Protection Associated with Intermittent Hypobaric Hypoxia Exposure in a Rat Model: Role of Oxidative Stress. *Int. J. Mol. Sci.* **2018**, *19*, 366. [CrossRef] [PubMed]
20. Chang, H.R.; Lien, C.F.; Jeng, J.R.; Hsieh, J.C.; Chang, C.W.; Lin, J.H.; Yang, K.T. Intermittent Hypoxia Inhibits Na^+-H^+ Exchange-Mediated Acid Extrusion Via Intracellular Na^+ Accumulation in Cardiomyocytes. *Cell Physiol. Biochem.* **2018**, *46*, 1252–1262. [CrossRef] [PubMed]
21. Lien, C.F.; Lee, W.S.; Wang, I.C.; Chen, T.I.; Chen, T.L.; Yang, K.T. Intermittent hypoxia-generated ROS contributes to intracellular zinc regulation that limits ischemia/reperfusion injury in adult rat cardiomyocyte. *J. Mol. Cell Cardiol.* **2018**, *118*, 122–132. [CrossRef]
22. Navarrete-Opazo, A.; Mitchell, G.S. Therapeutic potential of intermittent hypoxia: A matter of dose. *Am. J. Physiol. Regul. Integr. Comp. Physiol.* **2014**, *307*, R1181–1197. [CrossRef] [PubMed]
23. Yellon, D.M.; Hausenloy, D.J. Myocardial reperfusion injury. *N. Engl. J. Med.* **2007**, *357*, 1121–1135. [CrossRef] [PubMed]
24. Feissner, R.F.; Skalska, J.; Gaum, W.E.; Sheu, S.S. Crosstalk signaling between mitochondrial Ca^{2+} and ROS. *Front. Biosci.* **2009**, *14*, 1197–1218. [CrossRef] [PubMed]
25. Pinton, P.; Giorgi, C.; Siviero, R.; Zecchini, E.; Rizzuto, R. Calcium and apoptosis: ER-mitochondria Ca^{2+} transfer in the control of apoptosis. *Oncogene* **2008**, *27*, 6407–6418. [CrossRef] [PubMed]
26. Gu, S.; Hua, H.; Guo, X.; Jia, Z.; Zhang, Y.; Maslov, L.N.; Zhang, X.; Ma, H. PGC-1alpha Participates in the Protective Effect of Chronic Intermittent Hypobaric Hypoxia on Cardiomyocytes. *Cell Physiol. Biochem.* **2018**, *50*, 1891–1902. [CrossRef] [PubMed]
27. Neumann, A.K.; Yang, J.; Biju, M.P.; Joseph, S.K.; Johnson, R.S.; Haase, V.H.; Freedman, B.D.; Turka, L.A. Hypoxia inducible factor 1 alpha regulates T cell receptor signal transduction. *Proc. Natl. Acad Sci. USA* **2005**, *102*, 17071–17076. [CrossRef]
28. Formisano, L.; Guida, N.; Valsecchi, V.; Cantile, M.; Cuomo, O.; Vinciguerra, A.; Laudati, G.; Pignataro, G.; Sirabella, R.; Di Renzo, G.; et al. Sp3/REST/HDAC1/HDAC2 Complex Represses and Sp1/HIF-1/p300 Complex Activates ncx1 Gene Transcription, in Brain Ischemia and in Ischemic Brain Preconditioning, by Epigenetic Mechanism. *J. Neurosci.* **2015**, *35*, 7332–7348. [CrossRef]

29. Chen, T.I.; Hsu, Y.C.; Lien, C.F.; Lin, J.H.; Chiu, H.W.; Yang, K.T. Non-lethal levels of oxidative stress in response to short-term intermittent hypoxia enhance Ca^{2+} handling in neonatal rat cardiomyocytes. *Cell Physiol. Biochem.* **2014**, *33*, 513–527. [CrossRef]
30. Guo, H.C.; Guo, F.; Zhang, L.N.; Zhang, R.; Chen, Q.; Li, J.X.; Yin, J.; Wang, Y.L. Enhancement of Na/K pump activity by chronic intermittent hypobaric hypoxia protected against reperfusion injury. *Am. J. Physiol. Heart Circ. Physiol.* **2011**, *300*, 2280–2287. [CrossRef]
31. Ding, H.L.; Zhu, H.F.; Dong, J.W.; Zhu, W.Z.; Zhou, Z.N. Intermittent hypoxia protects the rat heart against ischemia/reperfusion injury by activating protein kinase C. *Life Sci.* **2004**, *75*, 2587–2603. [CrossRef] [PubMed]
32. Ma, H.J.; Li, Q.; Ma, H.J.; Guan, Y.; Shi, M.; Yang, J.; Li, D.P.; Zhang, Y. Chronic intermittent hypobaric hypoxia ameliorates ischemia/reperfusion-induced calcium overload in heart via Na/Ca^{2+} exchanger in developing rats. *Cell Physiol. Biochem.* **2014**, *34*, 313–324. [CrossRef] [PubMed]
33. Magalhaes, J.; Goncalves, I.O.; Lumini-Oliveira, J.; Marques-Aleixo, I.; Passos, E.; Rocha-Rodrigues, S.; Machado, N.G.; Moreira, A.C.; Rizo, D.; Viscor, G.; et al. Modulation of cardiac mitochondrial permeability transition and apoptotic signaling by endurance training and intermittent hypobaric hypoxia. *Int. J. Cardiol.* **2014**, *173*, 40–45. [CrossRef] [PubMed]
34. Hausenloy, D.J.; Yellon, D.M. Myocardial ischemia-reperfusion injury: A neglected therapeutic target. *J. Clin. Investig.* **2013**, *123*, 92–100. [CrossRef] [PubMed]
35. Wagner, S.; Rokita, A.G.; Anderson, M.E.; Maier, L.S. Redox regulation of sodium and calcium handling. *Antioxid. Redox Signal.* **2013**, *18*, 1063–1077. [CrossRef] [PubMed]
36. Becker, L.B. New concepts in reactive oxygen species and cardiovascular reperfusion physiology. *Cardiovasc. Res.* **2004**, *61*, 461–470. [CrossRef] [PubMed]
37. Estrada, J.A.; Williams, A.G., Jr.; Sun, J.; Gonzalez, L.; Downey, H.F.; Caffrey, J.L.; Mallet, R.T. delta-Opioid receptor (DOR) signaling and reactive oxygen species (ROS) mediate intermittent hypoxia induced protection of canine myocardium. *Basic Res. Cardiol.* **2016**, *111*, 17. [CrossRef]
38. Brieger, K.; Schiavone, S.; Miller, F.J., Jr.; Krause, K.H. Reactive oxygen species: From health to disease. *Swiss Med. Wkly.* **2012**, *142*, w13659. [CrossRef]
39. Sies, H. Hydrogen peroxide as a central redox signaling molecule in physiological oxidative stress: Oxidative eustress. *Redox Biol.* **2017**, *11*, 613–619. [CrossRef]
40. Ray, P.D.; Huang, B.W.; Tsuji, Y. Reactive oxygen species (ROS) homeostasis and redox regulation in cellular signaling. *Cell Signal.* **2012**, *24*, 981–990. [CrossRef]
41. Kalyanaraman, B.; Cheng, G.; Hardy, M.; Ouari, O.; Bennett, B.; Zielonka, J. Teaching the basics of reactive oxygen species and their relevance to cancer biology: Mitochondrial reactive oxygen species detection, redox signaling, and targeted therapies. *Redox Biol.* **2018**, *15*, 347–362. [CrossRef] [PubMed]
42. Kvandova, M.; Majzunova, M.; Dovinova, I. The role of PPARγ in cardiovascular diseases. *Physiol. Res.* **2016**, *65*, S343–S363. [PubMed]
43. Gonchar, O.A.; IMankovska, I.N. Time-dependent effect of severe hypoxia/reoxygenation on oxidative stress level, antioxidant capacity and p53 accumulation in mitochondria of rat heart. *Ukr. Biochem. J.* **2017**, *89*, 39–47. [CrossRef]
44. Wada, T.; Becskei, A. Impact of Methods on the Measurement of mRNA Turnover. *Int. J. Mol. Sci.* **2017**, *18*, 2723. [CrossRef] [PubMed]
45. Jia, W.; Jian, Z.; Li, J.; Luo, L.; Zhao, L.; Zhou, Y.; Tang, F.; Xiao, Y. Upregulated ATF6 contributes to chronic intermittent hypoxia-afforded protection against myocardial ischemia/reperfusion injury. *Int. J. Mol. Med.* **2016**, *37*, 1199–1208. [CrossRef] [PubMed]
46. Wang, Z.H.; Chen, Y.X.; Zhang, C.M.; Wu, L.; Yu, Z.; Cai, X.L.; Guan, Y.; Zhou, Z.N.; Yang, H.T. Intermittent hypobaric hypoxia improves postischemic recovery of myocardial contractile function via redox signaling during early reperfusion. *Am. J. Physiol. Heart Circ. Physiol.* **2011**, *301*, 1695–1705. [CrossRef]

6

Activation of Nrf2/HO-1 Pathway and Human Atherosclerotic Plaque Vulnerability: An In Vitro and In Vivo Study

Susanna Fiorelli [1], Benedetta Porro [1], Nicola Cosentino [1], Alessandro Di Minno [1], Chiara Maria Manega [1], Franco Fabbiocchi [1], Giampaolo Niccoli [2], Francesco Fracassi [2], Simone Barbieri [1], Giancarlo Marenzi [1], Filippo Crea [2], Viviana Cavalca [1,*], Elena Tremoli [1] and Sonia Eligini [1]

1 Centro Cardiologico Monzino, I.R.C.C.S., 20138 Milan, Italy; susanna.fiorelli@cardiologicomonzino.it (S.F.); benedetta.porro@ccfm.it (B.P.); nicola.cosentino@cardiologicomonzino.it (N.C.); alessandro.diminno@cardiologicomonzino.it (A.D.M.); chiara.manega@cardiologicomonzino.it (C.M.M.); franco.fabbiocchi@cardiologicomonzino.it (F.F.); simone.barbieri@cardiologicomonzino.it (S.B.); giancarlo.marenzi@cardiologicomonzino.it (G.M.); elena.tremoli@cardiologicomonzino.it (E.T.); sonia.eligini@cardiologicomonzino.it (S.E.)

2 Department of Cardiovascular & Thoracic Sciences, Fondazione Policlinico Universitario A. Gemelli, I.R.C.C.S., Università Cattolica del Sacro Cuore, 00168 Rome, Italy; gniccoli73@hotmail.it (G.N.); francesco.fracassi@yahoo.it (F.F.); filippo.crea@rm.unicatt.it (F.C.)

* Correspondence: viviana.cavalca@cardiologicomonzino.it

Abstract: Reactive oxygen species (ROS) induce nuclear factor erythroid 2–related factor 2 (Nrf2) activation as an adaptive defense mechanism, determining the synthesis of antioxidant molecules, including heme-oxygenase-1 (HO-1). HO-1 protects cells against oxidative injury, degrading free heme and inhibiting ROS production. HO-1 is highly expressed in macrophages during plaque growth. Macrophages are morpho-functionally heterogeneous, and the prevalence of a specific phenotype may influence the plaque fate. This heterogeneity has also been observed in monocyte-derived macrophages (MDMs), a model of macrophages infiltrating tissue. The study aims to assess oxidative stress status and Nrf2/HO-1 axis in MDM morphotypes obtained from healthy subjects and coronary artery disease (CAD) patients, in relation to coronary plaque features evaluated in vivo by optical coherence tomography (OCT). We found that MDMs of healthy subjects exhibited a lower oxidative stress status, lower Nrf2 and HO-1 levels as compared to CAD patients. High HO-1 levels in MDMs were associated with the presence of a higher macrophage content, a thinner fibrous cap, and a ruptured plaque with thrombus formation, detected by OCT analysis. These findings suggest the presence of a relationship between in vivo plaque characteristics and in vitro MDM profile, and may help to identify patients with rupture-prone coronary plaque.

Keywords: oxidative stress; nuclear factor erythroid 2–related factor 2; heme-oxygenase-1; macrophages; plaque vulnerability; optical coherence tomography

---

1. Introduction

The progression of coronary atherosclerotic plaque and its destabilization with plaque rupture and thrombus formation are the key mechanisms of acute myocardial infarction (AMI) [1,2]. Post-mortem reports have demonstrated that the vulnerable plaque is characterized by a large lipid/necrotic core, a thin fibrous cap, and a great amount of resident macrophages [3]. Macrophages are versatile cells and, in relation to microenvironmental stimuli, they respond by activating different signal transduction pathways, expressing several receptors, and acquiring specific phenotypes. At coronary atherosclerotic plaque level, macrophage population is also characterized by morphological and functional heterogeneity that may enhance plaque growth and/or rupture [4].

An increasing body of evidence suggests that oxidative stress is closely associated with the atherosclerotic process and plaque instability [5,6] through different pathological mechanisms, including endothelial dysfunction, lipid oxidation, expression of adhesion molecules, and monocyte recruitment [7–9]. In response to oxidative stress stimuli, cells implement several defense mechanisms and, among them, the activation of nuclear erythroid factor 2 – related factor 2 (Nrf2)/heme oxygenase-1 (HO-1) pathway was reported to be associated with atherosclerosis [10,11]. Under unstressed conditions, Nrf2 is constitutively expressed and sequestrated in the cytoplasm by Keap1 (Kelch-like erythroid cell-derived protein with cap 'n' collar homology-associated protein 1), inducing its proteasomal degradation. On the other hand, under an oxidative stress stimulus, the complex Keap1/Nrf2 dissociates itself, and Nrf2 translocates into the nucleus. At this level, Nrf2 binds the antioxidant responsive element and promotes the transcription of proteins with antioxidant activity. Among them, HO-1 plays a fundamental role in the antioxidant mechanism within the cell by degrading the prooxidant heme to carbon monoxide, biliverdin, and ferrous ion [11]. In addition, HO-1 induction partially inhibits nicotinamide adenine dinucleotide phosphate (NADPH) oxidase activity representing a mechanism of cytoprotection against oxidative stress [12].

HO-1 expression is induced by various proatherogenic stimuli and risk factors for cardiovascular diseases [13–15]. HO-1 is highly expressed in atherosclerotic plaques, mainly localized in macrophages and foam cells [16], where its antioxidant and anti-inflammatory properties could be fundamental to counteract the development of early stage lesions [17]. In particular, HO-1 reduces the immune cell recruitment and infiltration [18], regulates the macrophage polarization also driving a phenotypic shift towards an anti-inflammatory phenotype [19,20], and inhibits the maturation of dendritic cells [21], thus affecting lesion formation.

HO-1 may also affect plaque progression and its anti-atherogenic role was highlighted in in vitro and in vivo models. Indeed, the induction of HO-1 in co-cultures of human aortic endothelial cells and smooth muscle cells inhibited the monocyte oxidized low density lipoprotein (oxLDL)-dependent chemotaxis [22]. Accordingly, the evidence of an augmented atherogenesis after HO-1 inhibition and an attenuation in the development of atherosclerotic lesion after HO-1 induction confirms this protective role. Indeed, accelerated and more advanced atherosclerotic plaques were described in HO-1 knockout mice [23]. Moreover, the inhibition of HO-1 expression in hyperlipidemic rabbits [24] or in LDL-receptor deficient mice fed with high-fat diet [25] resulted in greater atherosclerotic lesions and increased plasma and tissue lipid peroxide levels.

Despite these preclinical studies, HO-1 levels, higher than those of healthy subjects, were observed in lymphocytes and monocytes isolated from coronary artery disease (CAD) patients. In particular, its expression was higher in patients with AMI than in those with stable angina (SA) [26]. More recently, Cheng et al. showed an increased HO-1 expression in carotid atherosclerotic lesions with a vulnerable phenotype. Of note, the HO-1 levels positively correlated with plaque macrophage and lipid content, and they inversely correlated with stable plaque features, like the presence of intra-plaque smooth muscle cells and collagen [16]. Overall, animal and human studies suggest that HO-1 reflects the severity of atherosclerosis, indicating that a high level of this protein in vulnerable plaque macrophages may represent an antioxidant response, aiming at counteracting the oxidative damage inside atherosclerotic plaque [16,27].

Since resident macrophages are not easily obtainable and manageable, macrophages obtained from a spontaneous differentiation of monocytes (MDMs) are considered to be a good in vitro model to study tissue macrophages. We have previously reported the co-existence of two main and different macrophage morphotypes (round and spindle cells) after 7-day culture of human monocytes isolated from healthy subjects [28]. Similarly to tissue macrophages heterogeneity [4], different MDM morphotypes showed different functional properties: in particular, round-shaped cells were reminiscent of M2 macrophage phenotype with anti-inflammatory and reparative characteristics. On the contrary, spindle-shaped cells showed a pro-inflammatory profile resembling M1 macrophages [28].

In our more recent study, we demonstrated that the peculiar morpho-phenotype profile of MDMs isolated from CAD patients is associated with the characteristics of coronary vulnerable plaque, as assessed by optical coherence tomography (OCT) [29]. This accurate intracoronary imaging technique allows the visualization and the characterization of the atherosclerotic plaque [30,31] providing its detailed architecture and highlighting the rupture prone plaques.

Currently, no data have been provided on the association between HO-1 levels and macrophage phenotype in CAD patients. In this work, we investigated HO-1 levels and the activation of Nrf2/HO-1 axis in different MDM morphotypes obtained from healthy subjects and CAD patients, also in relation to coronary plaque morphology and activity, as analyzed in vivo by OCT.

2. Materials and Methods

2.1. Study Population

Thirty consecutive CAD patients undergoing coronary angiography, due to SA or AMI, as their first manifestation of ischemic heart disease, were enrolled at Centro Cardiologico Monzino, Milan, Italy. SA was defined as angina on effort with a stable pattern of symptoms for at least the last six months prior to admission; AMI diagnosis encompassed patients presenting with non-ST-elevation (NSTEMI) or ST-elevation-acute myocardial infarction (STEMI). NSTEMI was defined as chest pain at rest in the last 48 h preceding the admission associated with evidence of transient ST-segment depression on 12-lead electrocardiogram and normal (unstable angina) or elevated (NSTEMI) serum troponin I (TnI) levels. The diagnosis of STEMI was based on typical symptoms lasting more than 30 min and new ST-segment elevation at the J point in ≥2 contiguous leads. Exclusion criteria were: Previous history of CAD; severe chronic heart failure, severe heart valve disease, acute and chronic infections, liver diseases, neoplasia, evidence of immunologic disorders, recent (<3 months) surgical procedures or trauma and use of anti-inflammatory or immunosuppressive drugs and antioxidant supplements. All AMI included patients underwent percutaneous coronary intervention, specifically within 24 h from admission for NSTEMI and within 12 h of symptom onset for STEMI patients. OCT assessment was performed both in SA and in AMI in order to investigate coronary plaque features. The control group consisted of 10 healthy subjects, with neither history of CAD, nor cardiovascular risk factors, nor inflammatory disorders, and specifically not taking any cardiovascular therapy. The ethics committee approved the study protocol, and all participants provided written informed consent to participate in the study. The study was performed according to the Declaration of Helsinki.

2.2. Monocyte Isolation and Culture

Venous blood samples were drawn from the antecubital vein of healthy subjects and CAD patients when fasting into tubes containing ethylenediaminetetraacetic acid (EDTA) (9.3 mM; Vacutainer Systems, Becton Dickinson, USA). Mononuclear cells were isolated by Ficoll-Paque Plus (GE Healthcare, Milan, Italy) density centrifugation and plated (2×10^6/mL) in 35 mm well plates (Primaria, Falcon, Como, Italy) as previously described [11]. After 90 min, non-adherent cells were removed and adherent cells were cultured over 7 days at 37 °C (5% CO2) in Medium 199 (Lonza, Milan, Italy) supplemented with 2 mM L-glutamine, 100 U/mL penicillin, 100 μg/mL streptomycin, and 10% autologous serum without replacement of the medium throughout the entire culture period. MDM morphology was examined by phase contrast microscopy (Axiovert 200 M; Zeiss, Milan, Italy) at 20× or 40× magnification. MDMs were defined spindle/elongated when a length > 70 μm and a width < 30 μm were detected, and round MDMs when width and length were similar and >30 to 40 μm. Cells, whose morphology and dimension did not satisfy the above criteria, were classified as undefined.

2.3. *Liquid Chromatography Tandem Mass Spectrometry (LC-MS/MS) Analysis*

For the determination of reduced (GSH) and oxidized (GSSG) glutathione levels, MDMs were washed with phosphate buffer saline (PBS) and detached by gentle scraping. After centrifugation (400× *g*, 10 min), the supernatant was removed and MDMs were lysed in PBS containing 0.1 μg of leupeptin, 0.2 M benzamidine, and 1 μg of trypsin inhibitor. Lysed cells were mixed in a 1:1 (*v*/*v*) ratio with 10% trichloroacetic acid (TCA) containing 1 mM EDTA. After centrifugation (10,000× *g*, 10 min) at room temperature (RT), the supernatant was diluted 1:20 with 0.1% formic acid. The analysis was performed using LC-MS/MS method as previously described [32]. Liquid chromatography was performed on Luna analytical PFP column (100 × 2.0 mm, 3 μm) using an Accela HPLC pump system (Thermo Fisher Scientific, Milan, Italy). Mass spectrometric analysis was performed using a TSQ Quantum Access (Thermo Fisher Scientific) triple quadrupole mass spectrometer coupled with electrospray ionization (ESI) operated in positive mode.

2.4. *Western Blot Analysis*

Western blot analysis was performed on MDMs total lysate, and on MDMs cytosolic and nuclear fractions.

To obtain MDMs total lysate, cells were harvested and lysated in a buffer composed by 20 mM Tris, 4% sodium dodecyl sulfate (SDS) and 20% glycerol, containing 1 mM sodium orthovanadate, 1 mM NaF, 1 μg/mL leupeptin hemisulfate, 1 mM benzamidine hydrochloride, 1 mM EDTA, 10 μg/mL soybean trypsin inhibitor, 0.5 mM pefabloc, 0.5 mM dithiothreitol (DTT) [33].

For the isolation of cytosolic fraction, MDMs were harvested in 50 μL of buffer containing 10 mM HEPES, pH 7.9, 1.5 mM $MgCl_2$, 10 mM KCl, 0.5 mM DTT and then Triton X-100 (0.2% final concentration) was added. Cells were centrifuged at 15,000× *g* for 1 min at 4 °C to separate cytosols from nuclei [34]. The supernatant containing cytosolic fraction was used for the analysis and the pellet was used for nuclear fraction extraction.

The pellet nuclear fraction was washed with a buffer (20 mM HEPES, pH 7.9, 400 mM NaCl, 25% glycerol, 1.5 mM $MgCl_2$, 0.2 mM EDTA) containing 0.5 mM PMSF, and 0.5 mM DTT and centrifuged at 15,000× *g* (10 min, 4 °C) [34]. The supernatant containing nuclear fraction was used for the analysis.

SDS-polyacrylamide gel electrophoresis (PAGE) was performed as previously described [33]. After blotting, membranes carrying MDMs total lysates were incubated overnight at 4 °C with primary antibodies directed against HO-1 (1:250) (catalogue number: ab13248; Abcam, Milan, Italy) or Nrf2 (1:200) (Santa Cruz Biotechnology; catalogue number: sc-722, Milan, Italy).

Membranes carrying MDMs cytosolic and nuclear fraction samples were incubated overnight at 4 °C with primary antibodies directed against Nrf2 (1:200) (Santa Cruz Biotechnology). After incubation with horseradish peroxidase-conjugated anti-mouse (1:10,000, catalogue number: 715-035-151) or anti-rabbit secondary antibody (1:5000, catalogue number: 111-035-003) (Jackson ImmunoResearch Labs Inc., Li StarFISH, Milan, Italy), as appropriate, for 1 h at RT, protein bands were detected by chemiluminescence. β-Actin was used as internal standard for control of protein load.

2.5. *Immunofluorescence Analysis*

Immunofluorescence was performed as previously described [35]. Fixed MDMs were incubated overnight at 4 °C with a monoclonal rabbit anti-human HO-1 antibody (1:100) (Abcam), or with a polyclonal rabbit anti-human Nrf2 antibody (1:200) (Santa Cruz Biotechnology). Detection was performed with Alexa Fluor 488 (1:200, catalogue number: A11034, 60 min at RT) (Life Technologies Italia, Monza, Italy). Nuclei were visualized by Hoechst 33258 (1:10,000, catalogue number: B2883) (Sigma-Aldrich, Milan, Italy). Negative control experiments were performed by omitting the primary antibodies. Fluorescence quantification was performed as previously described [33]. Data are expressed as mean ± SD of fluorescence intensity/μm^2 for each MDM morphotype, subtracted of the negative control value obtained in the absence of primary antibody. Multiple fields of view (at least three fields, 400× magnification) were captured for each culture.

2.6. OCT Image Acquisition and Analysis

OCT examination was performed in AMI patients at culprit lesion and in SA patients at the minimal lumen area (MLA) site. Some plaque characteristics were determined such as the measurement of the thickness of fibrous cap, lipid content, and macrophage accumulation. Fibrous cap thickness was defined as the minimum distance from the coronary artery lumen to the inner border of the lipid pool and a thin-cap fibroatheroma (TCFA) was defined as a minimal fibrous cap thickness ≤ 65 μm; thick-cap fibroatheroma was a plaque with a minimal fibrous cap thickness > 65 μm. The max lipid arc was measured on the frame with the largest lipid core by visual screening. A plaque showing two or more lipid containing quadrants was considered a lipid-rich plaque. A lipid plaque with fibrous cap discontinuity and cavity formation inside the plaque was defined as rupture plaque. A thrombus was defined as an irregular mass protruding into the lumen with a measured dimension > 250 μm.

The macrophage infiltration (MØI) in the analyzed lesions by OCT, has been assessed as previously reported [36], according to the International Working Group for Intravascular Optical Coherence Tomography (IWG-IVOCT) Consensus standards [37]. Briefly, macrophages were qualitatively identified on raw OCT data within a 300×125 μm^2 (lateral x axial) region of interest (ROI). In particular, macrophages have been visualized by OCT imaging as signal-rich, distinct, or confluent punctate regions that exceed the intensity of background speckle noise and generate a backward shadowing. For caps having a thickness < 125 μm^2, the depth of the ROI was matched to the cap thickness. Median filtering was performed with a 3×3 square kernel to remove speckle noise. In plaques with MØI, quantitative evaluation of macrophage content was obtained by measuring the normalized standard deviation (NSD) known to have a high degree of positive correlation with histological measurements of macrophage content, by using a dedicated software provided by S. Jude medical [38,39]. In particular, NSD was measured for each pixel within each cap using a 125 μm^2 window centered at the pixel location:

$$\text{NSD}(x,y) = [\sigma(x,y)125\ \mu\text{m}^2/(\text{Smax-Smin})] \times 100 \quad (1)$$

where NSD (x,y) is the normalized standard deviation of the OCT signal at pixel location (x,y), Smax is the maximum OCT image value, and Smin is the minimum OCT image value. Pixels within the (125×125) μm^2 window that did not overlap with the segmented cap were excluded.

2.7. Statistical Analysis

Continuous variables are presented as mean ± SD, variables not normally distributed are presented as median and interquartile ranges (IQR), and categorical variables as absolute numbers and percentages. Comparisons between 'healthy subjects' vs. 'CAD patients' groups were performed using independent samples *t*-Test for normally distributed variables and Wilcoxon rank-sum test for not normally distributed variables.

Comparisons between 'healthy subjects' vs. 'SA', 'NSTEMI', and 'STEMI' groups were performed using ANOVA Test for normally distributed variables and Kruskal–Wallis Test for not normally distributed variables. Categorical variables were compared using Fisher's exact test. Post-hoc testing of main effects was performed using Bonferroni adjustment for multiple comparisons (α/[number of comparisons]). Correlations between variables were determined using Spearman's rank test. Trends of variation from healthy subjects to STEMI patients were assessed by general linear models. All tests were two-sided, and a p value of less than 0.05 was required for statistical significance. All calculations were computed by using SAS software package v9.4 (SAS Institute Inc., Cary, NC, USA).

3. Results

3.1. Clinical Features

Demographic and clinical characteristics of the enrolled subjects are shown in Table 1. CAD patients had higher body mass index (BMI) and they were more frequently males. Moreover, as expected, in CAD group there was a prevalence of subjects with cardiovascular risk factors as smoking, diabetes, dyslipidemia, hypertension, and family history of cardiovascular disease. Despite there being more subjects with dyslipidemia among CAD patients, the LDL cholesterol values were similar between patients and healthy subjects, as the result of pharmacological treatment. Furthermore, CAD patients had higher levels of glycaemia and of C-reactive protein (hs-CRP), an inflammatory marker.

Of the 30 consecutive CAD patients, 10 (33.3%) had a diagnosis of SA, whereas 20 (66.6%) of AMI (10 NSTEMI (33.3%) and 10 STEMI (33.3%)). Among CAD patients, those with STEMI showed a higher BMI, higher levels of hs-CRP, creatinine, glycaemia, total cholesterol, triglycerides, TnI, and creatine phosphokinase-MB (CK-MB). In addition, lower high-density lipoprotein (HDL) cholesterol levels were observed in AMI patients. No difference in admission therapy was observed among CAD patients.

3.2. Oxidative Stress Status

The levels of GSH and GSSG, whose ratio is a recognized index of oxidative stress, were measured in MDMs obtained from the study population. The results are showed in Figure 1. The GSH/GSSG ratio was significantly lower in CAD patients as compared to healthy subjects. The analysis of the different clinical presentations of CAD revealed a progressive decrease of GSH/GSSG ratio in MDMs going from SA, NSTEMI, to STEMI patients ($p_{trend} < 0.005$).

3.3. HO-1 and Nrf2 Expression

MDMs of CAD patients displayed higher levels of HO-1 protein as compared to those observed in healthy subjects (0.12 ± 0.09 vs. 0.06 ± 0.02, $p < 0.05$) (Figure 2a). Moreover, the immunofluorescence analysis showed higher protein levels in spindle compared to round cells in all study patients' groups. (Figure 2b) with a significant increase in both MDM morphotypes of STEMI patients. A progressive increase was also shown going from healthy subjects to STEMI patients (p_{trend} round < 0.0002, p_{trend} spindle < 0.0001; Figure 2c).

This behavior was mirrored by the levels of the transcription factor Nrf2. Higher levels of total Nrf2 were detected in CAD patients as compared to healthy subjects (1.49 ± 0.73 vs. 0.51 ± 0.62, $p < 0.01$) (Figure 3a). In addition, the evaluation of Nrf2 levels in cytoplasmic and nuclear fractions demonstrated a significant increase of Nrf2 translocation into the nucleus in CAD patients as compared to those of healthy subjects (Figure 3b). The immunofluorescence analysis of the MDM morphotypes evidenced significantly higher Nrf2 protein levels in both spindle and round MDMs of AMI patients (NSTEMI and STEMI) as compared to those of healthy subjects (Figure 3c). Furthermore, an increasing trend in protein levels was detected in both MDM morphotypes paralleling the severity of the clinical presentations (p_{trend} round < 0.02, p_{trend} spindle $= 0.06$) (Figure 3d).

Table 1. Baseline clinical, laboratory, and angiographic characteristics of the study subjects.

Variables	Healthy Subjects (n = 10)	CAD (n = 30)	p Value Healthy Subjects vs. CAD °°	CAD: SA (n = 10)	CAD: NSTEMI (n = 10)	CAD: STEMI (n = 10)	ANOVA p Value °
Demographics							
Age (years)	61.5 ± 10	63.8 ± 12.1	0.5927	70.3 ± 7.2	61.0 ± 11.9	60.0 ± 14.8	0.1660
Male sex, n (%)	5 (50)	26 (86.7)	**0.0290** ‡	8 (80)	9 (90)	8 (80)	0.8179 ‡
Body mass index (kg/m^2)	23.5 ± 1.6	29.3 ± 4.6	**0.0004**	28.0 ± 4.5 *	28.1 ± 3.7 *	32.3 ± 4.6 *	**0.0002**
Clinical characteristics							
Current smoking, n (%)	0	18 (60)	**0.0010** ‡	7 (70)	6 (60)	5 (50)	0.5884 ‡
Diabetes mellitus, n (%)	0	16 (53.3)	**0.0030** ‡	5 (50)	5 (50)	6 (60)	0.6593 ‡
Dyslipidemia, n (%)	0	16 (53.3)	**0.0030** ‡	7 (70)	5 (50)	4 (40)	0.4704 ‡
Hypertension, n (%)	0	14 (46.7)	**0.0070** ‡	4 (40)	5 (50)	5 (50)	0.9004 ‡
Family history of CAD, n (%)	0	17 (56.7)	**0.0020** ‡	4 (40)	9 (90) #	4 (40)	**0.0149** ‡
LVEF (%)	NA	50.1 ± 8.8		48 ± 9.3	51.3 ± 9.8	51.0 ± 7.7	0.7510
Laboratory data							
WBC (× 10^9/L)	7.6 ± 2.9	9.2 ± 3.9	0.2537	8.9 ± 2.4	9.1 ± 5.1	9.7 ± 4.0	0.6852
RBC (× 10^{12}/L)	4.5 ± 0.8	5.1 ± 2.0	0.3917	4.6 ± 0.6	5.1 ± 0.6	5.8 ± 3.6	0.4330
Neutrophil count (× 10^9/L)	4.8 ± 2.3	6.2 ± 3.5	0.2585	5.8 ± 2.3	6.1 ± 4.7	6.9 ± 3.4	0.6249
Lymphocyte count (× 10^9/L)	2.1 ± 1.2	2.0 ± 0.9	0.7190	2.3 ± 1.2	1.9 ± 0.7	1.8 ± 0.9	0.6889
Eosinophil count (× 10^9/L)	0.1 ± 0.1	0.2 ± 0.2	0.6303	0.2 ± 0.1	0.1 ± 0.1	0.2 ± 0.4	0.6123
Monocyte count (× 10^9/L)	0.5 ± 0.2	0.6 ± 0.3	0.1398	0.6 ± 0.2	0.6 ± 0.4	0.7 ± 0.4	0.3176
Basophil count (× 10^9/L)	0.03 ± 0.02	0.01 ± 0.00	**0.0072**	0.01 ± 0.02	0.01 ± 0.03 *	0.01 ± 0.02	**0.0343**
Platelets (× 10^9/L)	248 ± 61.9	230.8 ± 83.4	0.5714	213.7 ± 49.8	252.2 ± 106.0	223.7 ± 85.6	0.6549
hs-CRP (mg/L)	1.9 (1.4–2.3)	4.9 (2.0–21.0)	**0.0141** †	2.1 (1.6–2.1) ǂ#	6.7 (1.6–17.0) *	38.6 (6.0–75.0) *	**0.0003** §
Creatinine (mg/dL)	1 ± 0.1	1.0 ± 0.5	0.9372	0.8 ± 0.3 #	0.8 ± 0.3 #	1.4 ± 0.5	**0.0015**
Glycaemia (mg/dL)	93.5 ± 12.2	140.2 ± 42.8	**0.0017**	116.4 ± 27.1 #	130.3 ± 33.4 #	178.8 ± 43.7 *	**<0.0001**
Total cholesterol (mg/dL)	187.7 ± 22.1	204.4 ± 42.6	0.2438	181.1 ± 34.1 #	207.9 ± 42.8	226.1 ± 41.7	**0.0417**
LDL (mg/dL)	112.6 ± 26	122.4 ± 41.9	0.4924	102.0 ± 23.5	130.4 ± 47.5	135.2 ± 46.2	0.1890
HDL (mg/dL)	41.1 ± 5.3	48.83 ± 14.9	**0.0242**	61.6 ± 16.7 *,ǂ,#	44.5 ± 10.5	43.3 ± 9.3	**0.0004**
Triglycerides (mg/dL)	143.8 ± 31.6	161.2 ± 55.5	0.3333	117.9 ± 42.7 ǂ#	176.7 ± 59.6	190.3 ± 32.2	**0.0034**
Peak TnI (µg/dL)	NA	1 (0.0–29.4)		NA	1.2 (0.5–1.4) #	29.7 (25.0–163.0)	**<0.0001** §
Peak CK-MB (µg/dL)	NA	11.1 (2.1–110.0)		2.1 (1.5–2.1)	12.3 (2.5–28.0) #	281 (110.0–521.0) #	**<0.0001** §
Angiographic data							
Culprit or treated vessel, n (%)							0.1489 ‡
LAD	NA	14 (46.7)		3 (30)	8 (80)	3 (30)	
LCX	NA	10 (30.3)		4 (40)	1 (10)	5 (50)	
RCA	NA	6 (20)		3 (30)	1 (10)	2 (20)	
Multivessel disease, n (%)	NA	17 (56.7)		8 (80)	4 (40)	5 (50)	0.3276 ‡
Admission therapy							
ASA, n (%)	0	11 (36.7)	**0.0380** ‡	3 (30)	5 (50)	3 (30)	0.3192 ‡
Beta-Blockers, n (%)	0	8 (26.7)	0.1650 ‡	2 (20)	5 (50)	1 (10)	0.2319 ‡
ACE-inhibitors, n (%)	0	9 (30)	0.0810 ‡	5 (50)	2 (20)	2 (20)	0.3192 ‡
Statins, n (%)	0	10 (30.3)	**0.0430** ‡	5 (50)	2 (20)	3 (30)	0.3459 ‡

SA: Stable angina; CAD: Coronary artery disease; LVEF: Left ventricular ejection fraction; WBC: White blood cells; RBC: Red blood cells; LDL: Low-density lipoprotein; HDL: High-density lipoprotein; hs-CRP: High-sensitive C-reactive protein: TnI: Troponin-I; CK-MB: Creatine phosphokinase-MB; LAD: Left anterior descending; LCX: Left circumflex; RCA: Right coronary artery; ASA: Aspirin; ACE-inhibitors, angiotensin-converting enzyme-inhibitors; NA: Not assessed. Data are expressed as mean ± SD or median and interquartile range. * $p < 0.05$ vs. healthy subjects; ǂ$p < 0.05$ vs. NSTEMI; # $p < 0.05$ vs. STEMI; ° by ANOVA test, except: § Kruskal–Wallis Test, ‡ Fisher exact test; °° by Independent t-test, except: † Wilcoxon rank-sum test, ‡ Fisher exact test.

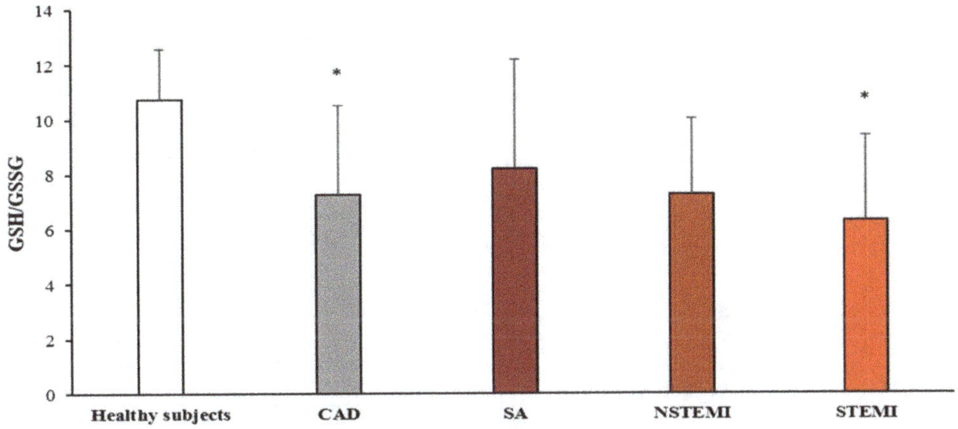

Figure 1. GSH/GSSG evaluation in monocyte-derived macrophages (MDMs). MDMs were obtained from CAD patients and healthy subjects. Data are expressed as mean ± SD and derive from independent cultures obtained from 10 healthy subjects and 30 CAD patients (SA $n = 10$; NSTEMI $n = 10$; STEMI $n = 10$. * $p < 0.05$ vs. healthy subjects.

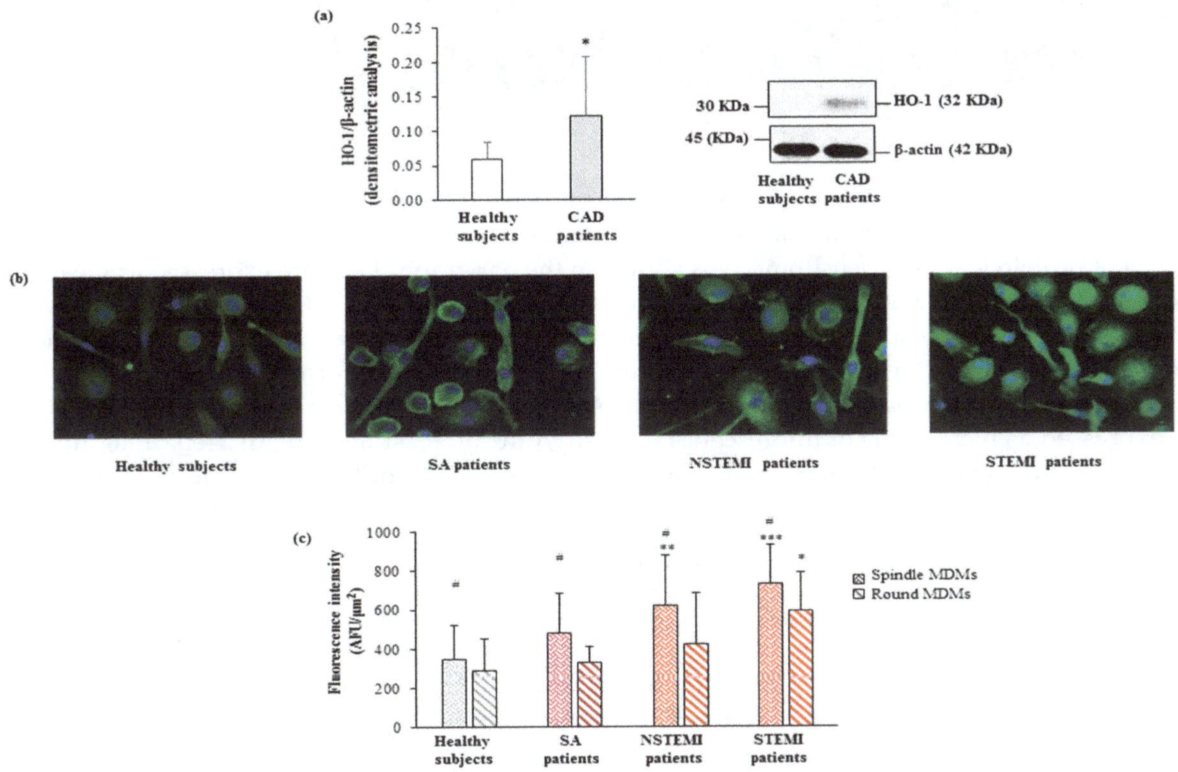

Figure 2. HO-1 levels in CAD patients and healthy subjects. (**a**) The protein levels of HO-1 were detected by western blot analysis. β-actin was used as a control of protein loading. Densitometry is shown in the bar graph. Data are expressed as mean ± SD and derive from MDMs obtained from 10 healthy subjects and 17 CAD patients. (**b**) Representative images of HO-1 in round and spindle MDMs obtained from healthy subjects and CAD patients (400× original magnification), nuclei were visualized by Hoechst 33258. (**c**) Quantitative analysis of HO-1 in round and spindle MDMs. Data are expressed as mean ± SD of fluorescence intensity/μm^2 (at least three fields, 400× magnification, were analyzed) and data derive from independent cultures obtained from 10 healthy subjects and 30 CAD patients (SA $n = 10$; NSTEMI $n = 10$; STEMI $n = 10$. $^{\#}$ $p < 0.05$ vs. round; * $p < 0.05$, ** $p < 0.01$, *** $p < 0.001$ vs. healthy subjects.

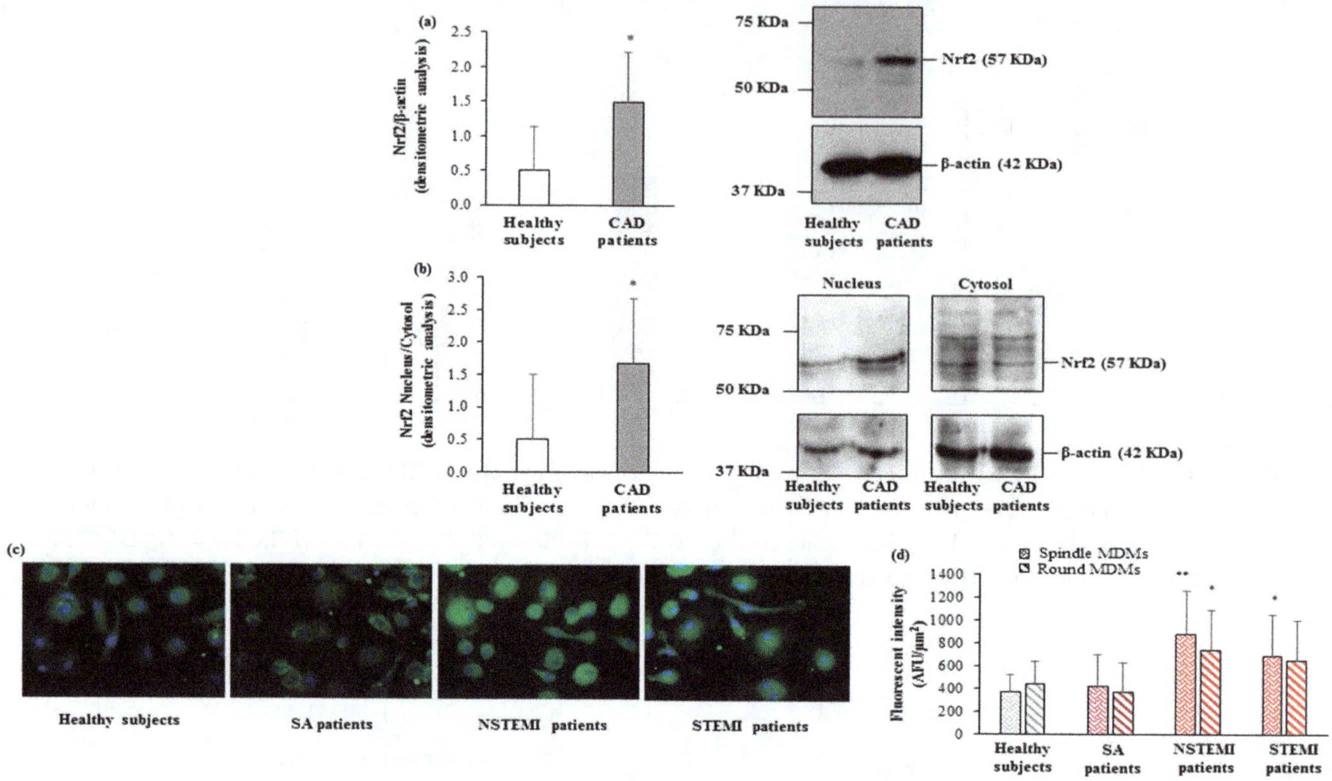

Figure 3. Nrf2 levels in CAD patients and healthy subjects. (**a**,**b**) Nrf2 levels in (**a**) total cellular lysate, (**b**) nuclear and cytosolic compartments were detected by western blot analysis. β-actin was used as a control of protein loading. Densitometry is shown in the bar graph. Data are expressed as mean ± SD and derive (**a**) from MDMs obtained from 10 healthy subjects and 17 CAD patients; (**b**) from MDMs obtained from 5 healthy subjects and 5 CAD patients. (**c**) Representative images of Nrf2 in round and spindle MDMs obtained from healthy subjects and CAD patients (400× original magnification), nuclei were visualized by Hoechst 33258. (**d**) Quantitative analysis of Nrf2 in round and spindle MDMs. Data are expressed as mean ± SD of fluorescence intensity/μm^2 (at least three fields, 400× magnification, were analyzed) and derive from independent cultures obtained from 10 healthy subjects and 30 CAD patients (SA $n = 10$; NSTEMI $n = 10$; STEMI $n = 10$). * $p < 0.05$, ** $p < 0.01$, vs. healthy subjects.

3.4. *Association Between In Vivo Plaque Morphology and HO-1 Levels in MDMs*

The coronary plaque characteristics in CAD patients are illustrated in Table 2.

Table 2. Optical coherence tomography features of coronary plaque in CAD patients.

Variables	CAD ($n = 30$)
Lipid plaque, n (%)	26 (86.67)
Fibrous plaque, n (%)	1 (3.33)
Calcific plaque, n (%)	3 (10)
Plaque rupture, n (%)	15 (50)
MLA, mm^2 (IQR)	1.70 (1.43–2.58)
TCFA, n (%)	15 (50)
Thrombus, n (%)	14 (46.67)
Lipid quadrants, n	2.70 ± 1.02
Lipid arc, degree ° (IQR)	163 (133.5–280)
Max lipid arc, degree °	206.37 ± 87.10
Macrophage infiltration detection, n (%)	21 (70)
Macrophage NSD	6.24 ± 1.16

MLA: Minimal lumen area; TCFA: Thin-cap fibroatheroma; NSD: Normalized standard deviation. Data are expressed as mean ± SD or median and IQR.

Patients with high levels of HO-1 in both MDM morphotypes more frequently displayed a TCFA ($p = 0.049$ and $p = 0.015$, spindle and round, respectively) (Figure 4a), a ruptured plaque ($p = 0.001$ and $p = 0.036$, spindle and round, respectively) (Figure 4b), and presence of thrombi ($p = 0.0005$ and $p = 0.028$, spindle and round, respectively) (Figure 4c).

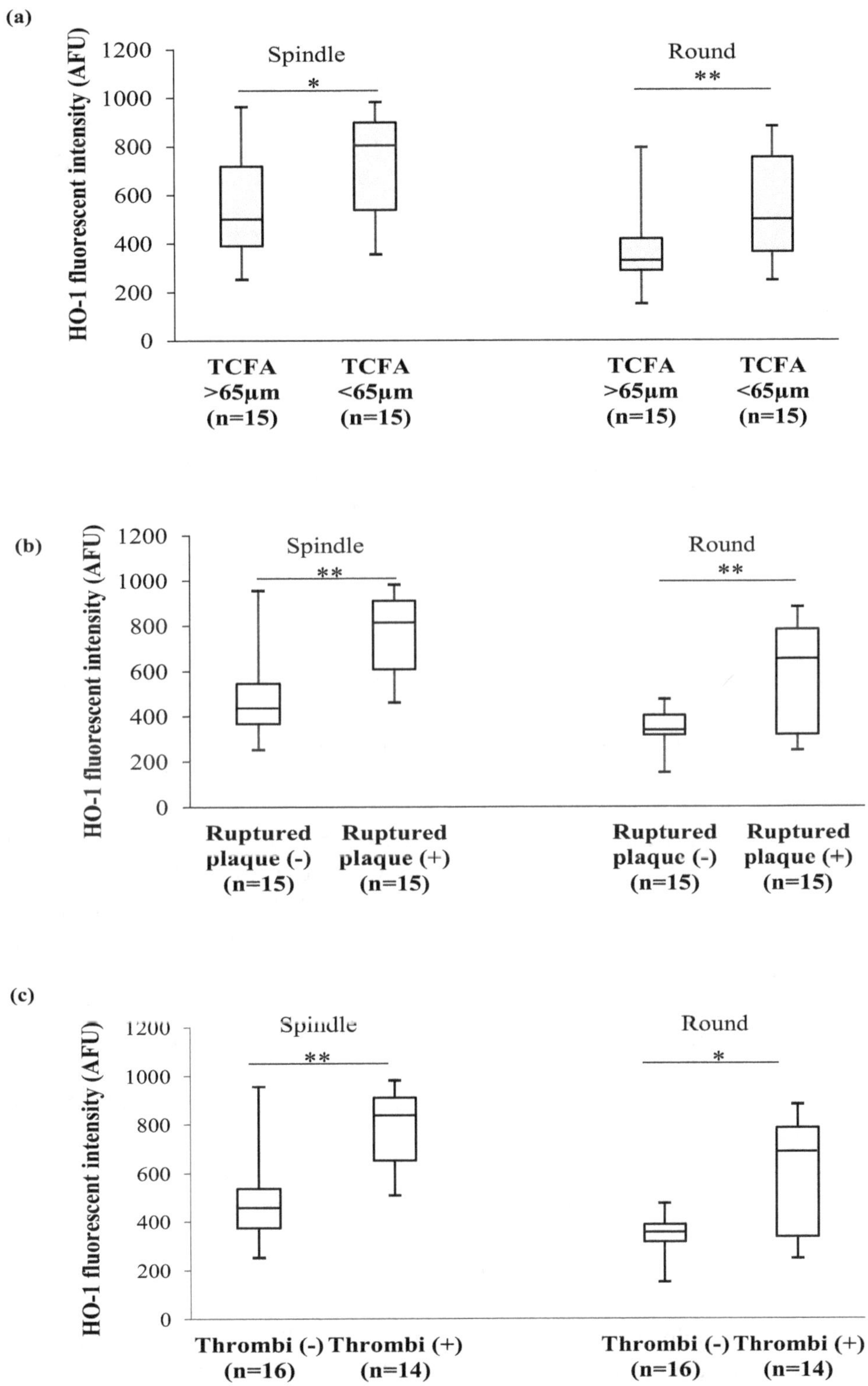

Figure 4. In vivo plaque features and HO-1 levels in MDM morphotypes. (**a**) Association between HO-1 levels and thin cap fibroatheroma (TCFA), (**b**) fibrous cap integrity, (**c**) presence of thrombi detected by means of optical coherence tomography (OCT). Data are expressed as median and IQR. * $p < 0.05$, ** $p < 0.01$.

In addition, in both MDM morphotypes we observed significant positive correlations between HO-1 levels and macrophage content (NSD) (spindle: r = 0.62, $p = 0.003$; round: r = 0.59, $p = 0.005$) (Figure 5a). Moreover, borderline positive correlations were observed between HO-1 levels and max lipid arc (spindle: r = 0.34, $p = 0.06$; round: r = 0.43, $p = 0.02$) (Figure 5b).

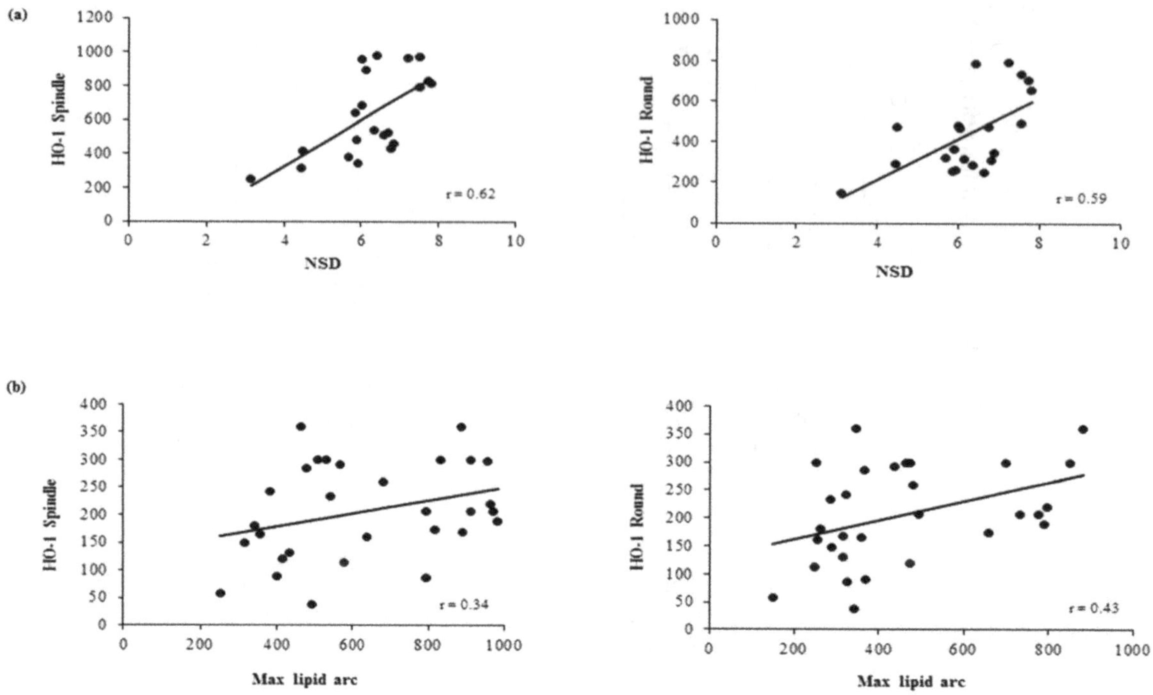

Figure 5. Correlations between in vivo plaque features and HO-1 levels in MDM morphotypes. Correlation between HO-1 levels in round and spindle MDMs and (**a**) macrophage content (NSD) and (**b**) max lipid arc, detected by means of OCT in in vivo plaque.

4. Discussion

In the present study we demonstrate higher HO-1 protein levels and the activation of its related Nrf2/HO-1 pathway in MDMs obtained from CAD patients as compared to those obtained from healthy subjects. For the first time, we analyzed this pathway in the different MDM morphotypes. Moreover, we show a positive association between HO-1 levels in MDMs obtained in vitro and the vulnerable coronary plaque features, as detected in vivo by OCT.

HO-1 is a stress response protein that is expressed in several cell types, including macrophages. It is induced by several stimuli inducing oxidative stress, such as cardiovascular risks factors [13–15], hypoxia [40], and GSH depletion [41]. In line with in animal models studies [42,43], our previous data obtained in whole blood from CAD patients demonstrated a reduction of GSH levels and a related imbalance of the GSH/GSSG ratio, that is commonly used as an index of oxidative status [44,45]. Here, we detect a progressive decrease of this marker in MDMs obtained from SA, NSTEMI, and STEMI patients, indicating an increase of oxidative stress status in patients with a worse prognosis.

Augmented oxidative stress levels activate the Nrf2/HO-1 pathway as one of the cellular protective mechanisms. The important protective role of HO-1 against human atherosclerosis has been highlighted in population genetic studies that evidenced a polymorphism in the promoter region of human HO-1 gene associated with atherosclerosis predisposition [46–48]. In the clinical setting HO-1 deficiency is a very rare condition, but the autopsy case report of HO-1 deficient boy showed hyperlipidemia, presence of foam cells in the liver, and of fatty streak and fibrous plaque in the aorta. These are several lines of evidence that outline the role of HO-1 against atherogenesis [49].

Nevertheless, in our experimental conditions, MDMs from CAD patients expressed a progressive increase of HO-1 levels in both morphotypes, going from healthy to STEMI patients. This is in line with previous studies that infer an adaptive defense role of HO-1 against atherosclerosis [50]. The mechanisms underlying the cardioprotective function of HO-1 involve the antioxidant action of its products [51,52], the reduction in leukocyte recruitment and infiltration, and the suppression of the pro-inflammatory response of immune cells [18].

The increase in HO-1 levels observed in our study CAD population goes in parallel with the increased levels of Nrf2, the transcription factor involved in cell protection from oxidant stressors. Several results strengthen the anti-atherogenic role of Nrf2 in preserving vascular integrity and endothelial function [53], potentially through the release of NO [54] and the protection from cell apoptosis [55]. Moreover, the activation of Nrf2 protects human coronary artery endothelial cells against oxidative challenge [56]. In the present study, we also demonstrate the activation of Nrf2 pathway by its translocation into the nucleus. Indeed, CAD MDMs show higher levels in the nucleus as compared to healthy MDMs. In turn, the latter exhibit very low and feeble protein levels in the nucleus, and high protein levels in the cytosol. This is in line with Nrf2 cytosolic localization in basal condition, where Nrf2 is associated to Keap-1 protein, which induces its ubiquitination and proteasomal degradation. The increase of the Nrf2 stability and its translocation into the nucleus in response to stress signals activates the antioxidant gene transcription [57].

This defense mechanism is evidenced also in atherosclerotic plaques by the presence of high levels of HO-1. Notably, its levels in both carotid [16] and coronary [58] lesions positively correlated with atherosclerotic grade and plaque vulnerability. In addition, Li et al. showed increased levels of HO-1 in lymphocytes and monocytes associated with the severity of the pathology [59]. Accordingly, in our study, patients with the highest levels of HO-1 in MDMs, more frequently showed a vulnerable coronary plaque featured by a TCFA, and an increase in macrophage and lipid content. In addition, in our study, higher levels of HO-1 are detected in MDMs of CAD patients that presented with ruptured plaque and with the presence of thrombus as compared to those with non-ruptured plaque and without thrombi formation. This finding may be explained in light of previous studies reporting a positive correlation between HO-1 and matrix metalloproteinase-9 levels and a negative correlation between the presence of smooth muscle cells and collagen deposition [16]. Moreover, it has been observed that MMP-9 is abundant in carotid plaques characterized by the presence of intraplaque hemorrhage [60].

Overall, these results highlight the activation of Nrf2/HO-1 pathway as an antioxidant response mechanism in MDMs from CAD patients and point out that HO-1 levels may reflect coronary plaque vulnerability. This association could help in identifying patients with rupture-prone plaque and suggests Nrf2/HO-1 pathway as a new potential therapeutic target to counteract plaque progression.

Author Contributions: S.E., V.C., F.C., G.M. and E.T. proposed and supervised the project. S.F., S.E. and V.C. designed the experiments. S.F., B.P., A.D.M. and C.M.M. performed the in vitro experiments. N.C. and G.M. were involved in clinical evaluation and subject enrolment. F.F. (Franco Fabbiocchi), F.F. (Francesco Fracassi) and G.N. performed the in vivo analyses. S.B. performed the statistical analysis. S.F., S.E., V.C., wrote the manuscript. S.F., B.P., A.D.M., S.E., V.C., G.N., F.F. (Franco Fabbiocchi), F.F. (Francesco Fracassi), N.C. and G.M. contributed to data analysis and interpretation of the results. All authors reviewed the manuscript.

Acknowledgments: The authors thank Fabrizio Veglia for his help in the statistical analysis revision. The authors also thank Alessandra Terragni for her assistance in English language editing.

References

1. Schaar, J.A.; Muller, J.E.; Falk, E.; Virmani, R.; Fuster, V.; Serruys, P.W.; Colombo, A.; Stefanadis, C.; Ward Casscells, S.; Moreno, P.R.; et al. Terminology for high-risk and vulnerable coronary artery plaques. Report of a meeting on the vulnerable plaque, June 17 and 18, 2003, Santorini, Greece. *Eur. Heart J.* **2004**, *25*, 1077–1082. [CrossRef] [PubMed]
2. Schwartz, S.M.; Galis, Z.S.; Rosenfeld, M.E.; Falk, E. Plaque rupture in humans and mice. *Arterioscler. Thromb. Vasc. Biol.* **2007**, *27*, 705–713. [CrossRef]

3. Vancraeynest, D.; Pasquet, A.; Roelants, V.; Gerber, B.L.; Vanoverschelde, J.L. Imaging the vulnerable plaque. *J. Am. Coll. Cardiol.* **2011**, *57*, 1961–1979. [CrossRef] [PubMed]
4. Chinetti-Gbaguidi, G.; Colin, S.; Staels, B. Macrophage subsets in atherosclerosis. *Nat. Rev. Cardiol.* **2015**, *12*, 10–17. [CrossRef]
5. Yokoyama, M. Oxidant stress and atherosclerosis. *Curr. Opin. Pharmacol.* **2004**, *4*, 110–115. [CrossRef] [PubMed]
6. Uno, K.; Nicholls, S.J. Biomarkers of inflammation and oxidative stress in atherosclerosis. *Biomark. Med.* **2010**, *4*, 361–373. [CrossRef] [PubMed]
7. Higashi, Y.; Matsuoka, H.; Umei, H.; Sugano, R.; Fujii, Y.; Soga, J.; Kihara, Y.; Chayama, K.; Imaizumi, T. Endothelial function in subjects with isolated low HDL cholesterol: Role of nitric oxide and circulating progenitor cells. *Am. J. Physiol. Endocrinol. Metab.* **2010**, *298*, E202–E209. [CrossRef] [PubMed]
8. Peluso, I.; Morabito, G.; Urban, L.; Ioannone, F.; Serafini, M. Oxidative stress in atherosclerosis development: The central role of LDL and oxidative burst. *Endocr. Metab. Immune Disord. Drug Targets* **2012**, *12*, 351–360. [CrossRef] [PubMed]
9. Lee, C.F.; Ullevig, S.; Kim, H.S.; Asmis, R. Regulation of Monocyte Adhesion and Migration by Nox4. *PLoS ONE* **2013**, *8*, e66964. [CrossRef]
10. Wang, L.J.; Lee, T.S.; Lee, F.Y.; Pai, R.C.; Chau, L.Y. Expression of heme oxygenase-1 in atherosclerotic lesions. *Am. J. Pathol.* **1998**, *152*, 711–720. [PubMed]
11. Morita, T. Heme oxygenase and atherosclerosis. *Arterioscler. Thromb. Vasc. Biol.* **2005**, *25*, 1786–1795. [CrossRef] [PubMed]
12. Luo, M.; Tian, R.; Lu, N. Nitric oxide protected against NADPH oxidase-derived superoxide generation in vascular endothelium: Critical role for heme oxygenase-1. *Int. J. Biol. Macromol.* **2019**, *126*, 549–554. [CrossRef] [PubMed]
13. Ndisang, J.F.; Zhao, W.; Wang, R. Selective regulation of blood pressure by heme oxygenase-1 in hypertension. *Hypertension* **2002**, *40*, 315–321. [CrossRef] [PubMed]
14. Fukano, Y.; Oishi, M.; Chibana, F.; Numazawa, S.; Yoshida, T. Analysis of the expression of heme oxygenase-1 gene in human alveolar epithelial cells exposed to cigarette smoke condensate. *J. Toxicol. Sci.* **2006**, *31*, 99–109. [CrossRef] [PubMed]
15. Ishii, T.; Itoh, K.; Ruiz, E.; Leake, D.S.; Unoki, H.; Yamamoto, M.; Mann, G.E. Role of Nrf2 in the regulation of CD36 and stress protein expression in murine macrophages: Activation by oxidatively modified LDL and 4-hydroxynonenal. *Circ. Res.* **2004**, *94*, 609–616. [CrossRef] [PubMed]
16. Cheng, C.; Noordeloos, A.M.; Jeney, V.; Soares, M.P.; Moll, F.; Pasterkamp, G.; Serruys, P.W.; Duckers, H.J. Heme oxygenase 1 determines atherosclerotic lesion progression into a vulnerable plaque. *Circulation* **2009**, *119*, 3017–3027. [CrossRef]
17. Araujo, J.A.; Zhang, M.; Yin, F. Heme oxygenase-1, oxidation, inflammation, and atherosclerosis. *Front. Pharmacol.* **2012**, *3*, 119. [CrossRef] [PubMed]
18. Durante, W. Protective role of heme oxygenase-1 against inflammation in atherosclerosis. *Front. Biosci. (Landmark Ed.)* **2011**, *16*, 2372–2388. [CrossRef]
19. Naito, Y.; Takagi, T.; Higashimura, Y. Heme oxygenase-1 and anti-inflammatory M2 macrophages. *Arch. Biochem. Biophys.* **2014**, *564*, 83–88. [CrossRef]
20. Zhang, M.; Nakamura, K.; Kageyama, S.; Lawal, A.O.; Gong, K.W.; Bhetraratana, M.; Fujii, T.; Sulaiman, D.; Hirao, H.; Bolisetty, S.; et al. Myeloid HO-1 modulates macrophage polarization and protects against ischemia-reperfusion injury. *JCI Insight* **2018**, *3*. [CrossRef] [PubMed]
21. Chauveau, C.; Remy, S.; Royer, P.J.; Hill, M.; Tanguy-Royer, S.; Hubert, F.X.; Tesson, L.; Brion, R.; Beriou, G.; Gregoire, M.; et al. Heme oxygenase-1 expression inhibits dendritic cell maturation and proinflammatory function but conserves IL-10 expression. *Blood* **2005**, *106*, 1694–1702. [CrossRef] [PubMed]
22. Ishikawa, K.; Navab, M.; Leitinger, N.; Fogelman, A.M.; Lusis, A.J. Induction of heme oxygenase-1 inhibits the monocyte transmigration induced by mildly oxidized LDL. *J. Clin. Investig.* **1997**, *100*, 1209–1216. [CrossRef] [PubMed]
23. Yet, S.F.; Layne, M.D.; Liu, X.; Chen, Y.H.; Ith, B.; Sibinga, N.E.; Perrella, M.A. Absence of heme oxygenase-1 exacerbates atherosclerotic lesion formation and vascular remodeling. *FASEB J.* **2003**, *17*, 1759–1761. [CrossRef] [PubMed]

24. Ishikawa, K.; Sugawara, D.; Goto, J.; Watanabe, Y.; Kawamura, K.; Shiomi, M.; Itabe, H.; Maruyama, Y. Heme oxygenase-1 inhibits atherogenesis in Watanabe heritable hyperlipidemic rabbits. *Circulation* **2001**, *104*, 1831–1836. [CrossRef] [PubMed]
25. Ishikawa, K.; Sugawara, D.; Wang, X.; Suzuki, K.; Itabe, H.; Maruyama, Y.; Lusis, A.J. Heme oxygenase-1 inhibits atherosclerotic lesion formation in LDL-receptor knockout mice. *Circ. Res.* **2001**, *88*, 506–512. [CrossRef] [PubMed]
26. Chen, S.M.; Li, Y.G.; Wang, D.M. Study on changes of heme oxygenase-1 expression in patients with coronary heart disease. *Clin. Cardiol.* **2005**, *28*, 197–201. [CrossRef] [PubMed]
27. Schaer, C.A.; Schoedon, G.; Imhof, A.; Kurrer, M.O.; Schaer, D.J. Constitutive endocytosis of CD163 mediates hemoglobin-heme uptake and determines the noninflammatory and protective transcriptional response of macrophages to hemoglobin. *Circ. Res.* **2006**, *99*, 943–950. [CrossRef]
28. Eligini, S.; Crisci, M.; Bono, E.; Songia, P.; Tremoli, E.; Colombo, G.I.; Colli, S. Human monocyte-derived macrophages spontaneously differentiated in vitro show distinct phenotypes. *J. Cell. Physiol.* **2013**, *228*, 1464–1472. [CrossRef] [PubMed]
29. Eligini, S.; Cosentino, N.; Fiorelli, S.; Fabbiocchi, F.; Niccoli, G.; Refaat, H.; Camera, M.; Calligaris, G.; De Martini, S.; Bonomi, A.; et al. Biological profile of monocyte-derived macrophages in coronary heart disease patients: Implications for plaque morphology. *Sci. Rep.*. under review.
30. Jang, I.K.; Bouma, B.E.; Kang, D.H.; Park, S.J.; Park, S.W.; Seung, K.B.; Choi, K.B.; Shishkov, M.; Schlendorf, K.; Pomerantsev, E.; et al. Visualization of coronary atherosclerotic plaques in patients using optical coherence tomography: comparison with intravascular ultrasound. *J. Am. Coll. Cardiol.* **2002**, *39*, 604–609. [CrossRef]
31. Jang, I.K.; Tearney, G.J.; MacNeill, B.; Takano, M.; Moselewski, F.; Iftima, N.; Shishkov, M.; Houser, S.; Aretz, H.T.; Halpern, E.F.; et al. In vivo characterization of coronary atherosclerotic plaque by use of optical coherence tomography. *Circulation* **2005**, *111*, 1551–1555. [CrossRef]
32. Squellerio, I.; Caruso, D.; Porro, B.; Veglia, F.; Tremoli, E.; Cavalca, V. Direct glutathione quantification in human blood by LC-MS/MS: comparison with HPLC with electrochemical detection. *J. Pharm. Biomed. Anal.* **2012**, *71*, 111–118. [CrossRef] [PubMed]
33. Eligini, S.; Colli, S.; Basso, F.; Sironi, L.; Tremoli, E. Oxidized low density lipoprotein suppresses expression of inducible cyclooxygenase in human macrophages. *Arterioscler. Thromb. Vasc. Biol.* **1999**, *19*, 1719–1725. [CrossRef] [PubMed]
34. Eligini, S.; Brambilla, M.; Banfi, C.; Camera, M.; Sironi, L.; Barbieri, S.S.; Auwerx, J.; Tremoli, E.; Colli, S. Oxidized phospholipids inhibit cyclooxygenase-2 in human macrophages via nuclear factor-kappaB/IkappaB- and ERK2-dependent mechanisms. *Cardiovasc. Res.* **2002**, *55*, 406–415. [CrossRef]
35. Eligini, S.; Brioschi, M.; Fiorelli, S.; Tremoli, E.; Banfi, C.; Colli, S. Human monocyte-derived macrophages are heterogenous: Proteomic profile of different phenotypes. *J. Proteomics* **2015**, *124*, 112–123. [CrossRef]
36. Scalone, G.; Niccoli, G.; Refaat, H.; Vergallo, R.; Porto, I.; Leone, A.M.; Burzotta, F.; D'Amario, D.; Liuzzo, G.; Fracassi, F.; et al. Not all plaque ruptures are born equal: An optical coherence tomography study. *Eur. Heart J. Cardiovasc. Imaging* **2017**, *18*, 1271–1277. [CrossRef] [PubMed]
37. Tearney, G.J.; Regar, E.; Akasaka, T.; Adriaenssens, T.; Barlis, P.; Bezerra, H.G.; Bouma, B.; Bruining, N.; Cho, J.M.; Chowdhary, S.; et al. Consensus standards for acquisition, measurement, and reporting of intravascular optical coherence tomography studies: A report from the International Working Group for Intravascular Optical Coherence Tomography Standardization and Validation. *J. Am. Coll. Cardiol.* **2012**, *59*, 1058–1072. [CrossRef] [PubMed]
38. Tearney, G.J.; Yabushita, H.; Houser, S.L.; Aretz, H.T.; Jang, I.K.; Schlendorf, K.H.; Kauffman, C.R.; Shishkov, M.; Halpern, E.F.; Bouma, B.E. Quantification of macrophage content in atherosclerotic plaques by optical coherence tomography. *Circulation* **2003**, *107*, 113–119. [CrossRef]
39. Di Vito, L.; Agozzino, M.; Marco, V.; Ricciardi, A.; Concardi, M.; Romagnoli, E.; Gatto, L.; Calogero, G.; Tavazzi, L.; Arbustini, E.; et al. Identification and quantification of macrophage presence in coronary atherosclerotic plaques by optical coherence tomography. *Eur. Heart J. Cardiovasc. Imaging* **2015**, *16*, 807–813. [CrossRef]
40. Neubauer, J.A.; Sunderram, J. Heme oxygenase-1 and chronic hypoxia. *Respir. Physiol. Neurobiol.* **2012**, *184*, 178–185. [CrossRef]
41. Ewing, J.F.; Maines, M.D. Glutathione depletion induces heme oxygenase-1 (HSP32) mRNA and protein in rat brain. *J. Neurochem.* **1993**, *60*, 1512–1519. [CrossRef]

42. Biswas, S.K.; Newby, D.E.; Rahman, I.; Megson, I.L. Depressed glutathione synthesis precedes oxidative stress and atherogenesis in Apo-E(-/-) mice. *Biochem. Biophys. Res. Commun.* **2005**, *338*, 1368–1373. [CrossRef]
43. Ozsarlak-Sozer, G.; Sevin, G.; Ozgur, H.H.; Yetik-Anacak, G.; Kerry, Z. Diverse effects of taurine on vascular response and inflammation in GSH depletion model in rabbits. *Eur. Rev. Med. Pharmacol. Sci.* **2016**, *20*, 1360–1372.
44. Cavalca, V.; Veglia, F.; Squellerio, I.; Marenzi, G.; Minardi, F.; De Metrio, M.; Cighetti, G.; Boccotti, L.; Ravagnani, P.; Tremoli, E. Glutathione, vitamin E and oxidative stress in coronary artery disease: Relevance of age and gender. *Eur. J. Clin. Investig.* **2009**, *39*, 267–272. [CrossRef]
45. Eligini, S.; Porro, B.; Lualdi, A.; Squellerio, I.; Veglia, F.; Chiorino, E.; Crisci, M.; Garlasche, A.; Giovannardi, M.; Werba, J.P.; et al. Nitric oxide synthetic pathway in red blood cells is impaired in coronary artery disease. *PLoS ONE* **2013**, *8*, e66945. [CrossRef]
46. Kaneda, H.; Ohno, M.; Taguchi, J.; Togo, M.; Hashimoto, H.; Ogasawara, K.; Aizawa, T.; Ishizaka, N.; Nagai, R. Heme oxygenase-1 gene promoter polymorphism is associated with coronary artery disease in Japanese patients with coronary risk factors. *Arterioscler. Thromb. Vasc. Biol.* **2002**, *22*, 1680–1685. [CrossRef]
47. Chen, Y.H.; Lin, S.J.; Lin, M.W.; Tsai, H.L.; Kuo, S.S.; Chen, J.W.; Charng, M.J.; Wu, T.C.; Chen, L.C.; Ding, Y.A.; et al. Microsatellite polymorphism in promoter of heme oxygenase-1 gene is associated with susceptibility to coronary artery disease in type 2 diabetic patients. *Hum. Genet.* **2002**, *111*, 1–8. [CrossRef]
48. Schillinger, M.; Exner, M.; Mlekusch, W.; Ahmadi, R.; Rumpold, H.; Mannhalter, C.; Wagner, O.; Minar, E. Heme oxygenase-1 genotype is a vascular anti-inflammatory factor following balloon angioplasty. *J. Endovasc. Ther.* **2002**, *9*, 385–394. [CrossRef] [PubMed]
49. Kawashima, A.; Oda, Y.; Yachie, A.; Koizumi, S.; Nakanishi, I. Heme oxygenase-1 deficiency: The first autopsy case. *Hum. Pathol.* **2002**, *33*, 125–130. [CrossRef]
50. Wu, M.L.; Ho, Y.C.; Yet, S.F. A central role of heme oxygenase-1 in cardiovascular protection. *Antioxid. Redox Signal.* **2011**, *15*, 1835–1846. [CrossRef]
51. Kawamura, K.; Ishikawa, K.; Wada, Y.; Kimura, S.; Matsumoto, H.; Kohro, T.; Itabe, H.; Kodama, T.; Maruyama, Y. Bilirubin from heme oxygenase-1 attenuates vascular endothelial activation and dysfunction. *Arterioscler. Thromb. Vasc. Biol.* **2005**, *25*, 155–160. [CrossRef]
52. Balla, G.; Jacob, H.S.; Balla, J.; Rosenberg, M.; Nath, K.; Apple, F.; Eaton, J.W.; Vercellotti, G.M. Ferritin: A cytoprotective antioxidant strategem of endothelium. *J. Biol. Chem.* **1992**, *267*, 18148–18153.
53. Zakkar, M.; Van der Heiden, K.; Luong le, A.; Chaudhury, H.; Cuhlmann, S.; Hamdulay, S.S.; Krams, R.; Edirisinghe, I.; Rahman, I.; Carlsen, H.; et al. Activation of Nrf2 in endothelial cells protects arteries from exhibiting a proinflammatory state. *Arterioscler. Thromb. Vasc. Biol.* **2009**, *29*, 1851–1857. [CrossRef] [PubMed]
54. Buckley, B.J.; Marshall, Z.M.; Whorton, A.R. Nitric oxide stimulates Nrf2 nuclear translocation in vascular endothelium. *Biochem. Biophys. Res. Commun.* **2003**, *307*, 973–979. [CrossRef]
55. Ungvari, Z.; Bagi, Z.; Feher, A.; Recchia, F.A.; Sonntag, W.E.; Pearson, K.; de Cabo, R.; Csiszar, A. Resveratrol confers endothelial protection via activation of the antioxidant transcription factor Nrf2. *Am. J. Physiol. Heart Circ. Physiol.* **2010**, *299*, H18–H24. [CrossRef] [PubMed]
56. Donovan, E.L.; McCord, J.M.; Reuland, D.J.; Miller, B.F.; Hamilton, K.L. Phytochemical activation of Nrf2 protects human coronary artery endothelial cells against an oxidative challenge. *Oxid. Med. Cell. Longev.* **2012**, *2012*, 132931. [CrossRef] [PubMed]
57. Nguyen, T.; Nioi, P.; Pickett, C.B. The Nrf2-antioxidant response element signaling pathway and its activation by oxidative stress. *J. Biol. Chem.* **2009**, *284*, 13291–13295. [CrossRef] [PubMed]
58. Song, J.; Sumiyoshi, S.; Nakashima, Y.; Doi, Y.; Iida, M.; Kiyohara, Y.; Sueishi, K. Overexpression of heme oxygenase-1 in coronary atherosclerosis of Japanese autopsies with diabetes mellitus: Hisayama study. *Atherosclerosis* **2009**, *202*, 573–581. [CrossRef] [PubMed]
59. Li, Y.G.; Wang, D.M.; Chen, S.M.; Tan, X.R.; Fang, X.Y.; Wu, J.W.; Zhang, G.H.; Mai, R.Q. Haem oxygenase-1 expression and coronary heart disease–association between levels of haem oxygenase-1 expression and angiographic morphology as well as the quantity of coronary lesions. *Acta Cardiol.* **2006**, *61*, 295–300. [CrossRef] [PubMed]
60. Choudhary, S.; Higgins, C.L.; Chen, I.Y.; Reardon, M.; Lawrie, G.; Vick, G.W., 3rd; Karmonik, C.; Via, D.P.; Morrisett, J.D. Quantitation and localization of matrix metalloproteinases and their inhibitors in human carotid endarterectomy tissues. *Arterioscler. Thromb. Vasc. Biol.* **2006**, *26*, 2351–2358. [CrossRef]

7

5′,8-Cyclopurine Lesions in DNA Damage: Chemical, Analytical, Biological and Diagnostic Significance

Chryssostomos Chatgilialoglu [1,2,*], Carla Ferreri [1], Nicholas E. Geacintov [3], Marios G. Krokidis [4], Yuan Liu [5,6,7], Annalisa Masi [1], Vladimir Shafirovich [3], Michael A. Terzidis [8] and Pawlos S. Tsegay [6]

1 Istituto per la Sintesi Organica e la Fotoreattività, Consiglio Nazionale delle Ricerche, Via P. Gobetti 101, 40129 Bologna, Italy; carla.ferreri@isof.cnr.it (C.F.); annalisa.masi@isof.cnr.it (A.M.)
2 Center for Advanced Technologies, Adam Mickiewicz University, 61-614 Poznań, Poland
3 Department of Chemistry, New York University, 31 Washington Place, New York, NY 10003-5180, USA; nicholas.geacintov@nyu.edu (N.E.G.); vs5@nyu.edu (V.S.)
4 Institute of Nanoscience and Nanotechnology, N.C.S.R. "Demokritos", 15310 Agia Paraskevi Attikis, Greece; m.krokidis@inn.demokritos.gr
5 Department of Chemistry and Biochemistry, Florida International University, 11200 SW 8th Street, Miami, FL 33199, USA; yualiu@fiu.edu
6 Biochemistry Ph.D. Program, Florida International University, Miami, FL 33199, USA; ptseg001@fiu.edu
7 Biomolecular Sciences Institute, Florida International University, Miami, FL 33199, USA
8 Laboratory of Chemical and Environmental Technology, Department of Chemistry, Aristotle University of Thessaloniki, 54124 Thessaloniki, Greece; mterzidi@gmail.com
* Correspondence: chrys@isof.cnr.it

Abstract: Purine 5′,8-cyclo-2′-deoxynucleosides (cPu) are tandem-type lesions observed among the DNA purine modifications and identified in mammalian cellular DNA in vivo. These lesions can be present in two diasteroisomeric forms, 5′*R* and 5′*S*, for each 2′-deoxyadenosine and 2′-deoxyguanosine moiety. They are generated exclusively by hydroxyl radical attack to 2′-deoxyribose units generating C5′ radicals, followed by cyclization with the C8 position of the purine base. This review describes the main recent achievements in the preparation of the cPu molecular library for analytical and DNA synthesis applications for the studies of the enzymatic recognition and repair mechanisms, their impact on transcription and genetic instability, quantitative determination of the levels of lesions in various types of cells and animal model systems, and relationships between the levels of lesions and human health, disease, and aging, as well as the defining of the detection limits and quantification protocols.

Keywords: reactive oxygen species; free radicals; DNA damage; cyclopurines; DNA and RNA polymerases; nucleotide excision repair; LC-MS/MS; xeroderma pigmentosum; cancer

---

1. Introduction

1.1. Reactive Oxygen Species (ROS)

Since the start of life on Earth, the human body has lived in an oxidative atmosphere due to the presence of molecular oxygen (dioxygen in its ground triplet state), which plays an immense role in biological processes [1,2]. Each individual consumes ca. 3.5 kg of molecular oxygen every day and 2.8% of it is utilized to generate free radicals. Superoxide dismutase (SOD) and nitric oxide synthase (NOS) are two classes of enzymes that control the production of superoxide radical anions ($O_2^{\bullet-}$) and induce the formation of nitric oxide ($^{\bullet}NO$) [2]. These two radicals are the progenitors of endogenous reactive oxygen species (ROS) and reactive nitrogen species (RNS). The ROS/RNS network includes molecules such as hydrogen peroxide (H_2O_2), hypochlorous acid (HOCl), and peroxynitrite ($ONOO^-$), as well as radicals such as hydroxyl radicals ($HO^\bullet$), nitrogen dioxide ($^\bullet NO_2$), and carbonate radical anions ($CO_3^{\bullet-}$). However, the ROS/RNS network also functions as an efficient cellular defense mechanism, being involved in the elimination of viral and microbial infections. The overproduction of ROS/RNS has been linked with the etiology of many diseases. The main processes that generate $HO^\bullet$ radicals are depicted in reactions (1)–(3): the Fenton reaction of H_2O_2, the reduction of HOCl by the superoxide radical anion, and the spontaneous decomposition of protonated $ONOO^-$, respectively [2,3].

$$H_2O_2 + Fe^{2+} \rightarrow Fe^{3+} + HO^- + HO^\bullet \quad (1)$$

$$HOCl + O_2^{\bullet-} \rightarrow O_2 + Cl^- + HO^\bullet \quad (2)$$

$$ONOO^- + H^+ \leftrightarrows ONOOH \rightarrow {}^\bullet NO_2 + HO^\bullet \quad (3)$$

1.2. Reactivity of Hydroxyl Radicals with DNA

Hydroxyl radicals ($HO^\bullet$) are known for their reactivity and ability to cause DNA strand breaks and chemical modifications of nucleobase. In addition to the induction of DNA damage through the metabolism of oxygen, DNA damage may be induced through other environmentally originated insults such as ionizing radiation, UV light, and chemical mutagens. Although the majority of DNA damage induced through oxidative metabolism are single lesions, there are also types of multiple lesions, such as tandem or clustered lesions and DNA/DNA or DNA/protein crosslinking, that may challenge the repair machinery of the cell. Indeed, enzymatic systems such as base excision repair (BER) and nucleotide excision repair (NER) are known to remove the majority of DNA lesions and safeguard the integrity of the genome [4]. However, the lesions may accumulate in tissues, mainly due to the progressive loss of protective systems and consequent poor repair, as it occurs in the aging process [5]. Also, enzymatic deficiencies can give rise to the accumulation of damage to cellular components that are linked to certain pathologies.

As far as the chemical mechanisms are concerned, the site of DNA attack by diffusible $HO^\bullet$ radicals are known to be either the hydrogen atom abstraction from the 2-deoxyribose units or the addition to the base moieties, the latter being the predominant one (accounting for 85–90% of sites attacked) [6]. Moreover, experimental evidence on direct strand scission suggests the transfer of the radical center from the base moieties to the sugar backbone [6]. The order of reactivity of $HO^\bullet$ radicals towards the various hydrogen atoms of the 2-deoxyribose moiety is generally accepted to follow that of the hydrogen atom exposure to solvent (i.e., $H5' > H4' > H3' \cong H2' \cong H1'$) [7,8]. The proportion of attacks at H5′ of DNA by $HO^\bullet$ radicals is estimated to be 55% for all possible sugar positions [9]. After abstraction of a hydrogen atom from 2-deoxyribose, the fate of the carbon-centered radical depends upon the environment. There are several studies focused on the selective generation of these species to obtain quantitative data [10–12]. The C5′ radical has a very peculiar behavior with respect to the other positions of 2-deoxyribose, since its corresponding peroxyl radical does not generate an abasic site and leads to the formation of unique cyclic base–sugar adducts, the purine 5′,8-cyclo-2′-deoxynucleosides (cPu). These tandem-type lesions can be observed in models of DNA modifications and have also been identified in mammalian cellular DNA in vivo [13].

1.3. Historical Background of 'Cyclopurines'

The chemistry of purine 5′,8-cyclo-2′-deoxynucleosides (cPu) has its origin in the literature of ionizing radiation [14] and goes back to 1968, when Keck discovered that the attack of $HO^\bullet$ radicals to adenosine-5′-monophosphate leads, among other products, to the production of 5′*S* and 5′*R* diastereoisomers of 5′,8-cycloadenosine-5′-monophosphate [15]. The abstraction of the H5′ atom of the 2-deoxyribose moiety by the $HO^\bullet$ radical was suggested as the initiating step of this reaction, followed by intramolecular cyclization of the C5′ radical onto C8, leading to the formation of a new covalent bond, C5′–C8 (Figure 1).

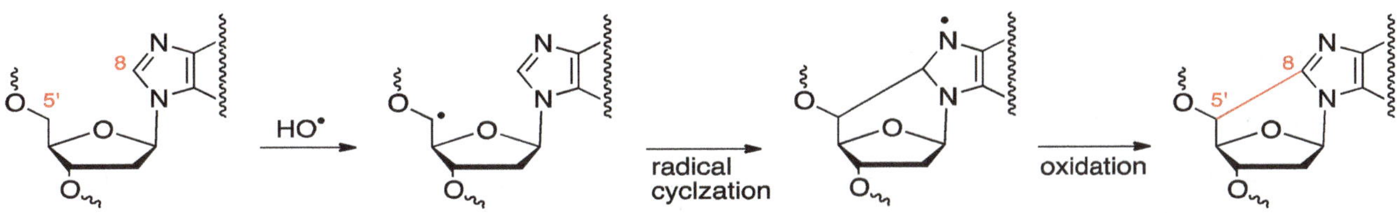

Figure 1. Purine 2′-deoxynucleotide reacts with a hydroxyl radical (HO•), yielding the purine 5′,8-cyclo-2′-deoxynucleotide via cyclization of the C5′ radical followed by oxidation.

In the late 1980s, Dizdaroglu and coworkers demonstrated that 5′,8-cyclo-2′-deoxyadenosine (cdA) and 5′,8-cyclo-2′-deoxyguanosine (cdG) exist in 5′*R* and 5′*S* diastereoisomeric forms (Figure 2A) and are generated by the reaction of HO• radicals with the genetic material via C5′ radical chemistry of purine moieties (Figure 1) [16,17]. The induction by ionizing radiation and identification of *R*-cdG and *S*-cdG in living cells were also reported in 1987 [18].

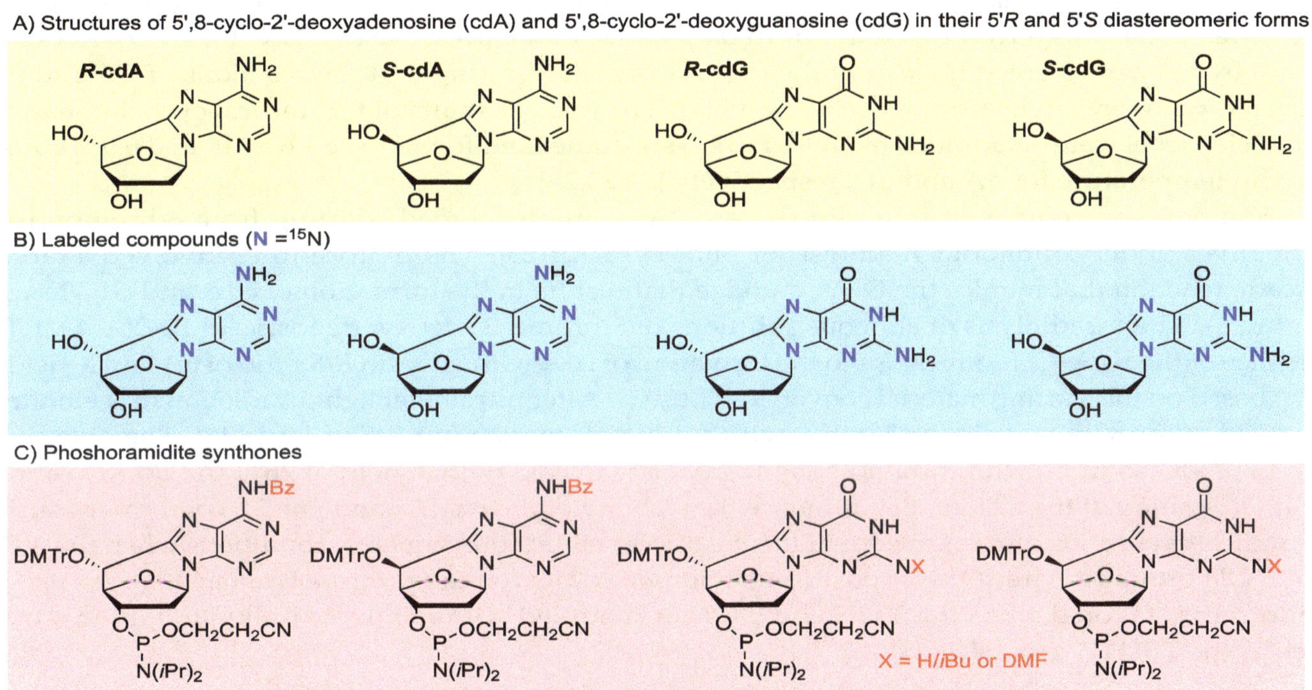

Figure 2. Libraries of purine 5′,8-cyclo-2′-deoxynucleosides (cPu).

The increasing interest in cPu lesions prompted research activity by several groups on DNA damage to investigate different areas of interest, including: nucleoside level, biomimetic models, DNA repair systems, structural investigations, biological effects, and relevance to human disease. Remarkable efforts have been made to improve the detection and quantification of the lesions in order to evaluate them in vivo and in vitro [13,19,20]. They are considered the 'smallest' tandem lesions and are the substrates of nucleotide excision repair (NER).

2. Synthesis of the cPu Library

Continuous efforts of the research community deepened our understanding of complex biological processes. When research efforts identify the structural features of specific small molecules that participate in these biological phenomena, it is of significant importance to develop synthetic procedures for generating large amounts of such molecules of interest for in-depth studies. The synthesis of molecules found in nature is also a challenge, compared to their natural synthesis, and bear important information for synthetic chemists concerning their physicochemical properties and reactivities.

Bioinspired synthetic strategies are indeed being applied and have been successful in providing novel insights.

2.1. 5′,8-Cyclo-2′-Deoxyadenosine (cdA) and 5′,8-Cyclo-2′-Deoxyguanosine (cdG)

The existence of cdA and cdG lesions in biological systems is due to the reactions of hydroxyl radicals with the natural DNA nucleotides. A full chemical understanding of the mechanisms of formation of these lesions has been obtained by extensive exploration of the fate of the C5′ purine nucleoside radicals.

The simplest models are well understood and are based on the scenarios of reactions of 2′-deoxyadenosine (dA) and 2′-deoxyguanosine (dG) with $HO^{\bullet}$ radicals in the absence or presence of molecular oxygen. The two diastereomeric forms, *S*-cdG and *R*-cdG, were identified as the products of γ-irradiation of a N_2O-saturated aqueous solution of dG; the overall reaction yields were 8–10% and the *R*/*S* product ratios were 8.3:1 (Figure 2A) [21]. In a similar experiment using dA instead of dG, two diastereomeric forms, *S*-cdA and *R*-cdA, were identified, with an *R*/*S* ratio of 6:1 and an overall yield of 10–11% [22,23]. The diastereomeric outcome is rationalized in terms of favorable hydrogen-bonded structures in the *pro*-(5′*R*) conformation. In both cases, in the presence of 2.7×10^{-4} M O_2, hydrated 5′-aldehydes were formed instead of the cyclopurines, indicating that the reactions of C5′ radicals occurred with oxygen instead of cyclization [24]. The rate constants of C5′ radical cyclization at the nucleoside level were determined by pulse radiolysis studies and found to be 1.6×10^5 and 6.9×10^5 s^{-1} at room temperature for dA and dG, respectively [22,23,25].

Synthetic procedures of four cPu (Figure 2A) were developed, starting from 8-bromopurine derivatives under continuous radiolysis or photolysis [21,23]. These procedures involved a radical cascade reaction that mimics the DNA damage that results in the formation of cdA and cdG lesions (Figure 3). The γ-radiolysis of aqueous solutions of 8-bromo-2′-deoxyadenosine (8-Br-2′-dA) in the presence of the $K_4Fe(CN)_6$ gives rise to the formation of cdA with an ratio *R*/*S* ratio of 6:1 and a yield of 67% (based on the starting material conversion) [22,23]. After ultraviolet light irradiation in acetonitrile, the yield of formation attained 65%, but with a diastereomeric ratio *R*/*S* of 1.7:1 [26]. The photolysis of 8-Br-2′-dG solution with ultraviolet light gave rise to cdG with a yield of 26% and an *R*/*S* ratio of 8:1 [21]. A study of the factors that influence the stability of the *pro*-5′R and *pro*-5′S conformers during the radical cyclization also revealed that the *pro*-5′R is stabilized in aqueous conditions when the 3′OH and 5′OH groups are free. The opposite behavior was observed in aprotic solvents where the *pro*-5′S conformer is favored when the 3′OH and 5′OH are coupled with relatively bulky lipophilic groups such as the TBDMS group [27,28].

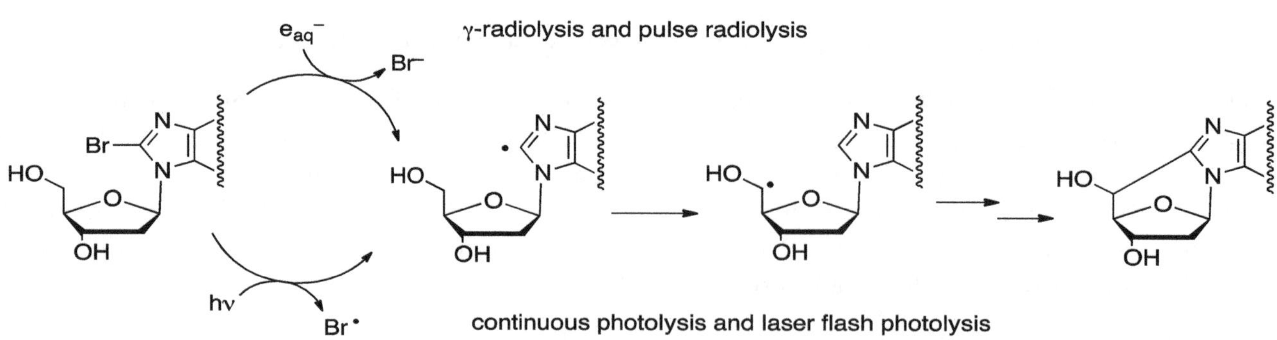

Figure 3. Bioinspired radical transformations for the synthesis of 5′,8-cyclopurines.

The abovementioned results regarding the chemical generation and fates of the C5′ radicals in purine nucleosides led to the development of shorter procedures that afforded the four diastereoisomers of cPu in good to very good yields. Overall, starting from 8-Br-2′-dA, both *R*-cdA and *S*-cdA were obtained in good yields when the reaction was performed in aprotic solvents in the absence of oxygen using ultraviolet light irradiation. Starting with 8-Br-2′-dG, the selective protection of the free hydroxyl groups with bulky lipophilic groups such as the TBDMS group was found to improve the radical cascade of cyclization yield in aprotic solvents and shifted the *R*/*S* ratio in favor of the *S* diastereoisomer. One-step deprotection afforded both lesions in good yield. The 8-bromo derivatives are accessible following standard bromination conditions that have been reported in the literature. Thus, the C5′ radicals can be easily generated in situ by a variety of radical methodologies affording the products in good to very good yields [27,28].

2.2. Isotopically Labeled Derivatives

The relatively low abundance of the cPu in cellular DNA demands the development of highly sensitive analytical methods for their identification and quantification in biological samples. One of the sensitive and reliable techniques utilized by various laboratories in the field is based on liquid chromatography isotope dilution tandem mass spectrometry (see below). The synthesis of the stable $^{15}N_5$ isotopes of the *R*-cdA, *S*-cdA, *R*-cdG, and *S*-cdG (Figure 2B) by the above-described methodologies for the synthesis of the natural lesions has been reported in the literature. In particular, photoirradiation of $^{15}N_5$-labeled 8-Br-2′-dA in acetonitrile with UV light at $\lambda = 254$ nm was shown to yield the $^{15}N_5$-labeled *R*-cdA and *S*-cdA products. The same procedure under similar conditions was also reported to give access to the $^{15}N_5$-labeled *R*-cdG and *S*-cdG products, starting from the corresponding $^{15}N_5$-labeled 8-Br-2′-dG [29,30]. Another synthetic strategy that involves two steps utilizes $^{15}N_5$-labeled 2′-deoxyadenosine and $^{15}N_5$-labeled 2′-deoxyguanosine triphosphate aqueous solutions, under γ-radiolysis conditions in the presence of nitrous oxide, to form the triphosphate 5′,8-cyclopurine intermediates. The latter are then subjected to enzymatic dephosphorylation, affording the $^{15}N_5$-labeled cdA and cdG lesions [31]. A more straightforward methodology has been utilized for generating both *R* and *S* $^{15}N_5$-labeled cdA lesions after γ-radiolysis of aqueous solution of $^{15}N_5$-labeled dA in the presence of nitrous oxide [30].

2.3. Phosphoramidite Synthones

Further studies of the biological characteristics of the cPu lesions require the design of more complex model systems that mimic biomaterials in their natural environments. Biochemical and biological experiments that can unveil important information regarding the formation of lesions and their repair in cells are based on synthetic DNA oligonucleotides that contain lesions incorporated at specific positions in the DNA sequences according to the requirements of the experimental design. These oligonucleotide strands can be synthesized according to standard automated DNA synthesis methods.

The phosphoramidite synthones of the cPu that are compatible with standard automated DNA synthesis conditions can be synthesized by multistep procedures [32]. In order to achieve the synthesis of protected phosphoramidites in an appropriate manner, the 3′OH and 5′OH protecting groups should be differentiated prior to cyclization. The four phosphoramidites of *S*-cdA, *R*-cdA, *S*-cdG, and *R*-cdG (Figure 2C) were obtained in a nine-step synthesis procedure starting from the 8-Br-2′-dA or 8-Br-2′-dG, respectively, in good overall yields. In the case of the cdA series, the bromo-derivative under UV-light irradiation cyclized to give the mixture of the two diastereoisomers in 55% yield (Figure 4a). The two diastereoisomers were separated by chromatography on silica and each of them was converted to the final products following standard procedures. In the cdG series, two different protecting groups were produced for 3′OH and 5′OH moieties using deprotection with the TBDMS group, monodeprotection, and reprotection with Et_3SiCl. This "key" derivative, cyclized under radical cascade conditions in the presence of Bu_3SnH and AIBN as radical initiators, affords a mixture of the two diastereoisomers in good yields (Figure 4b). The two diastereoisomers were separated by chromatography on silica gel and each of these steps was followed by the necessary steps to yield the final products.

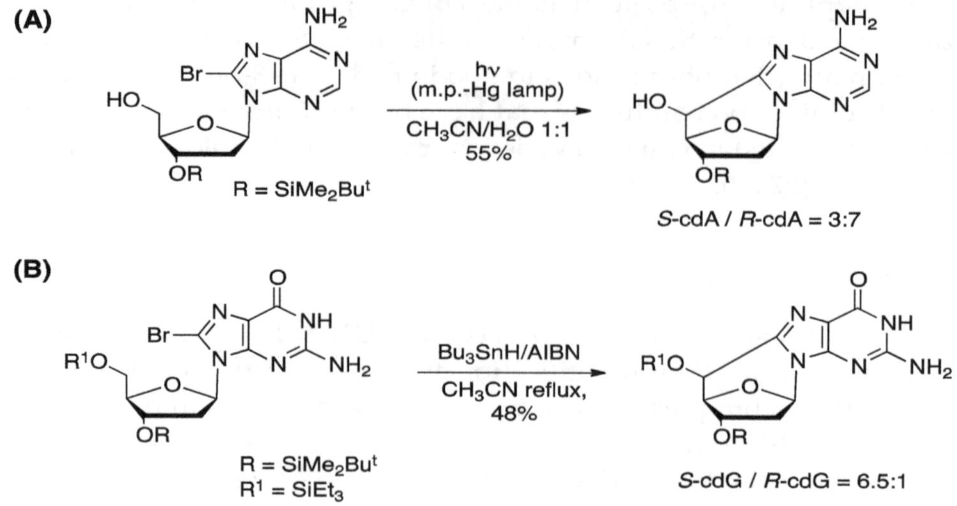

Figure 4. Synthesis of (**A**) 5′,8-cyclo-2′-deoxyadenosine (cdA) and (**B**) 5′,8-cyclo-2′-deoxyguanosine (cdG) in both diastereoisomeric forms, differentiating the two secondary 3′OH and 5′OH groups, by the radical cascade protocol. (m.p.: medium pressure).

The earlier syntheses of cPu phosphoramidites based on the protocol reported by Matsuda et al. [33] for the synthesis of the *S* diastereoisomer and the subsequent inversion of the C5′ configuration were characterized by lower overall yields [34,35]. At the present time, the companies Glen Research (Sterling, VA, USA) and Berry & Associates Inc. (Dexter, MI) produce phosphoramidites of *S*-cdA and *S*-cdG. These companies commercialize *S*-cdG protected with a 5′-tetrahydropyranyl (THP) group instead of the widely used dimethoxytrityl (DMT) group. Worth noting is that the procedure used to remove the THP group is more aggressive than the one used to remove the DMT group.

3. Synthesis of Oligonucleotides Containing Site-Specifically Inserted cPu

The phosphoramidite derivatives were incorporated (following standard procedures, in the 3′→5′ direction [34,35]) into specific sequences of oligodeoxynucleotides (ODNs). These site-specifically modified ODNs can serve as the simplest biomimetic models of oxidized DNA lesions for studies of DNA repair pathways of cPu lesions, their biological impact, insights into structure–function relationships, and relevance of repair to the onset of human diseases. These model systems are less complex than DNA in its natural environment, but nevertheless provide valuable insights into the biological characteristics of these DNA lesions. It is worth mentioning that ODNs can also be synthesized via the automated solid DNA synthesis approach following a 5′→3′ direction by using 3′-*O*-DMTr-5′-*O*-(*β*-cyanoethyl)-*N*,*N*,-diisopropylamino-2′-deoxyribonucleoside phosphoramidites [36].

Figure 5 shows the results of the synthetic procedures for the same oligonucleotide sequences containing the four cPu lesions [32]. The total yields of the ODNs containing *S*-cdA or *S*-cdG were always greater than those of the ODNs containing the corresponding *R* diastereoisomers. The stereochemistry of the cPu lesions were found to influence the coupling yield and thus the total yield of the full-length modified oligonucleotides.

It is well known that the double strands have a lower molar extinction coefficient at $\lambda = 260$ nm than the corresponding mixture of single strands because the absorbance of the aromatic base system is diminished due to base–base stacking interactions in the double helix (hypochromicity). When the temperature of the sample is raised above a temperature specific for the examined length and sequence, the duplexes dissociate into single strands and the intensity of UV absorbance increases. The melting point (Tm) of an oligonucleotide duplex is defined as the midpoint of the duplex-to-single-strand transition at $\lambda = 260$ nm. In order to evaluate the structural destabilization induced by the presence of each of the four diastereomeric lesions in different oligonucleotide sequences, melting studies were carried out, in the absence and presence of cPu lesions, by several research groups (Table 1). The thermodynamic destabilization of the duplexes containing the lesion was observed by the decrease of the melting points of modified duplex sequences with respect to unmodified ones. In the case of 17-mer double-stranded (ds) oligonucleotides (Table 1), it was demonstrated that all four cPu lesions destabilize identical duplex sequences, with cdA producing more destabilization than cdG. Shorter DNA duplexes are more easily destabilized by the lesions than longer ones.

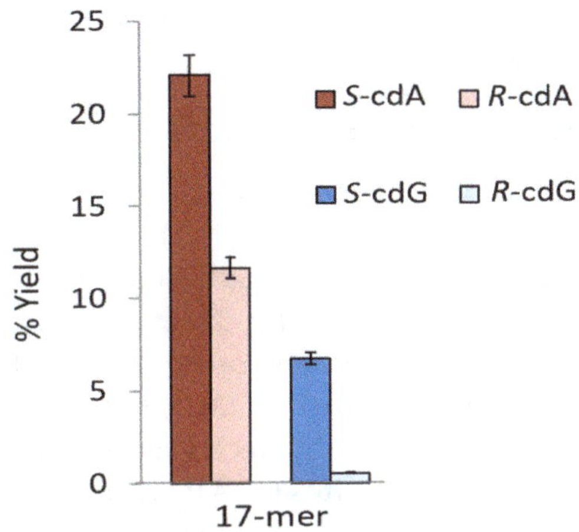

Figure 5. Comparison of the total yields of the 17-mer: 5′-d(CCA CCA AC**X** CTA CCA CC)-3′, where **X** = *S*-cdA, *R*-cdA, *S*-cdG, or *R*-cdG.

Table 1. Melting points (Tm) of a series of double-stranded (ds) oligonucleotide sequences.

Length	Sequence	Tm, °C	Reference
11-mer	5′-d(CGT AC**X** CAT GC)-3′		[37]
	3′-d(GCA TG**Y** GTA CG)-5′		
	X = dA, **Y** = T	49.5	
	X = *S*-cdA, **Y** = T	42.0	
12-mer	5′-d(GTG C**X**T GTT TGT)-3′		[38]
	3′-d(CAC G**Y**A CAA ACA)-5′		
	X = dG, **Y** = C	55.0	
	X = *S*-cdG, **Y** = C	46 ± 1	
14-mer	5′-d(ATC GTG **X**CT GAT CT)-3′		[24]
	3′-d(TAG CAC **Y**GA CTA GA)-5′		
	X – dA, **Y** – T	54 ± 1	
	X = *S*-cdA, **Y** = T	48 ± 1	
17-mer	5′-d(CCA CCA AC**X** CTA CCA CC)-3′		[39]
	3′-d(GGT GGT TG**Y** GAT GGT GG)-5′		
	X = dA, **Y** = T	65.2 ± 0.6	
	X = *R*-cdA, **Y** = T	58.9 ± 0.6	
	X = *S*-cdA, **Y** = T	60.5 ± 0.6	
	X = dG, **Y** = C	66.2 ± 0.7	
	X = *R*-cdG, **Y** = C	63.4 ± 1.0	
	X = *S*-cdG, **Y** = C	63.5 ± 0.6	
23-mer	5′-d(GCA GAC ATA TCC TAG AG**X** CAT AT)-3′		[40]
	3′-d(CGT CTG TAT AGG ATC TC**Y** GTA TA)-3′		
	X = dA, **Y** = T	60.0 ± 0.3	
	X = *R*-cdA, **Y** = T	59.0 ± 0.2	
	X = *S*-cdA, **Y** = T	58.0 ± 0.3	

The presence of the C5′–C8 bond in cPu lesions causes substantial structural changes in DNA. These structural changes include displacement of the purine base, an unusual sugar pucker, deformation of the sugar–phosphate backbone, and alterations in the base stacking with adjacent nucleotides in DNA, especially the 5′ side relative to the lesion [37–39,41,42]. Molecular dynamics simulations were also used by several groups to evaluate the structural distortions and dynamics induced by cPu lesions in comparison with unmodified DNA. It was found that the *R* diastereomers cause a greater backbone distortion and base stacking impairment than the *S* diastereomers [37–39,43]. The differences in the structural perturbation effects, as caused by the presence of the *R* and *S* diastereoisomers, provide some rationale for the observation that the *R* isomers are more efficiently recognized by the nucleotide excision repair (NER) system [37,39,44].

4. Base Excision Repair (BER) and Nucleotide Excision Repair (NER) Pathways

The stability of the human genome subjected to oxidative stress is maintained by the repair of oxidatively generated DNA lesions in cellular DNA [45]. Non-bulky oxidatively generated DNA lesions are typically repaired by BER mechanisms [46] that are highly conserved in all three domains of life: Bacteria, Archaea, and Eukarya [47,48]. The BER proteins bind to the damaged nucleotide and induce cleavage of the *N*-glycosyl bond, thus forming abasic sites. In the case of monofunctional glycosylases, abasic sites are cleaved by an apurinic (AP) human endonuclease (APE1) that results in the formation of fragments with 3′-OH and 5′-deoxyribose phosphate (5′-dRP) at the ends [49]. In turn, bifunctional glycosylases cleave abasic sites by AP lyase activities that result in the formation of single-strand breaks containing either a phosphate (P) group (β, δ-elimination), or an α,β-unsaturated aldehyde (PUA, β-elimination) at the 3′-end [50,51]. Thus, the BER mechanism results in the excision of the damaged nucleotide and the formation of single-strand breaks that can be detected by gel electrophoresis methods [52]. Brooks et al. [36] have shown that cdA lesions are not repaired by BER pathways in adult rat brain nuclear extracts [53], although other known BER substrates [54] were efficiently removed. Furthermore, none of the cPu lesions *R*-cdA, *S*-cdA, or *S*-cdG were found to be substrates of DNA glycosylases in HeLa cell extracts [44]. Pande et al. [55] also reported that the seven purified BER proteins (*E. coli* Fpg, Endo III, Endo V, and Endo VIII; human OGG1; human NEIL1; and NEIL2) neither bind to nor excise *S*-cdA or *S*-cdG lesions from double-stranded (ds) oligonucleotide duplexes. In summary, these results clearly demonstrate that the cPu lesions are not substrates of BER pathways.

The human NER machinery typically excises bulky DNA lesions, such as those derived from the binding of metabolically activated polycyclic aromatic hydrocarbons (PAH) to DNA [56]. Bulky DNA lesions are recognized by the DNA damage sensor and NER factor XPC-RAD23B, a heterodimeric protein complex. After binding to the DNA lesion, the ten-protein NER factor TFIIH, XPA, and the endonucleases XPF-ERCC1 and XPG are recruited to the XPC–TFIIH–DNA lesion complex. The two endonucleases incise the damaged strand on the two sides of the bulky DNA lesion, thus excising the characteristic ~24–30 nucleotide (nt) dual incision products that contain the lesion and are the hallmarks of successful NER [57,58].

More recently, it was demonstrated that certain non-bulky DNA lesions, such as the oxidatively generated diastereomeric spiroiminodihydantoin (Sp) and 5-guanidinohydantoin (Gh), are the substrates of both BER and NER pathways in human cell extracts [59] and in intact human cells [60]. In contrast, cPu lesions are repaired exclusively by the NER pathway [13,19,61].

Employing the host cell reactivation assay, Brooks et al. [36] found that the *S*-cdA lesions strongly block gene expression in Chinese hamster ovary (CHO) cells and in SV40-transformed human fibroblasts. Furthermore, the repair of cdA lesions was significantly suppressed in NER-deficient CHO cells and in human cells from patients who are carriers of XP complementation group A (XPA) mutations that are associated with neurodegeneration [36]. NER dual incision products were also detected after incubation of plasmid DNA harboring *R*-cdA or *S*-cdA lesions in HeLa cell extracts [44]. The NER incision of *R*-cdA was more efficient than in the case of *S*-cdA. In cell extracts supplemented with antiserum against XPA, the NER dual incision efficiency was significantly reduced; in turn, NER activity was restored by the addition of purified XPA protein [44]. The time course of dual incisions from 136 bp DNA duplexes harboring *S*-cdA and *S*-cdG lesions in HeLa cell extracts was compared to the rate of excision of *cis-anti*-B[*a*]PDE-dG adducts (an excellent substrate of human NER [62]) embedded in an identical sequence context in HeLa cell extracts; it was shown that the *cis-anti*-B[*a*]PDE-dG adduct (where B[*a*]PDE-dG denotes the product of reaction of the B[*a*]P diol epoxide metabolite (+)-7*R*,8*S*-dihydrodiol, 9*S*,10*R*-epoxy-tetrahydrobenzo[*a*]pyrene, BPDE) with the exocyclic amino group of guanine in double-stranded DNA was repaired more efficiently than the *S*-cdG lesion. In turn, the latter was a better NER substrate than the *S*-cdA lesion embedded in the same sequence context under identical conditions [55].

The relative NER efficiencies of all four cPu in the same sequence context were measured and compared in human HeLa cell extracts for the first time under identical conditions [39]. The cdA and cdG lesions were excised with similar efficiencies, but the NER excision rates measured for both *R* diastereoisomers were greater by a factor of ~2 than in the case of the *S* lesions. Molecular modeling and molecular dynamics simulations revealed the structural and energetic origins of this difference in NER-incision efficiencies. The characteristic C5′–C8 bond in both kinds of diastereoisomeric cPu lesions causes a greater local distortion of the DNA backbone and a greater disruption of local van der Waals stacking interactions in the case of the *R* than the *S* diastereoisomeric cdA and cdG lesions. Molecular dynamic simulations indicate that the local structural dynamic fluctuations are more pronounced in the case of the diastereoisomeric *R* than the *S* cPu lesions [39]. Therefore, the greater local dynamics and destabilization of stacking interactions associated with the *R* diastereoisomers appear to be correlated with their higher susceptibilities to NER compared to the *S*-cdA and *S*-cdG lesions.

The relative NER efficiencies at the level of chromatin or nucleosomes, the primary subunits of chromatin, can be significantly reduced to those of free or naked DNA because DNA lesions embedded in nucleosomes are less accessible to NER repair protein factors [63–65]. To explore the effects of nucleosome environments on the relative NER efficiencies, all four cPu were embedded at the *In* or *Out* rotational setting near the dyad axis in nucleosome core particles reconstituted either with native histones extracted from HeLa cells (HeLa-NCP) or with recombinant histones (Rec-NCP), as shown in Figure 6 [66]. These experiments showed that while the cPu and B[*a*]PDE-dG lesions in free DNA are good NER substrates, the non-bulky cdA and cdG lesions embedded at either the *In* or *Out* rotational setting of native HeLa histone-derived nucleosomes are completely resistant to NER in human cell extracts (Figure 7).

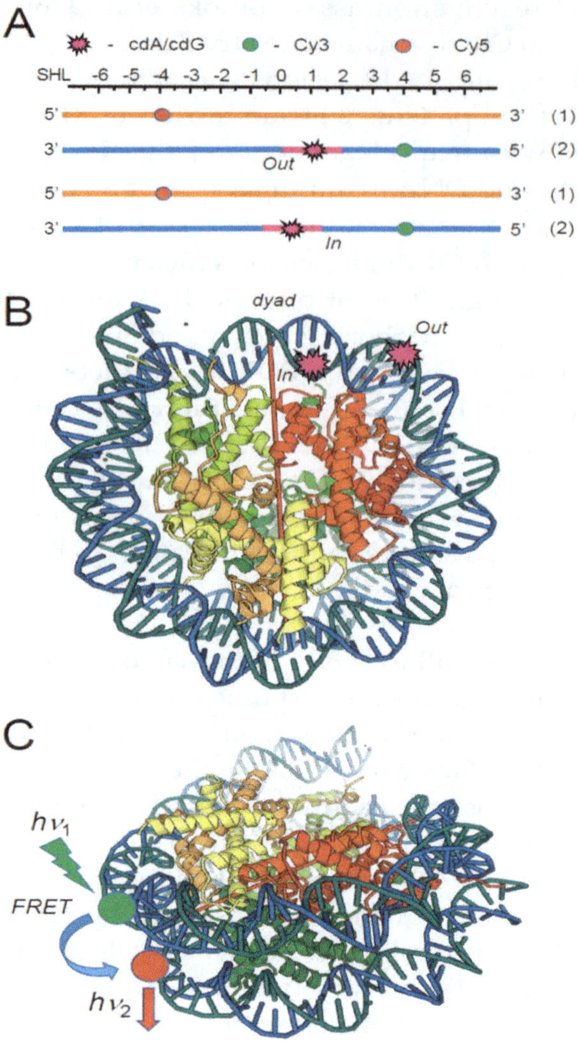

Figure 6. (**A**) Schematic illustrations of the placements of the DNA lesions and the Cy3 donor and Cy5 acceptor molecules at the "*Out*" and "*In*" rotational settings in the 147-mer 601 DNA duplexes used in the Förster Resonance Energy Transfer (FRET) experiments. (**B**) The crystal structure of the 601 nucleosome core particles (PDB 3LZ0) [67]. The dyad axis is indicated by the red line. (**C**) Positions of the Cy3 donor and Cy5 acceptor molecules in the nucleosome FRET experiments. The lesions were positioned at the 66th or 70th nucleotide (nt) counted from the 5′-end of the 147-mer, corresponding to the *Out* or *In* rotational settings. The internal Cy3 and Cy5 labels were positioned at nucleotides 43 and 39 counted from the dyad axis in opposite strands. Reproduced from reference [66].

By contrast, the relative excision rates of the *trans*- and *cis*-B[*a*]PDE-dG adducts that are excised at different rates in free DNA are reduced by the same factor of ~2.2 in HeLa nucleosomes and by the much greater factor of ~11 in recombinant histone nucleosomes. Molecular dynamics simulations showed that the *cis-anti*-B[*a*]PDE-dG adduct is more dynamic and more destabilizing than the smaller and more constrained cdG lesions, suggesting more facile access to the bulkier *cis-anti*-B[*a*]PDE-dG lesion [68]. By contrast to the bulky B[*a*]PDE-dG adducts, the cPu lesions embedded in the same sequence contexts in either post-translationally modified HeLa histone-derived nucleosomes or unmodified recombinant histone nucleosome core particles are fully resistant to NER in human cell extracts.

The NER response of the B[*a*]PDE-dG adducts in HeLa-NCPs is not directly correlated with the observed differences in the thermodynamic destabilization of HeLa NCPs, the Förster resonance energy transfer (FRET) values, or hydroxyl radical footprint patterns and is weakly dependent on the rotational settings [66]. These and other observations suggest that NER is initiated and limited by the binding of the DNA damage-sensing NER factor XPC-RAD23B to a transiently opened B[*a*]PDE-dG-modified DNA sequence in HeLa histone nucleosome particles that corresponds to the known footprint of XPC–DNA–RAD23B complexes (≥30 base pairs). These observations are consistent with the hypothesis that post-translation modifications and the dimensions and properties of the DNA lesions are the major factors that have an impact on the dynamics and initiation of NER in nucleosomes. These results further suggested the hypothesis that the non-bulky cPu DNA lesions do not sufficiently perturb the dynamics of nucleosomes and are therefore resistant to NER in human cell extracts [66].

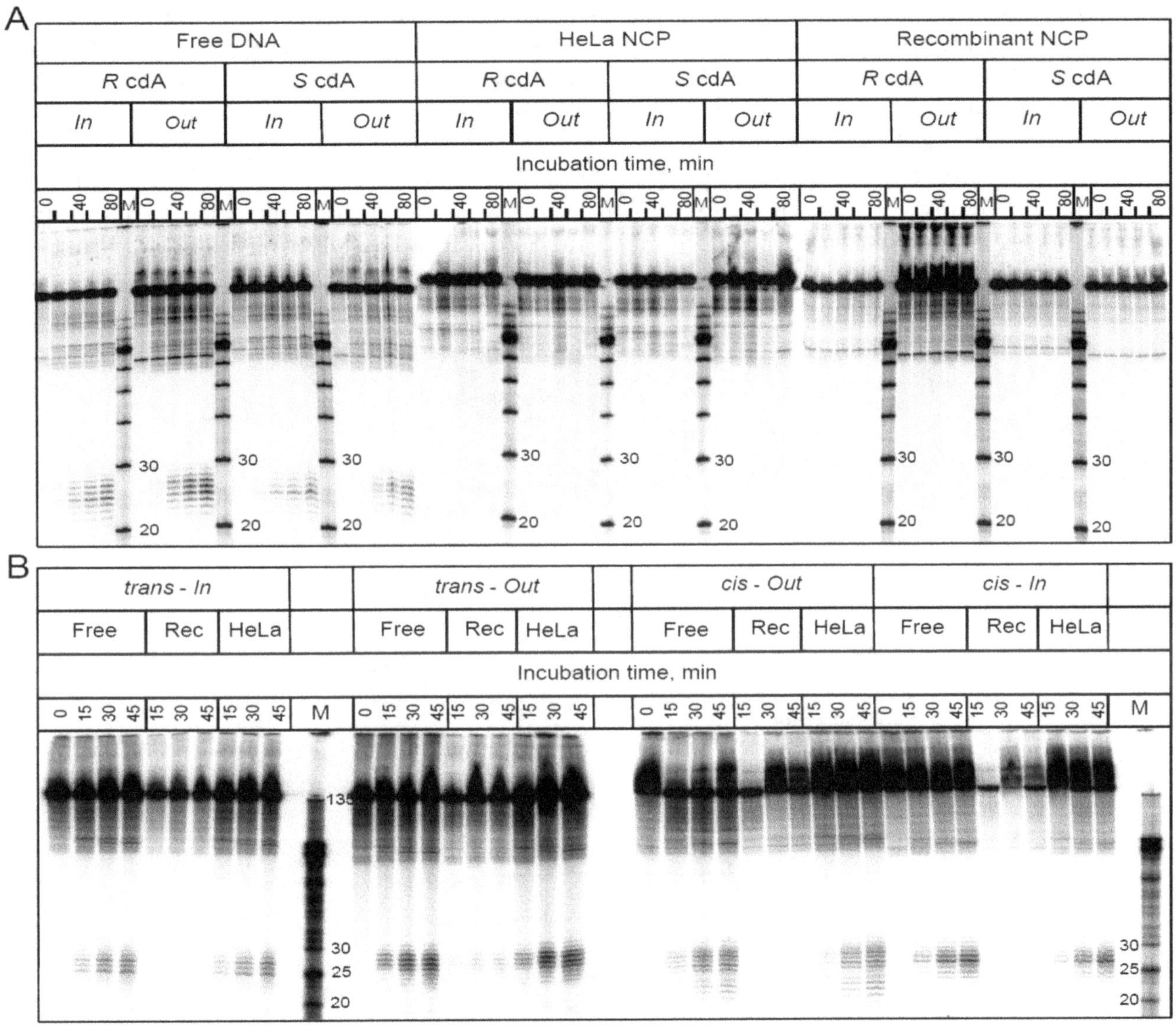

Figure 7. Representative autoradiographs of denaturing gels of the results of nucleotide excision experiment assays in HeLa cell extracts. (**A**) The substrates were either free 147-mer 601[cP] DNA sequences containing single cPu lesions or nucleosomes assembled with histone octamers derived from recombinant (Rec) histones or native, post-translationally modified histones extracted from HeLa cells. (**B**) Analogous nucleotide excision repair (NER) experiments with *cis*- and *trans*-B[a]PDE-dG adducts positioned at the same *In* and *Out* superhelical locations. Three separate gels are depicted in panels A and B. Reproduced from reference [66].

It remains unknown whether these observations carried out with post-translationally modified and recombinant histone-derived nucleosome particles in vitro are also relevant to the same lesions embedded in chromatin in intact human cells and tissues. It is therefore important to extend such studies to the more challenging in vivo environments in order to determine whether the physically smaller cPu lesions are also more resistant to repair in their natural chromatin settings than lesions derived from the binding of metabolites of bulky polycyclic compounds to native DNA in vivo.

5. Bypassing of cPu Lesions by DNA and RNA Polymerases and the Resulting Biological Consequences

As already discussed, unlike other oxidized DNA base lesions, cPu lesions cannot be repaired by the BER pathway. This results in the accumulation of the lesions in the genomic DNA and distorts the DNA backbone, initiating the NER pathway. On the other hand, NER repairs cPu lesions at low efficiency compared to its repair of other bulky DNA lesions, thereby resulting in the accumulation of cPu lesions in DNA. When DNA and RNA polymerases encounter the lesions during DNA replication and repair and gene transcription, they have to bypass the lesions to complete the biological processes (Figure 8). Studies have shown that repair DNA polymerases such as DNA polymerase β (pol β) and translesion DNA polymerases, including pol η and ι, and ζ, can bypass cdA [69–71]. Also, cdA lesions can be bypassed by an RNA polymerase [36,72]. A study from the Kuraoka group has found that *E. coli* polymerase I (pol I) can incorporate *R*-cdA and *S*-cdA into DNA [73].

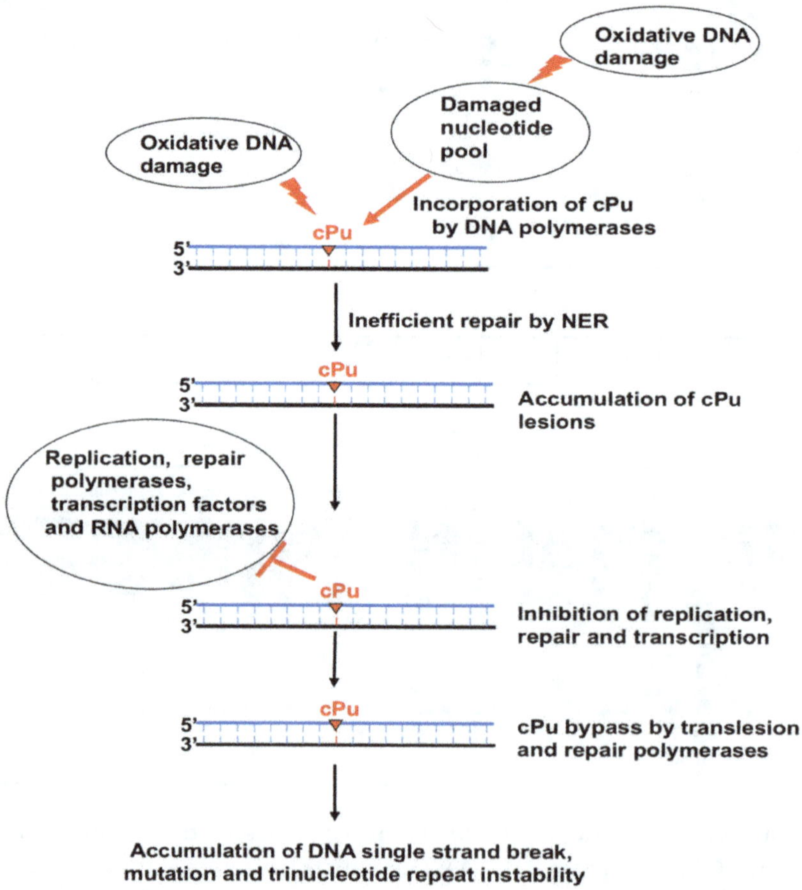

Figure 8. Accumulation of cPu disrupts DNA replication, repair, and gene transcription, leading to lesion bypass, mutations, and genome instability.

5.1. cPu Incorporation by a Replicative DNA Polymerase

The Kuraoka group found that the *E. coli* DNA polymerase I (pol I) large protein fragment, the Klenow fragment, which lacks 5′–3′ exonuclease activity [74], can incorporate 5′*R* and 5′*S* diastereoisomers of cdATP at different efficiencies [73]. The incorporation of *R*-cdATP and *S*-cdATP into a base pair with a dTMP by the pol I fragment is about 17,000-fold and 750-fold less efficient than that of dATP, respectively. The rate of the incorporation of *S*-cdATP and *R*-cdATP opposite to a dTMP by the Klenow fragment is 25.6 μM^{-1} min^{-1} and 1.13 μM^{-1} min^{-1}, respectively. The extension of a cdA by the pol I fragment is only slightly inhibited by the two isomers of cdA, although *S*-cdATP is more readily incorporated and extended by the Klenow fragment [73]. The pol I Klenow fragment incorporates *S*-cdATP more efficiently than *R*-cdATP. This is because the active site of the Klenow fragment binds to *R*-cdA and *S*-cdA with a different affinity due to the stereospecific difference between the two isomers. By superimposing *R*-cdATP and *S*-cdATP on the incoming dATP, the interaction between the Klenow fragment and the cdA lesions is revealed. It is found that the 5′-phosphate group of the *R*-cdA is turned away from the active site of the polymerase. In contrast, the 5′-phosphate of the *S*-cdA turns toward the active site [73]. Thus, compared to *S*-cdATP, the *R*-cdATP can barely form a hydrogen bond with dTMP through the transition from an opened to a closed conformation in the active site of the polymerase, thereby leading to low efficiency of its incorporation. Interestingly, both stereoisomers of cdATP can also be incorporated by the polymerase to base-pair with dCMP. However, *R*-cdATP preferentially base-pairs with dCMP rather than dTMP, suggesting that *R*-cdATP is more error-prone than *S*-cdATP.

5.2. Inhibition of Gene Transcription by a cPu Lesion and Its Bypass by an RNA Polymerase

A cdA lesion can also inhibit the binding of TATA box binding protein (TBP) [75] and RNA polymerase II to the CMV promoter that regulates the luciferase reporter gene [36], resulting in reduced synthesis of RNA that further decreases the luciferase reporter gene expression [36,72,75]. It has been shown that XP cells transfected with plasmids containing a single *S*-cdA lesion located at the second A in the TATA box of the CMV promoter exhibit 75% reduced luciferase gene expression [75]. In a study that has also tested the effects of *S*-cdA on the transcriptional activity of RNA polymerase II, XP cells were transfected with a plasmid carrying a single *S*-cdA lesion in the transcribed region of the luciferase reporter gene. The results show that XP cells transfected with plasmids containing the lesion still exhibit 20–30% of the luciferase activity of the XP cells transfected with the plasmids without a lesion [36]. This indicates that the presence of an *S*-cdA does not entirely abolish the activity of RNA polymerase II, further suggesting that RNA polymerase can partially bypass an *S*-cdA in the template, and the bypass of a cdA by RNA pol II can result in full-length transcribed products. It has been found that yeast RNA pol II bypasses a cdA by preferentially incorporating UTP that base-pairs with a cdA, although it also misincorporates rA, rG, and rC to base-pair with the lesion with low efficiency [72]. In the presence of ATP alone, yeast RNA pol II can efficiently incorporate it to base-pair with a cdA, but with a much lower rate than its incorporation of UTP. To continue to extend the nucleotide that base-pairs with a cdA, RNA pol II can incorporate an rA opposite a dA next to the lesion. Besides, the transcription initiation/elongation factor TFIIF can stimulate the activity of RNA pol II of bypassing a cdA lesion without affecting its fidelity, indicating that the *cis* and *trans* factors can also affect the efficiency of the bypassing of a cPu lesion by RNA pol II [72].

5.3. Inhibition of DNA Polymerase Activities by a cPu Lesion and Its Bypass by DNA Polymerases

A cPu lesion can alter the activity of DNA polymerases. It has been reported that the DNA synthesis activities of calf thymus replicative DNA polymerases, such as pol δ, and the bacterial phage DNA polymerase T7 DNA polymerase are inhibited completely by *R*-cdA and *S*-cdA lesions, resulting in replication fork stalling [44]. The primer extension activity of calf thymus pol δ is abolished at an *R*-cdA or *S*-cdA lesion site in the template strand, whereas T7 DNA polymerase can manage to extend the primer at the cPu lesions [44]. This indicates that the DNA synthesis of the pol δ ceases before the cPu lesions, while T7 DNA polymerase can bypass *R*-cdA or *S*-cdA by incorporating an additional nucleotide. T7 DNA polymerase bypasses an *R*-cdA more efficiently than an *S*-cdA. The results indicate that cPu lesions can lead to DNA replication stalling by inhibiting the activities of replication DNA polymerases, further suggesting that the lesions have to be bypassed by translesion DNA synthesis in cells to resolve the stalled replication fork and restart DNA replication.

An *S*-cdA lesion located in double-stranded DNA can also inhibit pol β DNA synthesis and BER in cell extracts during the repair of an abasic (AP) site at the complementary strand [76]. The inhibitory effect on the DNA synthesis activity is determined by the location of the AP site relative to the *S*-cdA lesion. The DNA synthesis at the AP site located at the 5th or 8th nucleotide upstream (−8 or −5) or downstream (+5 or +8) of *S*-cdA is not significantly affected [76]. However, the polymerase only can exhibit minimal incorporation of a nucleotide at the AP site located at the 1 nucleotide upstream (−1) and downstream (+1) of and opposite to (0) the *S*-cdA [76]. These indicate that an *S*-cdA lesion close to an abasic site can inhibit the DNA synthesis by the DNA polymerase during BER, whereas the lesion located at ≥5 nt away from an AP site does not affect its DNA synthesis. It is suggested that an *S*-cdA induces a geometric alteration of the DNA surrounding the AP site, and this may inhibit the binding of pol β to the DNA, thereby inhibiting its DNA synthesis.

The Basu group has further identified a translesion DNA polymerase that can bypass a cPu lesion in *E. coli*. They demonstrate that *S*-cdA and *S*-cdG strongly block *E. coli* replicative and repair DNA polymerases, including pol II, Klenow fragment, pol IV, and pol V, as well as *S. solfataricus* P2 DNA polymerase IV (Dpo4) [77,78]. Through the gene knockout of pol II, pol IV, and pol V in the SOS-induced or uninduced *E. coli* strains, the group has found that pol V is the one that is responsible for bypassing of *S*-cdA or *S*-cdG inserted in a plasmid through translesion DNA synthesis in *E. coli*, whereas pol II and pol IV do not play a role in bypassing of the lesions. This is further supported by the results showing that the *E. coli* strain with pol V deficiency that bears the plasmids containing an *S*-cdA or *S*-cdG cannot survive [77,78]. This demonstrates that pol V is required for the bypass of *S*-cdA or *S*-cdG in *E. coli*. However, in vitro biochemical characterization has shown that the *E. coli* Klenow fragment, pol IV, and Dpo4 can incorporate nucleotides to base-pair with *S*-cdA or *S*-cdG [78]. Specifically, the Klenow fragment preferentially incorporates dTTP and dCTG to base-pair with *S*-cdA or *S*-cdG, respectively. On the other hand, biochemical characterization indicates that pol IV can incorporate dTTP and dCTP to base-pair with *S*-cdA and dCTP to base-pair with *S*-cdG. Dpo4 can insert dTTP and dGTP to base-pair with *S*-cdA. However, it preferentially inserts dTTP over dCTP to base-pair with *S*-cdG. This may be due to a more opened and less rigid active site in Dpo4 polymerase that can tolerate distorted and bulky DNA lesions. A study conducted by Xu et al. has also shown that the bypass of *S*-cdG by *S. solfataricus* Dpo1 and Dpo4 is decreased by 140- and 65-fold compared to their bypass of dG [79]. The authors further demonstrate that Dpo1 and Dpo4 preferentially incorporate dCTP opposite to *S*-cdG. In addition, the polymerases can misincorporate dATP and dTTP opposite to the lesion [79]. The results of Dpo4 differ from those reported by Basu's group, which showed that Dpo4 preferentially incorporates dTTP opposite to *S*-cdG [78]. The inefficient bypass of *S*-cdG by Dpo1 and Dpo4 appears to result from their poor DNA synthesis at the lesion, with the nucleotide incorporation reduced by 200- and 3000-fold [79]. On the other hand, the binding affinity of Dpo1 and Dpo4 to a substrate containing *S*-cdG is either not affected or moderately reduced by only 6-fold [79]. The structural base of the DNA polymerases has been revealed by the results from a crystal structure containing a Dpo4:*S*-cdG:dCTP complex and showing that *S*-cdG is shifted to the major groove of the DNA substrate. This separates the α-phosphate dCTP from the C3′ atom of the primer terminal dideoxy C (C^{dd}) by 9.2 Å, thereby preventing nucleotidyl transfer activity of the polymerase for catalyzing the DNA synthesis [79]. In a crystal structure of the Dpo4:*S*-cdG:dTTP complex, the formation of a looped template has been also identified. This suggests that Dpo1 and Dpo4 can skip the lesion by looping out the template strand. This may further lead to sequence deletion, thus providing the structural evidence for the findings about a cPu-induced repeat sequence instability reported by the Liu laboratory [69]. These results indicate that the Klenow fragment, pol IV, Dpo1, and Dpo4 can manage to bypass cPu lesions. The results further indicate that pol IV, Dpo1, and Dpo4 are more error-prone than the Klenow fragment by performing nucleotide misincorporation to bypass a cPu lesion.

In eukaryotic cells, cPu lesions can also be readily bypassed by several human and yeast translesion DNA polymerases, pol η, pol ι, and pol ζ, but not pol κ. These DNA polymerases can incorporate different nucleotides to base-pair with an *S*-cdA and *S*-cdG [70]. For example, pol ι can incorporate dTTP, dGTP, and dATP to base-pair with *S*-cdA, whereas it incorporates dCTG, dATP, and dGTP to base-pair with *S*-cdG. In contrast, human pol η usually incorporates a correct nucleotide to base-pair with *S*-cdA or *S*-cdG, whereas yeast pol η can also misincorporate dTTP opposite to *S*-cdG [80]. Among these DNA polymerases, pol η and pol ζ can extend the nucleotides incorporated opposite an *S*-cdA or *S*-cdG lesion [70,80]. Human pol η can only extend a matched nucleotide that base-pairs with a cPu lesion. However, yeast pol η can extend both matched and mismatched nucleotides that base-pair with the lesions [80]. It is proposed that human pol η, pol ι, and pol ζ cooperate to bypass cPu lesions during DNA replication and repair. Cell-based mutation analysis has further demonstrated that the bypass of cPu lesions through these translesion DNA polymerase results in a wide spectrum of mutations, indicating the misincorporation of nucleotides through the bypass of cPu lesions in cells [70].

A recent study from the Yang group has further revealed the molecular basis underlying the bypass of an *S*-cdA by human pol η using the cocrystals of pol η and the DNA substrates containing an *S*-cdA lesion [81]. The crystal structures indicate that the C5′–C8 covalent bond of cdA distorts the backbone of the DNA template by shifting the sugar toward the minor groove. This further results in the change of the width of duplex DNA and pushes the adenine of the cdA to be tilted toward the 5′–direction, leading to the disruption of the base stacking between the damaged nucleotide and the adjacent base. The structural study further reveals that ~60% of adenines from the damaged nucleotide are shifted into the major groove, thereby preventing the formation of a hydrogen bond between the damaged nucleotide and dTTP. In addition, the presence of different type of metal ions can also alter the configuration of the active site, either facilitating the incorporation of dTTP opposite cdA or preventing the formation of the hydrogen bonds between cdA and dTTP. The effect is also mediated through the opening of the finger domain of pol η to accommodate the DNA backbone distortion. Since cdA is shifted to the major groove, this protects it from forming the hydrogen bonds with dT at the 3′-end. Instead, this allows cdA to make only van der Waals interaction with the dT, preventing the formation of a new base pair with the incoming nucleotide and primer extension. Thus, as a result, pol η fails to extend the dT opposite the cdA lesion. The study provides novel insights into the structural basis underlying the nucleotide incorporation by pol η in bypassing a cdA lesion [81].

Similar to the Y family translesion DNA polymerases, DNA polymerase β (pol β), a central component of BER [82,83], can also bypass a cdA lesion [84]. Pol β can readily bypass both *R*-cdA and *S*-cdA located in the substrates mimicking DNA replication and BER intermediates [84]. It has been shown that pol β wild-type mouse embryonic fibroblast (MEF) cell extracts can generate a significant amount of DNA synthesis products resulting from the bypass of an *R*-cdA and *S*-cdA lesion located in an open template, i.e., a DNA substrate containing a 1-nt gap or 1-nt gap with a sugar–phosphate residue [84]. However, pol β-knockout MEF cell extracts generate only a small amount of the lesion bypass products on all the substrates [84]. The results suggest that pol β also plays an important role in bypassing a cPu lesion during DNA replication and repair in mammalian cells. Further biochemical analysis has shown that pol β mainly incorporates a dT to base-pair with a *R*-cdA, but can also misincorporate dA, dG, and dC to base-pair with the damaged nucleotide at low efficiency. Pol β only inserts a dT opposite an *S*-cdA lesion. Moreover, the polymerase can readily extend the dT opposite an *R*-cdA, but fails to extend the dT opposite an *S*-cdA, indicating that pol β stalls at an *S*-cdA following its incorporation of a dT. This further inhibits the ligation of the nick by DNA ligase I (LIG I), allowing flap endonuclease 1 (FEN1) to cleave nucleotides. Subsequently, this results in the gaps and accumulation of single-strand DNA break intermediates [84]. Thus, pol β bypass of an *R*-cdA can lead to nucleotide misincorporation causing mutations, whereas its bypass of an *S*-cdA can cause the accumulation of DNA strand break intermediates that in turn results in recombination and genome instability [84].

5.4. Repeat Sequence Instability Through the Bypass of cdA by Pol β

Interestingly, although pol β stalls at an *S*-cdA that is located in the random sequences [84], it can efficiently extend a dT opposite to an *S*-cdA located in trinucleotide repeats, such as CAG repeats. This further results in CTG repeat deletion through BER [69]. It has been found that this is because both

R and *S* diasteroisomers of cdA located on the template strand induce the formation of a CAG repeat loop in the template that mimics the intermediates formed during maturation of the lagging strand and BER [69]. This is likely due to the distortion of the backbone of the repeats, which subsequently induces the G:C self-base pair in the CAG repeats. Since pol β preferentially skips over a hairpin or loop structure [85,86], the loop structure in CAG repeats induced by a cdA lesion can also be readily bypassed by pol β, thereby leading to the displacement of the downstream repeat strand into a flap during DNA lagging strand maturation and BER [69]. Subsequently, the flap is captured and cleaved efficiently by FEN1, resulting in CTG repeat deletion (Figure 9). This is further supported by the fact that the locations of a gap relative to that of a cdA lesion in the CAG repeat template can govern the deletion of CTG repeats. A gap that is located upstream of or opposite to the lesion can result in CTG repeat deletion through the pol β bypass of a CAG repeat loop structure. However, a gap located downstream of the cdA that does not involve pol β loop bypass fails to cause repeat deletion [69]. These findings further demonstrate the essential role of pol β bypass of a loop structure containing a cPu lesion in mediating trinucleotide repeat deletion.

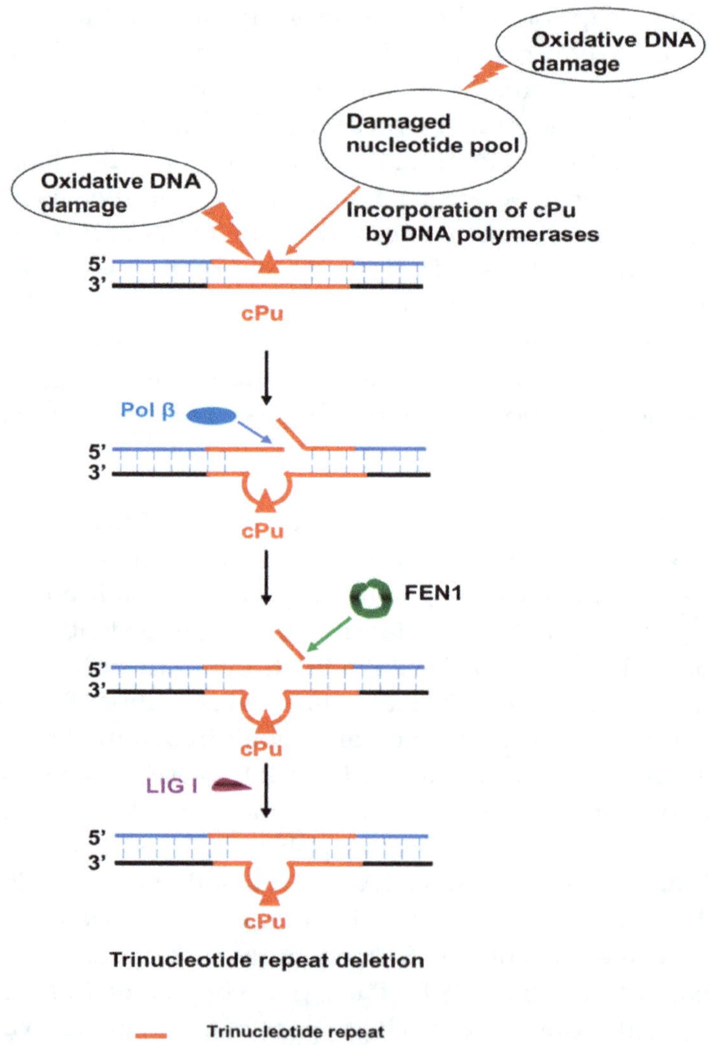

Figure 9. A cdA located at DNA repeat sequences induces repeat instability through pol β bypass of a loop structure. Pol β: DNA polymerase β; LIG I: DNA ligase I; FEN1: flap endonuclease 1.

Although a study provides a mechanistic insight into the incorporation of cdATP by the Klenow fragment using superimposing modeling, crystallography-based structural studies on the Klenow fragment and other replication and repair polymerases and translesion DNA polymerases, especially eukaryotic DNA polymerases, are needed to understand the molecular mechanisms underlying cPu incorporation in DNA. Because DNA replication and repair polymerases often coordinate with their cofactors during DNA replication and repair, the effects of the coordination on the bypass of a cPu lesion and their impact on cellular function remain to be elucidated. Moreover, the effects of cPu lesions on the instability of repeated DNA sequences, including mono-, di-, tri-, tetra- and hexanucleotide repeats, through DNA replication and repair and the crosstalk among different DNA metabolic pathways and the underlying mechanisms need to be explored.

6. Quantification of cPu Lesions in DNA Samples

6.1. *^{32}P-Postlabeling and Enzyme-Linked* Immunosorbent Assays

A variety of analytical techniques for the quantification of oxidative DNA lesions in material derived from various types of cells and organisms have been developed, including radiolabeling approaches, enzyme-linked immunosorbent assays (ELISA), and chromatographic and mass spectrometry-based approaches, while the description of DNA damage (mitochondrial vs. nuclear) remains an important challenge [87]. It should be noticed that data obtained in different laboratories gave rise to different results, mainly due to deviations among experimental conditions used in various laboratories or inconsistent quantification of lesions because of oxidation artifacts [88,89]. Accurate quantification of cPu lesions is necessary to understand the biological relevance, and therefore analytical strategies must be set up in order to deliver reproducible and reliable results. Large discrepancies in the levels of the lesions determined by different groups are likely due to variations in analytical procedures, mainly in the steps of DNA extraction, hydrolysis of the nucleobases, and derivatization [30]. Each quantification technique is characterized by its own advantages, counterbalanced by some limitations for applicability to specific fields. The ^{32}P-postlabeling approach for the measurement of cdA lesions developed by some research groups was important for increasing the detection limit to 1–5 lesions per 10^{10} unmodified nucleosides [90,91]. This methodology was then implemented for the identification and measurement of structurally diverse DNA carcinogen adducts [92], 8-oxo-dG in rat tissues [93], and oxidative DNA products, emerging as a suitable approach for detection of bulky lesions [94]. The levels of cdA in rat samples were significantly elevated in normal newborn samples compared to those in fetuses [90]. ^{32}P-labeled adducts could be visualized by screen-enhanced autoradiography and quantified by Phosphorimager, Instant Imager, or Scintillation Analyzer [91,95].

Immunoassay methodologies are simple, fast, cost-effective, and reproducible for the identification of oxidatively induced DNA adducts within individual cells or tissues, while the combination of this approach with immunofluorescence, along with the presence of a monoclonal antibody specific for the lesion, enhanced the effectiveness of the method [96]. In the latter case, artifacts from interferences were eliminated, such as endogenous antibodies and cross-reactants, increasing the specificity and the sensitivity of this approach. The enzyme-linked immunosorbent assay (ELISA) has been shown to be successful in a wide variety of biological matrices, such as DNA, plasma, and urine, especially for the measurement of 8-oxo-dG, the most commonly investigated lesion as a biomarker of oxidatively induced DNA damage [97]. Recently, an immunoassay using a novel monoclonal antibody (CdA-1) specific for cdA in single-stranded DNA was generated [98]. The levels of cdA lesions measured through this ELISA indicated an accumulation of cdA with age in brain tissues of Xpa (-/-) mice compared with wild-type (wt) mice, and significantly, similar evidence was also revealed in liver and kidney tissues among these mice at 6, 24, and 29 months of age [99].

6.2. *Liquid Chromatography Tandem Mass Spectrometry (LC–MS/MS)*

Although ELISA and ^{32}P-postlabeling methods represent applicable approaches with widespread use, owing to not requiring specialist experience and expensive equipment, lack of structural specificity seems to be the main problem of lesion overestimation compared with the concentrations measured by chromatography-based techniques coupled with isotope dilution mass spectrometric methodologies [100]. Indeed, the *m/z* values for parent and fragment ions contain important information about the structure of DNA lesions and the molecular composition, and when used in these approaches, are sufficiently specific for accurate quantification [101]. Gas chromatography/mass spectrometry (GC/MS) has been utilized to identify *R*-cdG and *S*-cdG in DNA, including a derivatization step by trimethylsilylation after the hydrolysis of the DNA sample [16,31,102]. However, this derivatization step may render the implementation of GC/MS methodology inapplicable due to artifacts at several stages of the analysis, likely due to the oxidation of the nucleobases present in acidic hydrolysis of DNA [103].

Liquid chromatography coupled with modern highly sensitive mass detectors, which follows a top-down approach, starting from the genetic material and going down to the single nucleoside level, is of great interest and extensively used due to its advantages in specificity and repeatability. On the contrary to immunoassays and the overestimation of specific DNA adducts [97], liquid chromatography–tandem mass spectrometry (LC–MS/MS) analysis ascertains accurate quantification of DNA adducts, providing lower limits the detection (LOD) and quantification (LOQ), thus improving the sensitivity of the methodology. Furthermore, the use of isotopically labeled reference compounds for the lesions enhances the reliability of the process, increasing the reproducibility and recovery of the quantification to a great extent. All these considerations make the LC–MS/MS approach an extremely important tool for the identification and estimation of oxidatively produced DNA products or adducts in relation to inflammatory and age-dependent disorders. The ion–current profiles of the corresponding mass transitions for *S*-cdA and *S*-cdG, which are acquired during LC–MS/MS analysis of the samples, are given in Figure 10. The ions of *m/z* 164 and 180 (together with 169 and 185 for the labeled cdA and cdG) result from the cleavage of both the *N*-glycosidic and the C4′–C5′ bonds of cPu. The stepwise procedure for the quantification of DNA lesions includes the isolation and purification of DNA from the biological specimen (biological fluid, cells, tissues, etc.) and the hydrolysis to single nucleosides by an enzymatic cocktail containing nucleases, followed by the analysis and quantification by liquid chromatography coupled with tandem mass spectrometry for the determination of the modified nucleosides. One of the problems for the exact quantification is the contamination of the genetic material before the enzymatic hydrolysis steps or by side products, such as those generated via oxidation of the unmodified nucleosides or/and degradation of the lesions (e.g., via oxidation of the lesions themselves), during the procedure. In order to avoid these sample workup artifacts, the presence of argon, metal chelators, and radical scavengers are added as part of the method [29,104].

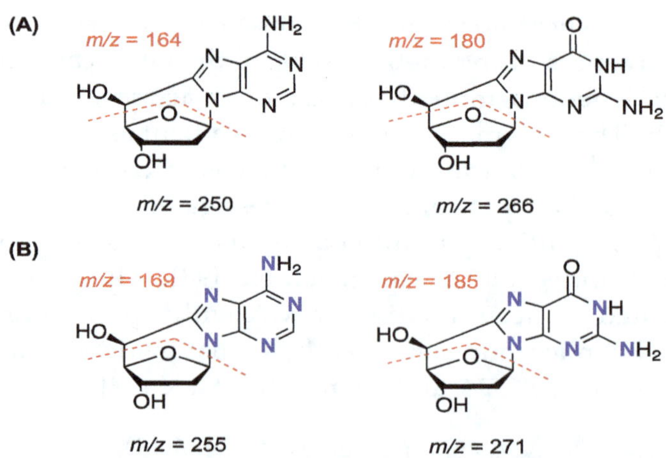

Figure 10. (**A**) MS/MS fragmentation spectra (ESI–MS/MS of the $[M + H]^+$ ion) of *S*-cdA (*m/z* 250→164) and *S*-cdG (*m/z* 266→180) lesions. Similar fragment ions were observed for *R*-cdA and *R*-cdG lesions. (**B**) MS/MS fragmentation spectra (ESI–MS/MS of the $[M + H]^+$ ion) of isotopically labeled (**N** = ^{15}N) *S*-cdA (*m/z* 255→169) and *S*-cdG (*m/z* 271→185) lesions.

Two different approaches have been reported to date for the quantification of cPu via isotope dilution LC–MS/MS analysis. One approach is the direct injection of the sample into the LC–MS/MS system right after the enzymatic digestion step. The advantage of this methodology is the rapidity, as the unmodified nucleosides and the lesions are quantified together in one single run. However, in this approach, the samples do not pass through a purification step, and therefore materials that are incompatible with the mass detector materials, such as buffer ions carried from the enzymatic hydrolysis procedure, can create problems. Moreover, the unfiltered sample which is directly injected into the LC–MS/MS system comprises the unmodified nucleosides along with the lesions, resulting in a concentration which can be up to ~60-fold in excess with respect to the lesions [105,106]. Another important consequence of the latter resulting in poor sensitivity is the limited solubility of the crude sample. A second approach for the quantification of DNA lesions is executed in two independent steps. Firstly, the sample is analyzed by an HPLC–UV system coupled with a sample collector, where the quantification of the unmodified nucleosides takes place based on their absorbance at 260 nm, whereas at the elution time-windows of the modified lesions, they are collected, pooled, and concentrated. In such an approach, the buffer ions carried from the enzymatic digestion step are separated from the analytes and the lesions from the unmodified nucleosides [29,30]. The sample containing the lesions can be further concentrated prior to LC–MS/MS analysis and each can be quantified by LC–MS/MS, increasing the overall sensitivity of the quantification method. An ameliorated enzymatic digestion protocol was developed based on the cost-effective genetically modified nuclease benzonase and the nuclease P1, which are necessary for the quantitative release of the cPu lesions [30]. Figure 11 illustrates a schematic representation of the isotope dilution LC–ESI–MS/MS chromatographic profile and the fragmentation pattern in the mass detector.

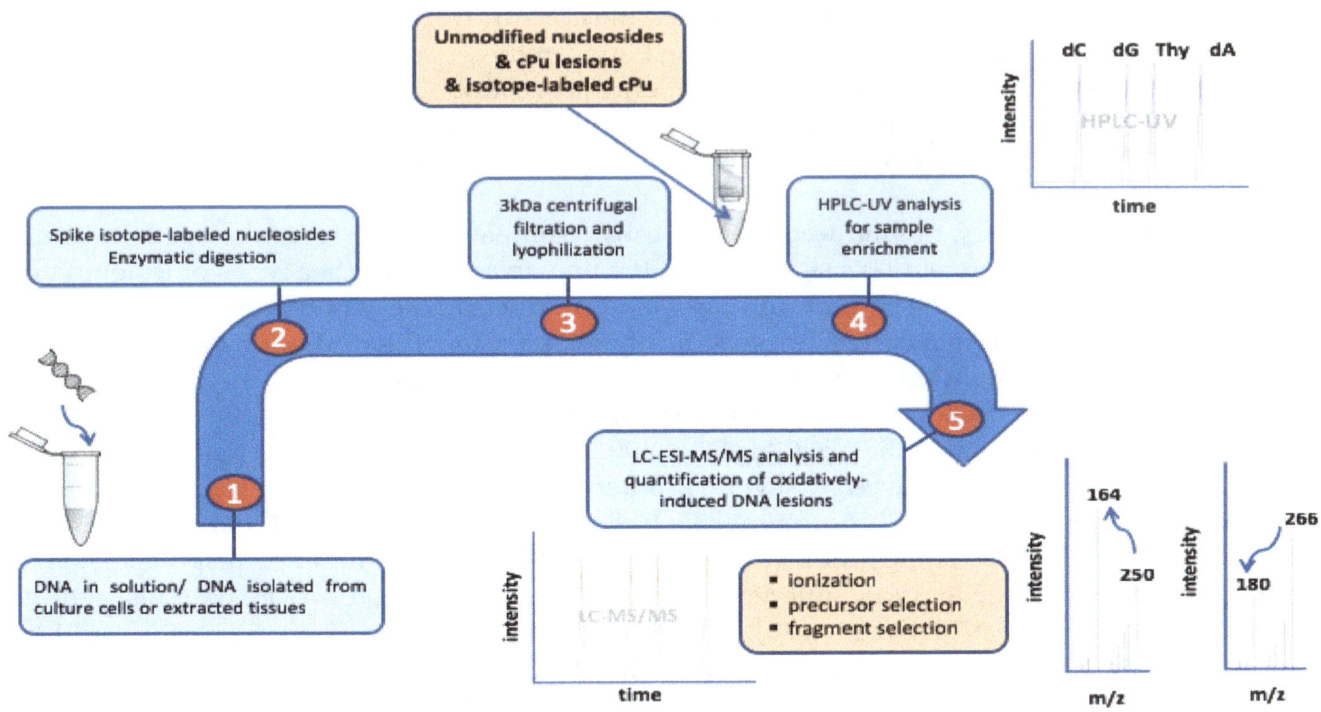

Figure 11. Workflow showing protocol steps for the quantification of cPu lesions via isotope dilution LC–ESI–MS/MS (the fragmentation pattern of the lesions is shown); dC: 2′-deoxycytidine; dG: 2′-deoxyguanosine; Thy: thymidine; dA: 2′-deoxyadenosine.

Comparison of cPu measured in an irradiated sample of calf thymus DNA by two different LC–MS/MS methods has been reported [107]. Data are obtained from the applied dose within the range of 0–60 Gy, and a linear dependence of lesions/10^6 nucleosides vs. dose has been reported. The level of lesions/10^6 nucleosides/Gy in γ-irradiated samples of calf thymus DNA in aqueous solutions, within the range of 0–60 Gy, was found to be 0.23 for cdA and 0.35 for cdG in ds-DNA using the method reported in Figure 11 [30], whereas the values of similar experiments were reported to be 14.2 for cdA and 20.1 cdG [105] using direct injection after the enzymatic digestion step without internal standards (i.e., spiked with known quantities of the isotopically enriched analytes), which is a ~60-fold excess for both cdA and cdG.

7. Biological Studies for Health Applications

7.1. Neurological Diseases

Xeroderma pigmentosum (XP) is a rare disorder of defective UV radiation-induced damage repair that is characterized by photosensitivity and cancer of the UV-exposed areas of the skin and mucous membranes of the eyes and mouth at an early age. The skin is normal at birth, and with the onset of the disorder, all patients develop freckle-like pigmentary changes in sun-exposed areas, which eventually appear as poikiloderma [108,109]. XP can result from mutations in any of eight genes, denoted as XPA–XPG and XPV. Cells from the patients in complementation groups A through G are defective in nucleotide excision repair (NER) [110]. There is a relatively higher rate of XPA mutation in the Japanese population [111,112], which results in severe loss of neurons in multiple regions of the brain and spinal cord, with an age of onset between 7 and 13 years. The signs and symptoms of XP neurological disease include peripheral neuropathy, sensory neural deafness, microcephaly, cerebral dysfunction, ventricular dilation, cortical atrophy, and basal ganglia and cerebellar disturbances [113]. However, how the molecular defects in the NER pathway lead to the XP disease is far from being fully understood.

The first observations of the relationship between defective NER and XP neurological disease were provided by Robbins and coworkers, who examined the cell survival after exposure to UV radiation in patient-derived cells. They observed a correlation between the poor cell survival after UV exposure and the severity of neurodegeneration in the patients [114]. Robbins and colleagues hypothesized that since UV light cannot reach into the human brain, XP neurological disease results from specific endogenous DNA damages caused by free radicals, a kind of damage that is normally repaired by the NER pathway [115].

In the of NER deficiency, this induced damage can accumulate causing neuronal death by blocking transcription of essential genes. This study paved the way to consider other types of lesions caused by UV light, such as pyrimidine (6-4) pyrimidine photoproducts and cyclobutane pyrimidine dimers (CPDs) that block transcription, which are specifically repaired by NER and not by any other human enzymatic repair process. In this scenario, the cPu lesions can fulfill many of the criteria expected in neurodegenerative DNA lesions in XP. Specifically, these lesions are: chemically stable [116,117]; formed endogenously in mammalian cellular DNA [90,118]; repaired by the NER pathway and not by any other known process, as proven by several studies [39,44,55,119]; able to block strongly, but not completely, the transcription by RNA polymerase II that occurs in cells from XP patients [36,120].

These observations led some groups to propose that these lesions might be responsible, at least in part, for the neurodegeneration suffered by xeroderma pigmentosum (XP) patients, who lack the capacity to carry out NER [61,111,120]. Elucidating the mechanism by which cPu could have an implication in XP remains the main goal of several research groups. The studies carried out so far suggest three principal directions: (i) If the lesions completely block transcription by RNAPII, multiple genes are inactivated, and consequently, the levels of encoded proteins are reduced, causing neuronal death; (ii) If the lesions partially block transcription by RNAPII, mutant RNA transcripts can be produced, and this can result in mutant proteins that play a role in neuronal death [121]. In addition, the presence of cPu-induced distortion in the DNA structure prevents the binding of transcription factors (TF) [75,122] to DNA, causing reduced or dysregulated gene expression, in turn resulting in neuronal death; (iii) Other possible implications of the presence of cPu lesions in XP have been hypothesized by Arczewska et al. [123], who suggest that cPu in XP may impair the ability of BER proteins or other repair proteins to bind to cPu lesions, block transcription, and trigger transcriptomic reprogramming.

Recent studies have demonstrated the ability of poly(ADP-ribose) polymerase 1 (PARP-1), responsible for maintaining the integrity of the genome, to recognize and bind DNA sequences containing *S*-cdA and *R*-cdA [40]. DNA recognition and repair proteins other than NER may bind to such lesions but fail to repair them. This may be relevant in understanding the potential role of these lesions in a number of neurological diseases.

A growing body of evidence [124] indicates that NER-defective diseases, besides presenting the accumulation of nuclear DNA lesions, present high ROS levels, mitochondrial dysfunction, and increased reliance on glycolysis. Recent evidence [125] indicates that mitochondrial dysfunction is a neglected but important component of the DNA repair-defective syndromes. Fang et al. found that the neurodegeneration of XPA patients may be associated with mitochondrial and mitophagic dysfunction through PARP-1 hyperactivation and NAD*/SIRT1 reduction [126]. Moreover, this study suggested that the dysfunction was in common with other neurodegenerative disorders, such as ataxia–telangiectasia (AT) and Cockayne syndrome (CS). However, it is possible to consider that because of the differences in neurodegeneration in XPA, AT, and CS, that different mechanisms are involved in these diseases. CS is a rare autosomal recessive disease caused by mutation of either of two genes (CSA or CSB) critical for a subpathway of NER, termed transcription-coupled NER (TC-NER), which preferentially deals with transcription-blocking (typically bulky) lesions within active regions of the genome. The neurological abnormalities that affect CS patients are qualitatively different from those seen in NER-deficient XP patients. In XP, the neurons are primarily affected. In contrast, in CS, the disease involves primarily the white matter of the brain and features brain calcification as well as vascular abnormalities, neither of which are ever observed in XP neurologic disease [61,127]. It was found that CSA is involved in the repair of *S*-cdA and 8-oxo-dG [128].

The differences between CS and XP neurological diseases reflect fundamentally different underlying mechanisms [61,129]. Mitochondrial dysfunction might be a primary or a secondary causative effect in these diseases, depending on the specific DNA repair defect; further evidences will be needed to prove that the accumulation of endogenous cPu lesions in the brain plays a causative role in XP neurological disease [99] and is involved mitochondrial and mitophagic dysfunction.

7.2. Cancer and Aging

Carcinogenesis is a complex process characterized by a progression of abnormalities over time before a cell becomes malignant, while a variety of DNA mutations play crucial roles in the process development. Increased rates of formation of ROS and their interaction with genetic material lead to high yields of oxidatively induced nucleoside modifications [130]. Inability to efficiently repair these harmful species in the genome may affect DNA integrity, causing genetic instability and enhancing cancer risk [131,132]. The levels of cdA and cdG, in their *R* and *S* diastereoisomeric forms, have been identified in estrogen receptor-alpha (ER-α) MCF-7 and triple-negative MDA-MB-231 breast cancer cells upon exposure to two distinct conditions of radical stress, i.e., 5 Gy of ionizing radiation and 300 μM hydrogen peroxide, followed by an interval period that allows DNA repair [133]. The isotope dilution LC–ESI–MS/MS analysis revealed that ER-α MCF-7 and MDA-MB-231 cells are highly susceptible to radiation-induced DNA damage (Figure 12). Estrogen-induced ROS formation can stimulate oxidative DNA damage, while differences in NER efficiency of the breast tissue strongly support the association between modified DNA nucleobases and breast cancer repair capacity [134,135]. This was the first study reporting the simultaneous quantification of the four cPu lesions of DNA in human breast cancer cell lines before and after treatment with DNA-damaging agents. Upon exposure to a dose of 5 Gy, it was found that *S*-cdG was the most abundant lesion, with detected levels in the range of 0.03–0.14 to 0.17–0.23/10^6 nucleosides in both cell lines, whereas *R*-cdG in MCF-7 and *R*-cdA in MCF-7 and MDA-MB-231 cells were indicated as being significantly increased with respect to untreated cell cultures. Exposure of cells to hydrogen peroxide revealed that the *S*-cdG was still observed as the most abundant, and statistically significant differences were obtained at the levels of *R*-cdA in both cell lines along with the levels of *R*-cdG in MDA-MB-231 cells [133].

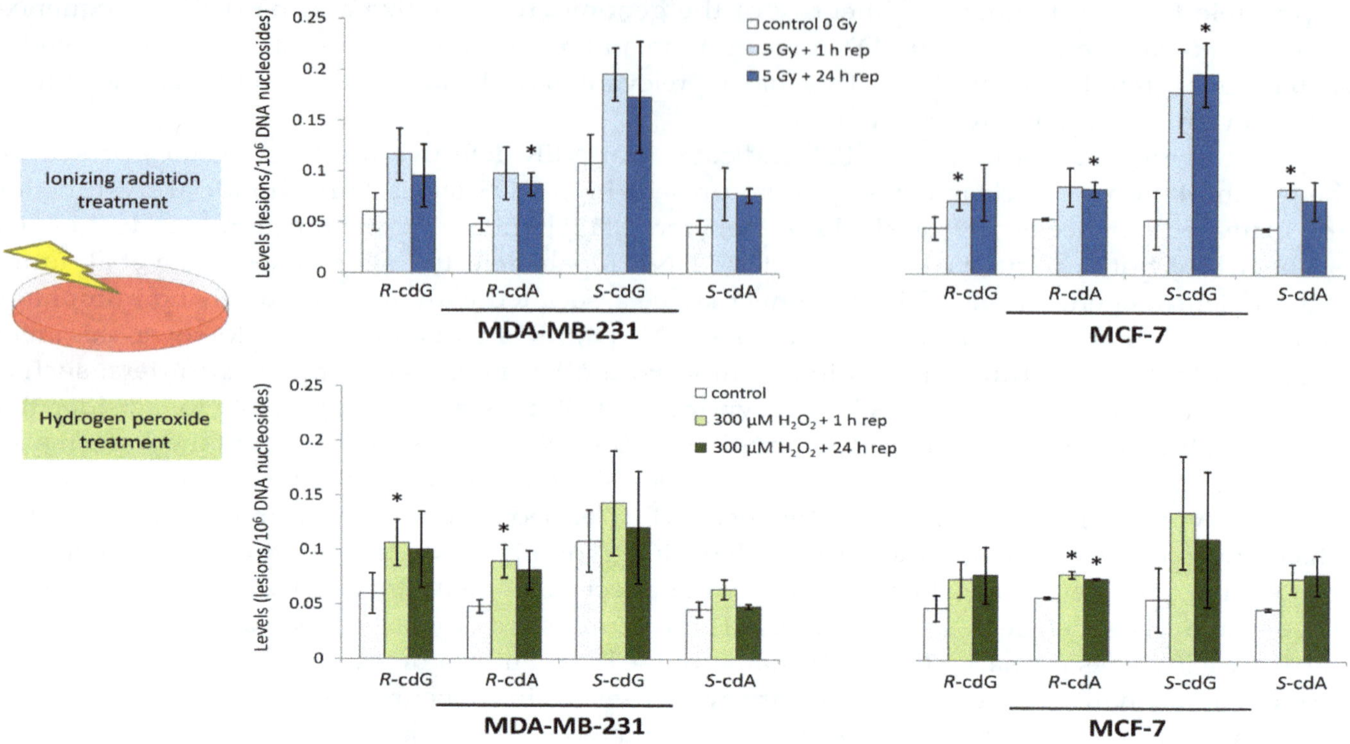

Figure 12. Levels of cPu lesions/10^6 nucleosides measured by LC–MS/MS from breast cancer MDA-MB-231 and MCF-7 cells with or without exposure to γ-irradiation (upper part) and hydrogen peroxide (lower part). The light bars (blue and green, respectively) represent samples exposed to 5 Gy or 300 μM H_2O_2 followed by a 1 h repair period, and the dark bars (blue and green, respectively) represent samples exposed to 5 Gy or 300 μM H_2O_2 followed by a 24 h repair period. The white bars represent untreated samples (control). The asterisks denote a statistically significant difference ($p < 0.05$) between the untreated controls and the treated samples. From reference [133].

Previous studies indicated the elevated levels of *S*-cdA in malignant HCC1937 and MCF-7 breast cancer cells after exposure to hydrogen peroxide (0.04–0.06/10^6 nucleosides) [136]. Moreover, the levels of *R*-cdG, *S*-cdG, and *S*-cdA were found to be increased in cultured lymphoblasts of women with BRCA1 mutations after exposure to 5 Gy of ionizing radiation followed by a subsequent 1 h period of cellular repair of this damage (0.2–0.3/10^6 nucleosides, 0.6–0.8/10^6 nucleosides, and 0.2–0.7/10^6 nucleosides, respectively) [137]. Increased levels of cdA and cdG have been observed in organs of prdx1$^{+/+}$ and prdx1$^{-/-}$ mice in a model study of the tumor susceptibility in prdx1$^{-/-}$ animals and the link with oncogenes, such as c-Myc and DNA damage [138]. Furthermore, elevated cPu levels have been measured in the skin of red Mc1r$^{e/e}$ mice compared with albino Mc1r$^{e/e}$ mice in a study of the mechanism of UV-independent carcinogenesis aiming to ascertain whether ROS-mediated oxidative DNA damage is influenced by the pheomelanin synthesis pathway [139].

Recently, the accumulation of both the *R* and *S* diastereoisomers of cPu was followed up in organs of tumor-bearing severe combined immunodeficient (SCID) mice of different ages (4 and 17 weeks old) in comparison with the corresponding control SCID mice, providing evidence of increased oxidatively induced DNA damage occurring during tumor progression [140]. Identification and quantification by isotope dilution LC–ESI–MS/MS in two distinct tissues (the liver and kidney) revealed that tumor-bearing SCID mice had 1.1–1.4-fold higher levels of the four cyclopurines than the control SCID mice, while *S*-cdG was the most abundant lesion, with levels of 0.28–0.38/10^6 nucleosides in the liver and 0.24–0.25/10^6 nucleosides in the kidney, respectively (Figure 13). Moreover, in 17-week-old tumor-bearing animals, the cPu lesion levels were elevated compared to 4-week-old tumor-bearing mice, in agreement with previously reported data among cancer and age-dependent processes [141]. The total amount of cPu lesions in both tissues for all groups are reported in Table 2. Young and old control SCID mice presented similar levels of cPu (0.70/10^6 nucleosides in the liver and 0.67/10^6 nucleosides in the kidney). Among young mice, the diseased ones showed 1.1-fold higher cPu levels. Between the early and latest stages of the lifespan of tumor-bearing SCID mice, increased damage is observed at the levels of 1.04/10^6 nucleosides in the liver and 0.9/10^6 nucleobases in the kidney of 17-week-old diseased mice. Important evidence also can be provided by the ratios of *R*/*S* for both cdG and cdA in different tissue compartments. The *R*/*S* ratios were found to be almost the same in the liver and kidney for all five experimental animal models for cdG (0.7–0.9). On the contrary, the ratio for cdA was approximately 1.3–1.9 in the kidney and 1.5–2.3 in the liver, being twice than that of cdG [140].

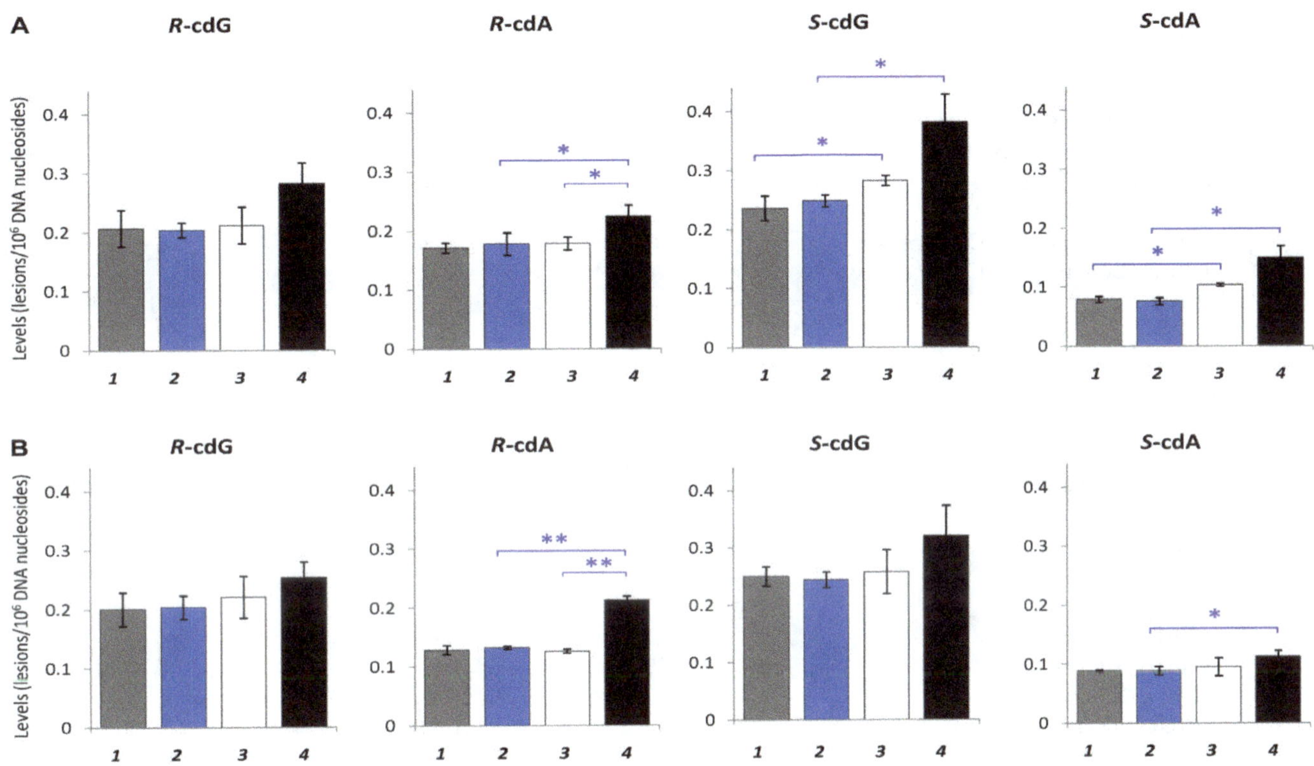

Figure 13. Levels of cPu lesions/10^6 nucleosides measured by LC–MS/MS in control severe combined immunodeficient (SCID) and tumor-bearing SCID mice. (**A**) Levels of *R*-cdG, *S*-cdG, *R*-cdA, and *S*-cdA lesions in genomic DNA isolated from the liver of control SCID mice (group ***1***: 4 weeks old; group **2**: 17 weeks old) and tumor-bearing SCID mice (group **3**: 4 weeks old; group **4**: 17 weeks old). (**B**) Levels of *R*-cdG, *S*-cdG, *R*-cdA, and *S*-cdA lesions in genomic DNA isolated from the kidney of control SCID mice (groups ***1*** and **2**) and tumor-bearing SCID mice (groups **3** and **4**). * Denotes a statistically significant difference ($p < 0.05$) and ** denotes a statistically significant difference $p < 0.005$ between the animal groups. From reference [140].

In Table 2, the diastereoisomer ratios for the different organs of control SCID and tumor-bearing SCID mice are summarized. The diastereomeric ratio (*R*/*S*) attracts interest since it can inform on mechanistic issues. It is worth underlining that the *S* form is always more abundant than the *R* form in cdG, whereas in cdA, the *R* form is always more abundant than the *S* form. These cyclopurine isomer ratios can be explained by the influence of at least two factors: (i) the local conformations of the supramolecular organization of DNA are taken at the reactive sites prior to C5′ radical cyclization, which make one type of diastereoisomer more prevalent, and (ii) the NER efficiency for the repair of the diastereoisomers. Indeed, previous work in human HeLa cell extracts indicated that the cdA and cdG lesions are excised with similar efficiency by NER and that the *R*-diastereoisomers of both cdA and cdG cause greater distortion of the DNA backbone and are better substrates of NER than the corresponding *S* ones [39].

Table 2. Diastereomeric ratios (*R*/*S*) of cdG and cdA lesions in animal tissues by isotope dilution liquid chromatography–tandem mass spectrometry #.

DNA Source	*R*-cdG/*S*-cdG	*R*-cdA/*S*-cdA	Reference
Liver			[140]
normal Swiss mice 4w	0.85	1.81	
normal Swiss mice 17w	0.79	1.77	
control SCID mice 4w	0.87	2.16	
control SCID mice 17w	0.82	2.34	
tumor-bearing SCID mice 4w	0.75	1.73	
tumor-bearing SCID mice 17w	0.74	1.51	
Kidney			
normal Swiss mice 4w	0.86	2.03	
normal Swiss mice 17w	0.77	1.72	
control SCID mice 4w	0.80	1.45	
control SCID mice 17w	0.84	1.49	
tumor-bearing SCID mice 4w	0.85	1.56	
tumor-bearing SCID mice 17w	0.79	1.88	
Liver			[118,142]
wild-type mice 10w	0.65	0.95	
wild-type mice 21w	0.96	1.64	
$ERCC^{-/\Delta}$ mice 10w	1.20	2.65	
$ERCC^{-/\Delta}$ mice 21w	1.30	2.05	
Kidney			
wild-type mice 10w	0.39	0.77	
wild-type mice 21w	0.22	0.38	
$ERCC^{-/\Delta}$ mice 10w	0.53	1.42	
$ERCC^{-/\Delta}$ mice 21w	0.26	0.40	
Brain			
wild-type mice 10w	0.45	0.64	
wild-type mice 21w	0.29	0.67	
$ERCC^{-/\Delta}$ mice 10w	0.32	0.81	
$ERCC^{-/\Delta}$ mice 21w	0.44	0.90	
Liver			[29,143]
LEA rats 3m	0.68	0.91	
$LEC^{+/-}$ rats 3m	0.77	1.29	
$LEC^{+/-}$ rats 12m	0.50	0.86	
$LEC^{-/-}$ rats 1m	0.80	1.75	
$LEC^{-/-}$ rats 3m	0.79	0.96	
$LEC^{-/-}$ rats 6m	0.65	1.6	
Brain			
LEA rats 3m	0.88	1.70	
$LEC^{+/-}$ rats 3m	0.81	2.17	
$LEC^{+/-}$ rats 12m	1.61	2.80	
$LEC^{-/-}$ rats 1m	1.00	1.60	
$LEC^{-/-}$ rats 3m	0.73	1.33	
$LEC^{-/-}$ rats 6m	1.21	2.15	
Skin			[139]
red-$Mc1r^{e/e}$ mice	0.40	0.86	
albino-$Mc1r^{e/e}$ mice	0.42	1.20	

#w: weeks; SCID: severe combined immunodeficient; ERCC: Excision repair cross-complementing; LEA: Long–Evans Agouti; LEC: Long–Evans Cinnamon.

Increased levels of cPu lesions were observed in animal tissues and biological fluids at background levels or at increased levels in age-related disorders as well as in other inflammation-related pathologies. Accumulation of cPu lesions with aging in a tissue-specific manner (liver > kidney > brain) of wild-type and DNA repair-deficient progeroid ERCC1$^{-/\Delta}$ mice is also observed, indicating the significant higher levels of cdA and cdG in ERCC1$^{-/\Delta}$ mice compared to age-matched wt mice [118,142]. NanoLC–ESI–MS/MS analysis of liver tissues of Long–Evans Cinnamon (LEC) rats, as an animal model of human Wilson's disease, and Long–Evans Agouti (LEA) healthy rats demonstrated significant elevated levels of cdA and cdG in both *R* and *S* isoforms in diseased animals (up to 4-fold excess for liver tissues of LEC$^{+/-}$ and brain tissues of LEC$^{-/-}$) [29,143]. In Table 2, the *R*/*S* levels of cdG and cdA lesions in tissues of selected animal models associated with age-related processes are also reported. It is worth recalling that the *R*/*S* ratios can be used to support further biological implications in the formation and/or repair of these lesions, as discussed above. However, comparing all the data reported in Table 2, in some studies, a clear scenario of the *R*/*S* formation cannot be drawn, probably due to the analytical performance and uncertainty of the measurements. It would be recommended to reach uniformity of protocols and quantitative methods in order to diminish these pitfalls. It is worth mentioning that previous studies using different analytical tools, such as GC/MS or LC/MS, indicated the accumulation of *R*-cdA and *S*-cdA in liver DNA of neil1$^{-/-}$ mice not exposed to exogenous oxidative stress, revealing that NEIL1 plays a role in the cellular repair of *R*-cdA and *S*-cdA [144]. Previous studies by the same group on this direction highlighted elevated levels of *S*-cdA in organs of csb$^{-/-}$ mice compared to wild-type mice, providing the evidence that CSB is implicated in the repair of the DNA helix-distorting tandem lesion *S*-cdA [145] as well in tissues of prdx1$^{-/-}$ mice with approximately identical *R*-cdG/*S*-cdG and *R*-cdA/*S*-cdA ratios (for cdG, it was found to be 0.23, 0.40, and 0.17, while for cdA, 0.11, 0.08, and 0.23 in the brain, liver, and spleen, respectively) [138]. Lastly, increased levels of *S*-cdA were found in polymorphonuclear leukocytes of familial Mediterranean fever (FMF) patients when compared to control subjects [146], while *R*-cdA and *S*-cdA were found to be elevated in urine samples of prediabetic patients [147].

7.3. Addendum

During the revision process of this review, a "Letter to the editor" was published in *Free Radical Research* by Cadet et al. [148] criticizing the work produced by several laboratories [29,89,133,138,142,148] and described here. In particular, Cadet et al. [148] questioned the effectiveness of the published protocol on DNA samples shown in Figure 11 and of the LC–MS/MS methodologies used for detection and quantification of cPu as biomarkers of DNA oxidation reactions. In their Letter, the authors did not present any new experimental work, but only gave a "revisited analysis" of DNA damage detection based on some unsuccessful attempts (Cadet et al., unpublished) made in 2000 to measure the presence of cPu lesions in DNA by LC–ESI–MS/MS, using either the brain of NER-knockout mice or gamma-irradiated human cells exposed to 1 kGy radiation. Therefore, based on a unique unsuccessful study of almost twenty years ago, in their review, the authors started to cast doubts on peer-reviewed work published in the last ten years from several important laboratories around the world and advanced the hypothesis that cPu formation can be an experimental artifact. Apart from the fact that the isotope dilution LC–MS/MS methodology is a sensitive methodology that nowadays overcomes problems of low-sensitivity methods faced twenty years ago, the artifact formation in the analysis of biological samples cannot be underestimated, especially after the well-known debate of some years ago on similar problems raised with 8-oxo-dG lesion detection. Indeed, this review describes the experience acquired in the last years for new and effective protocols of digestion and LC–MS/MS analysis of DNA samples, highlighting the importance of quantitation sensitivity, accuracy, precision, linearity, reproducibility, and recovery, along with the DNA enzymatic digestion efficiency. The synthesis of ^{15}N isotopically labeled compounds for each of the four cPu diastereoisomers was also an important step that allowed to establish an analytical protocol with an HPLC cleanup and enrichment of the samples (Figure 11). Therefore, to answer critical questions referring to supposed experimental artifacts, the effort and time that must be spent is such that only a few labs can afford. By using control experiments and DNA analyses of samples before and after irradiation, the criticisms on artifacts could be successfully overcome, and the resulting robust analytical methodology can be nowadays used to gain information on the importance of cPu as biomarkers of hydroxyl radical damage in aging and diseases. This addendum is made to trace the evolution of cPu research over time and also to offer some arguments for debating how the analytical protocols can be developed and ameliorated over time.

8. Conclusions

In this review, we have summarized the recent work on purine 5′,8-cyclo-2′-deoxynucleosides (cPu), a subject of growing interest in studies on DNA damage and impact on human health and disease. The cPu lesions are forms of oxidative DNA damage that can be repaired by NER with low efficiencies. The accumulation of cPu lesions in the genome inhibits the DNA replication by diminishing the activities of repair polymerases, RNA polymerases, and the DNA binding of transcription factors. This results in severe adverse effects on cellular functions, including replication fork stalling, deficient DNA repair and gene transcription, mutagenesis, and genomic instability.

The use of these lesions as candidate biomarkers of DNA damage is increasingly appreciated because the cPu DNA lesions do not suffer from stability issues and artifacts of other oxidatively generated DNA lesions. Moreover, an interesting feature of the cPu lesions is that they exist in four diastereoisomeric forms in different ratios. The presence of these DNA lesions in different tissue-types and biological environments is well-documented. However, the impact of the stereoisomeric cPu lesions on various pathologies is poorly understood, and thus new knowledge needs to be developed in order to achieve deeper insights into these phenomena. The state of the art described in this review indicates that important steps have been made for clarifying many of the of biochemical and biological processes involving cPu lesions. These achievements will certainly facilitate further extended analytical studies of the detection of these modified nucleosides in vivo and the assessment of their impact on genome integrity, human health, and aging.

Acknowledgments: The support given by the EU COST Action CM1201 "Biomimetic Radical Chemistry" is kindly acknowledged.

Abbreviations

BER	base excision repair
CS	Cockayne syndrome
cdA	5′,8-cyclo-2′-deoxyadenosine
cdG	5′,8-cyclo-2′-deoxyguanosine
cPu	purine 5′,8-cyclo-2′-deoxynucleoside
dA	2′-deoxyadenosine
dG	2′-deoxyguanosine
LC–MS/MS	liquid chromatography tandem mass spectrometry
LEA	Long–Evans Cinnamon
NER	nucleotide excision repair
ODNs	oligodeoxynucleotides
ROS	reactive oxygen species
SCID	severe combined immunodeficient

Pol	DNA polymerase
Dpo4	DNA polymerase IV of *Sulfolobus solfataricus*
LIG I	DNA ligase I
TBDMS	tert-butyldimethylsilyl
XP	xeroderma pigmentosum

References

1. McCord, J.M.; Fridovich, I. The reduction of cytochrome c by milk xanthine oxidase. *J. Biol. Chem.* **1968**, *243*, 5753–5760. [PubMed]
2. Winterbourn, C.C. Biological chemistry of reactive oxygen species. In *Encyclopedia of Radicals in Chemistry, Biology and Materials*; Chatgilialoglu, C., Studer, A., Eds.; Wiley: Chichester, UK, 2012; Volume 3, pp. 1260–1281.
3. Geacintov, N.E.; Shafirovich, V. Reactions of small reactive species with DNA. In *Encyclopedia of Radicals in Chemistry, Biology and Materials*; Chatgilialoglu, C., Studer, A., Eds.; Wiley: Chichester, UK, 2012; Volume 3, pp. 1284–1317.
4. Friedberg, E.C.; Walker, G.C.; Siede, W. *DNA Repair Mutagenesis*; ASM Press: Washington, WA, USA, 1995.
5. Maynard, S.; Fang, E.F.; Scheibye-Knudsen, M.; Croteau, D.L.; Bohr, V.A. DNA Damage, DNA Repair, Aging, and Neurodegeneration. *Cold Spring Harb Perspect. Med.* **2015**, *5*, 1–18. [CrossRef] [PubMed]
6. Von Sonntag, C. *Free-Radical-Induced DNA Damage and Its Repair: A Chemical Persective*; Springer: Berlin, Germany, 2006.
7. Balasubramanian, B.; Pogozelski, W.K.; Tullius, T.D. DNA strand breaking by the hydroxyl radical is governed by the accessible surface areas of the hydrogen atoms of the DNA backbone. *Proc. Natl. Acad. Sci. USA* **1998**, *95*, 9738–9743. [CrossRef]
8. Chan, W.; Chen, B.; Wang, L.; Taghizadeh, K.; Demott, M.S.; Dedon, P.C. Quantification of the 2-deoxyribonolactone and nucleoside 5′-aldehyde products of 2-deoxyribose oxidation in DNA and cells by isotope-dilution gas chromatography mass mpectrometry: Differential effects of γ-radiation and Fe^{2+} –EDTA. *J. Am. Chem. Soc.* **2010**, *132*, 6145–6153. [CrossRef] [PubMed]
9. Aydogan, B.; Marshall, D.T.; Swarts, S.G.; Turner, J.E.; Boone, A.J.; Richards, N.G.; Bolch, W.E. Site-specific OH attack to the sugar moiety of DNA: A comparison of experimental data and computational simulation. *Radiat. Res.* **2002**, *157*, 38–44. [CrossRef]
10. Chatgilialoglu, C. Reactivity of nucleic acid sugar radicals. In *Radical and Radical Ion Reactivity in Nucleic Acid Chemistry*; Greenberg, M.M., Ed.; Wiley: Hoboken, NJ, USA, 2009; Chapter 4; pp. 99–133.
11. Gimisis, T.; Chatgilialoglu, C. Oxidatively formed sugar radicals in nucleic acids. In *Encyclopedia of Radicals in Chemistry, Biology and Materials*; Chatgilialoglu, C., Studer, A., Eds.; Wiley: Chichester, UK, 2012; Volume 3, pp. 1345–1370.
12. Greenberg, M.M. Reactivity of Nucleic Acid Radicals. *Adv. Phys. Org. Chem.* **2016**, *50*, 119–202. [PubMed]
13. Chatgilialoglu, C.; Ferreri, C.; Terzidis, M.A. Purine 5′,8-cyclonucleoside lesions: Chemistry and biology. *Chem. Soc. Rev.* **2011**, *40*, 1368–1382. [CrossRef] [PubMed]
14. Von Sonntag, C. *The Chemical Basis of Radiation Biology*; Taylor & Francis: London, UK, 1987.
15. Keck, K. Bildung von cyclonudleotiden bei bestrahlung wassiriger losungen von purinenucleotiden. *Z. Naturforsch.* **1968**, *23b*, 1034–1043. [CrossRef]
16. Dizdaroglu, M. Free-radical-induced formation of an 8,5′-cyclo-2′-deoxyguanosine moiety in deoxyribonucleic acid. *Biochem. J.* **1986**, *238*, 247–254. [CrossRef]
17. Dirksen, M.L.; Blakely, W.F.; Holwitt, E.; Dizdaroglu, M. Effect of DNA conformation on the hydroxyl radical-induced formation of 8,5′-cyclopurine 2′-deoxyribonucleoside residues in DNA. *Int. J. Radiat. Biol.* **1988**, *54*, 195–204. [CrossRef]
18. Dizdaroglu, M.; Dirksen, M.L.; Jiang, H.X.; Robbins, J.H. Ionizing-radiation-induced damage in the DNA of cultured human cells. Identification of 8,5′-cyclo-2-deoxyguanosine. *Biochem. J.* **1987**, *241*, 929–932. [CrossRef] [PubMed]
19. Jaruga, P.; Dizdaroglu, M. 8,5′-Cyclopurine-2′-deoxynucleosides in DNA: Mechanisms of formation, measurement, repair and biological effects. *DNA Repair* **2008**, *7*, 1413–1425. [CrossRef] [PubMed]
20. Dizdaroglu, M.; Jaruga, P. Mechanisms of free radical-induced damage to DNA. *Free Radic. Res.* **2012**, *46*, 382–419. [CrossRef] [PubMed]

21. Chatgilialoglu, C.; Bazzanini, R.; Jimenez, L.B.; Miranda, M.A. (5′*S*)- and (5′*R*)-5′,8-Cyclo-2′-deoxyguanosine: Mechanistic insights on the 2′-deoxyguanosin-5′-yl radical cyclization. *Chem. Res. Toxicol.* **2007**, *20*, 1820–1824. [CrossRef] [PubMed]
22. Flyunt, R.; Bazzanini, R.; Chatgilialoglu, C.; Mulazzani, Q.G. Fate of the 2′-deoxyadenosin-5′-yl radical under anaerobic conditions. *J. Am. Chem. Soc.* **2000**, *122*, 4225–4226. [CrossRef]
23. Chatgilialoglu, C.; Guerra, M.; Mulazzani, Q.G. Model studies of DNA C5′ radicals. Selective generation and reactivity of 2′-deoxyadenosin-5′-yl radical. *J. Am. Chem. Soc.* **2003**, *125*, 3839–3848.
24. Boussicault, F.; Kaloudis, P.; Caminal, C.; Mulazzani, Q.G.; Chatgilialoglu, C. The fate of C5′ radicals of purine nucleosides under oxidative conditions. *J. Am. Chem. Soc.* **2008**, *130*, 8377–8385. [CrossRef]
25. Chatgilialoglu, C.; D'Angelantonio, M.; Kciuk, G.; Bobrowski, K. New insights into the reaction paths of hydroxyl radicals with 2′-deoxyguanosine. *Chem. Res. Toxicol.* **2011**, *24*, 2200–2206. [CrossRef]
26. Jimenez, L.B.; Encinas, S.; Miranda, M.A.; Navacchia, M.L.; Chatgilialoglu, C. The photochemistry of 8-bromo-2′-deoxyadenosine. A direct entry to cyclopurine lesions. *Photochem. Photobiol. Sci.* **2004**, *3*, 1042–1046. [CrossRef]
27. Navacchia, M.L.; Chatgilialoglu, C.; Montevecchi, P.C. C5′-adenosinyl radical cyclization. A stereochemical investigation. *J. Org. Chem.* **2006**, *71*, 4445–4452. [CrossRef]
28. Terzidis, M.A.; Chatgilialoglu, C. Radical cascade protocol for the synthesis of (5′S)- and (5′R)-5′,8-cyclo-2′-deoxyguanosine derivatives. *Aust. J. Chem.* **2013**, *66*, 330–335. [CrossRef]
29. Wang, J.; Yuan, B.; Guerrero, C.; Bahde, R.; Gupta, S.; Wang, Y. Quantification of oxidative DNA lesions in tissues of Long-Evans Cinnamon rats by capillary high-performance liquid chromatography-tandem mass spectrometry coupled with stable isotope-dilution method. *Anal. Chem.* **2011**, *83*, 2201–2209. [CrossRef] [PubMed]
30. Terzidis, M.A.; Chatgilialoglu, C. An ameliorative protocol for the quantification of purine 5′,8-cyclo-2′-deoxynucleosides in oxidized DNA. *Front. Chem.* **2015**, *3*, 47. [CrossRef] [PubMed]
31. Jaruga, P.; Birincioglu, M.; Rodriguez, H.; Dizdaroglu, M. Mass spectrometric assays for the tandem lesion 8,5′-cyclo-2′-deoxyguanosine in mammalian DNA. *Biochemistry* **2002**, *41*, 3703–3711. [CrossRef] [PubMed]
32. Chatgilialoglu, C.; Ferreri, C.; Masi, A.; Sansone, A.; Terzidis, M.A.; Tsakos, M. A problem solving approach for the diastereoselective synthesis of (5′S)- and (5′R)-5′,8-cyclopurine lesions. *Org. Chem. Front.* **2014**, *1*, 698–702. [CrossRef]
33. Matsuda, A.; Tezuka, M.; Niizuma, K.; Suiyama, E.; Ueda, T. Synthesis of carbon-bridged 8,5′-cyclopurine Nucleosides: Nucleosides and Nucleotides. *Tetrahedron* **1978**, *34*, 2633–2637. [CrossRef]
34. Romieu, A.; Gasparutto, D.; Molko, D.; Cadet, J. Site-Specific Introduction of (5′S)-5′,8-Cyclo-2′-deoxyadenosine into oligodeoxyribonucleotides. *J. Org. Chem.* **1998**, *63*, 5245–5249. [CrossRef]
35. Romieu, A.; Gasparutto, D.; Cadet, J. Synthesis and characterization of oligonucleotides containing 5′,8-cyclopurine 2′-deoxyribonucleosides: (5′R)-5′,8-cyclo-2′-deoxyadenosine, (5′S)-5′,8-cyclo-2′-deoxyguanosine, and (5′R)-5′,8-cyclo-2′-deoxyguanosine. *Chem. Res. Toxicol.* **1999**, *12*, 412–421. [CrossRef] [PubMed]
36. Brooks, P.J.; Wise, D.S.; Berry, D.A.; Kosmoski, J.V.; Smerdon, M.J.; Somers, R.L.; Mackie, H.; Spoonde, A.Y.; Ackerman, E.J.; Coleman, K.; et al. The oxidative DNA lesion 8,5′-(S)-cyclo-2′-deoxyadenosine is repaired by the nucleotide excision repair pathway and blocks gene expression in mammalian cells. *J. Biol. Chem.* **2000**, *275*, 22355–22362. [CrossRef]
37. Zaliznyak, T.; Lukin, M.; de los Santos, C. Structure and stability of duplex DNA containing (5′S)-5′,8-cyclo-2′-deoxyadenosine: An oxidatively generated lesion repaired by NER. *Chem. Res. Toxicol.* **2012**, *25*, 2103–2111. [CrossRef]
38. Huang, H.; Das, R.S.; Basu, A.K.; Stone, M.P. Structure of (5′S)-8,5′-cyclo-2′-deoxyguanosine in DNA. *J. Am. Chem. Soc.* **2011**, *133*, 20357–20368. [CrossRef] [PubMed]
39. Kropachev, K.; Ding, S.; Terzidis, M.A.; Masi, A.; Liu, Z.; Cai, Y.; Kolbanovskiy, M.; Chatgilialoglu, C.; Broyde, S.; Geancitov, N.E.; et al. Structural basis for the recognition of diastereomeric 5′,8-cyclo-2′-deoxypurine lesions by the human nucleotide excision repair system. *Nucl. Acids Res.* **2014**, *42*, 5020–5032. [CrossRef] [PubMed]
40. Masi, A.; Sabbia, A.; Ferreri, C.; Manoli, F.; Lai, Y.; Laverde, E.; Liu, Y.; Krokidis, M.G.; Chatgilialoglu, C.; Faraone Mennella, M.R. Diastereomeric recognition of 5′,8-cyclo-2′-deoxyadenosine lesions by human Poly(ADP-ribose) polymerase 1 in a biomimetic model. *Cells* **2019**, *8*, 116. [CrossRef] [PubMed]

41. Haromy, T.P.; Raleigh, J.; Sundaralingam, M. Enzyme-bound conformations of nucleotide substrates. X-ray structure and absolute configuration of 8,5′-cycloadenosine monohydrate. *Biochemistry* **1980**, *19*, 1718–1722. [CrossRef]
42. Huang, H.; Das, R.S.; Basu, A.K.; Stone, M.P. Structures of (5′S)-8,5′-cyclo-2′-deoxyguanosine mismatched with dA or dT. *Chem. Res. Toxicol.* **2012**, *25*, 478–490. [CrossRef] [PubMed]
43. Karwowski, B.T. The role of (5′R) and (5′S) 5′,8-cyclo-2′-deoxyadenosine in ds-DNA structure: A comparative QM/MM theoretical study. *Comput. Theor. Chem.* **2013**, *1010*, 34–44. [CrossRef]
44. Kuraoka, I.; Bender, C.; Romieu, A.; Cadet, J.; Wood, R.D.; Lindahl, T. Removal of oxygen free-radical-induced 5′,8-purine cyclodeoxynucleosides from DNA by the nucleotide excision-repair pathway in human cells. *Proc. Natl. Acad. Sci. USA* **2000**, *97*, 3832–3837. [CrossRef] [PubMed]
45. Kidane, D.; Chae, W.J.; Czochor, J.; Eckert, K.A.; Glazer, P.M.; Bothwell, A.L.; Sweasy, J.B. Interplay between DNA repair and inflammation, and the link to cancer. *Crit. Rev. Biochem. Mol. Biol.* **2014**, *49*, 116–139. [CrossRef]
46. Wallace, S.S.; Murphy, D.L.; Sweasy, J.B. Base excision repair and cancer. *Cancer Lett.* **2012**, *327*, 73–89. [CrossRef]
47. Lindahl, T.; Wood, R.D. Quality control by DNA repair. *Science* **1999**, *286*, 1897–1905. [CrossRef]
48. Krokan, H.E.; Bjøras, M. Base excision repair. *Cold Spring Harb. Perspect. Biol.* **2013**, *5*, a012583. [CrossRef] [PubMed]
49. Doetsch, P.W.; Cunningham, R.P. The enzymology of apurinic/apyrimidinic endonucleases. *Mutat. Res.* **1990**, *236*, 173–201. [CrossRef]
50. Bailly, V.; Verly, W.G. Escherichia coli endonuclease III is not an endonuclease but a beta-elimination catalyst. *Biochem. J.* **1987**, *242*, 565–572. [CrossRef] [PubMed]
51. Bailly, V.; Derydt, M.; Verly, W.G. Delta-elimination in the repair of AP (apurinic/apyrimidinic) sites in DNA. *Biochem. J.* **1989**, *261*, 707–713. [CrossRef] [PubMed]
52. Porello, S.L.; Leyes, A.E.; David, S.S. Single-turnover and pre-steady-state kinetics of the reaction of the adenine glycosylase MutY with mismatch-containing DNA substrates. *Biochemistry* **1998**, *37*, 14756–14764. [CrossRef]
53. Brooks, P.J.; Marietta, C.; Goldman, D. DNA mismatch repair and DNA methylation in adult brain neurons. *J. Neurosci.* **1996**, *16*, 939–945. [CrossRef]
54. McCullough, A.K.; Dodson, M.L.; Lloyd, R.S. Initiation of base excision repair: Glycosylase mechanisms and structures. *Annu. Rev. Biochem.* **1999**, *68*, 255–285. [CrossRef]
55. Pande, P.; Das, R.S.; Shepard, C.; Kow, Y.W.; Basu, A.K. Repair efficiency of (5′S)-8,5-cyclo-2′-deoxyguanosine and (5′S)-8,5′-cyclo-2′-deoxyadenosine depends on the complementary base. *DNA Repair* **2012**, *11*, 926–931. [CrossRef]
56. Geacintov, N.E.; Broyde, S. Repair-resistant DNA lesions. *Chem. Res. Toxicol.* **2017**, *30*, 1517–1548. [CrossRef]
57. Naegeli, H.; Sugasawa, K. The xeroderma pigmentosum pathway: Decision tree analysis of DNA quality. *DNA Repair* **2011**, *10*, 673–683. [CrossRef]
58. Gillet, L.C.; Schärer, O.D. Molecular mechanisms of mammalian global genome nucleotide excision repair. *Chem. Rev.* **2006**, *106*, 253–276. [CrossRef] [PubMed]
59. Shafirovich, V.; Kropachev, K.; Anderson, T.; Liu, Z.; Kolbanovskiy, M.; Martin, B.D.; Sugden, K.; Shim, Y.; Chen, X.; Min, J.H.; et al. Base and nucleotide excision repair of oxidatively generated guanine lesions in DNA. *J. Biol. Chem.* **2016**, *291*, 5309–5319. [CrossRef] [PubMed]
60. Shafirovich, V.; Kropachev, K.; Kolbanovskiy, M.; Geacintov, N.E. Excision of oxidatively generated guanine lesions by competing base and nucleotide excision repair mechanisms in human cells. *Chem. Res. Toxicol.* **2019**, *32*, 753–761. [CrossRef] [PubMed]
61. Brooks, P.J. The 8,5′-cyclopurine-2′-deoxynucleosides: Candidate neurodegenerative DNA lesions in xeroderma pigmentosum, and unique probes of transcription and nucleotide excision repair. *DNA Repair* **2008**, *7*, 1168–1179. [CrossRef] [PubMed]
62. Hess, M.T.; Gunz, D.; Luneva, N.; Geacintov, N.E.; Naegeli, H. Base pair conformation-dependent excision of benzo[a]pyrene diol epoxide-guanine adducts by human nucleotide excision repair enzymes. *Mol. Cell. Biol.* **1997**, *17*, 7069–7076. [CrossRef]
63. Hara, R.; Mo, J.; Sancar, A. DNA damage in the nucleosome core is refractory to repair by human excision nuclease. *Mol. Cell. Biol.* **2000**, *20*, 9173–9181. [CrossRef]

64. Hara, R.; Sancar, A. The SWI/SNF chromatin-remodeling factor stimulates repair by human excision nuclease in the mononucleosome core particle. *Mol. Cell. Biol.* **2002**, *22*, 6779–6787. [CrossRef]
65. Hara, R.; Sancar, A. Effect of damage type on stimulation of human excision nuclease by SWI/SNF chromatin remodeling factor. *Mol. Cell. Biol.* **2003**, *23*, 4121–4125. [CrossRef]
66. Shafirovich, V.; Kolbanovskiy, M.; Kropachev, K.; Liu, Z.; Cai, Y.; Terzidis, M.A.; Masi, A.; Chatgilialoglu, C.; Amin, S.; Dadali, A.; et al. Nucleotide excision repair and impact of site-specific 5′,8-cyclopurine and bulky DNA lesions on the physical properties of nucleosomes. *Biochemistry* **2019**, *58*, 561–574. [CrossRef]
67. Vasudevan, D.; Chua, E.Y.; Davey, C.A. Crystal structures of nucleosome core particles containing the '601' strong positioning sequence. *J. Mol. Biol.* **2010**, *403*, 1–10. [CrossRef]
68. Cai, Y.; Kropachev, K.; Terzidis, M.A.; Masi, A.; Chatgilialoglu, C.; Shafirovich, V.; Geacintov, N.E.; Broyde, S. Differences in the access of lesions to the nucleotide excision repair machinery in nucleosomes. *Biochemistry* **2015**, *54*, 4181–4185. [CrossRef] [PubMed]
69. Xu, M.; Lai, Y.; Jiang, Z.; Terzidis, M.A.; Masi, A.; Chatgilialoglu, C.; Liu, Y. A 5′, 8-cyclo-2′-deoxypurine lesion induces trinucleotide repeat deletion via a unique lesion bypass by DNA polymerase β. *Nucleic Acids Res.* **2014**, *42*, 13749–13763. [CrossRef] [PubMed]
70. You, C.; Swanson, A.L.; Dai, X.; Yuan, B.; Wang, J.; Wang, Y. Translesion synthesis of 8,5′-cyclopurine-2′-deoxynucleosides by DNA polymerases η, ι, and ζ. *J. Biol. Chem.* **2013**, *288*, 28548–28556. [CrossRef] [PubMed]
71. Kuraoka, I.; Robins, P.; Masutani, C.; Hanaoka, F.; Gasparutto, D.; Cadet, J.; Wood, R.D.; Lindahl, T. Oxygen free radical damage to DNA. Translesion synthesis by human DNA polymerase η and resistance to exonuclease action at cyclopurine deoxynucleoside residues. *J. Biol. Chem.* **2001**, *276*, 49283–49288. [CrossRef] [PubMed]
72. Walmacq, C.; Wang, L.; Chong, J.; Scibelli, K.; Lubkowska, L.; Gnatt, A.; Brooks, P.J.; Wang, D.; Kashlev, M. Mechanism of RNA polymerase II bypass of oxidative cyclopurine DNA lesions. *Proc. Natl. Acad. Sci. USA* **2015**, *112*, E410–E419. [CrossRef]
73. Kamakura, N.; Yamamoto, J.; Brooks, P.J.; Iwai, S.; Kuraoka, I. Effects of 5′,8-cyclodeoxyadenosine triphosphates on DNA synthesis. *Chem. Res. Toxicol.* **2012**, *25*, 2718–2724. [CrossRef]
74. Klenow, H.; Henningsen, I. Selective elimination of the exonuclease activity of the deoxyribonucleic acid polymerase from Escherichia coli B by limited proteolysis. *Proc. Natl. Acad. Sci. USA* **1970**, *65*, 168–175. [CrossRef]
75. Marietta, C.; Gulam, H.; Brooks, P.J. A single 8,5′-cyclo-2′-deoxyadenosine lesion in a TATA box prevents binding of the TATA binding protein and strongly reduces transcription in vivo. *DNA Repair* **2002**, *1*, 967–975. [CrossRef]
76. Karwowski, B.T.; Bellon, S.; O'Neill, P.; Lomax, M.E.; Cadet, J. Effects of (5′*S*)-5′,8-cyclo-2′-deoxyadenosine on the base excision repair of oxidatively generated clustered DNA damage. A biochemical and theoretical study. *Org. Biomol. Chem.* **2014**, *12*, 8671–8682. [CrossRef]
77. Jasti, V.P.; Das, R.S.; Hilton, B.A.; Weerasooriya, S.; Zou, Y.; Basu, A.K. (5′S)-8,5′-cyclo-2′-deoxyguanosine is a strong block to replication, a potent pol V-dependent mutagenic lesion, and is inefficiently repaired in Escherichia coli. *Biochemistry* **2011**, *50*, 3862–3865. [CrossRef]
78. Pednekar, V.; Weerasooriya, S.; Jasti, V.P.; Basu, A.K. Mutagenicity and genotoxicity of (5′S)-8,5′-cyclo-2′-deoxyadenosine in Escherichia coli and replication of (5′S)-8,5′-cyclopurine-2′-deoxynucleosides in vitro by DNA polymerase IV, exo-free Klenow fragment, and Dpo4. *Chem. Res. Toxicol.* **2014**, *27*, 200–210. [CrossRef] [PubMed]
79. Xu, W.; Ouellette, A.M.; Wawrzak, Z.; Shriver, S.J.; Anderson, S.M.; Zhao, L. Kinetic and structural mechanisms of (5′*S*)-8,5′-cyclo-2′-deoxyguanosine-induced DNA replication stalling. *Biochemistry* **2015**, *54*, 639–651. [CrossRef] [PubMed]
80. Swanson, A.L.; Wang, J.; Wang, Y. Accurate and efficient bypass of 8,5′-cyclopurine-2′-deoxynucleosides by human and yeast DNA polymerase η. *Chem. Res. Toxicol.* **2012**, *25*, 1682–1691. [CrossRef] [PubMed]
81. Weng, P.J.; Gao, Y.; Gregory, M.T.; Wang, P.; Wang, Y.; Yang, W. Bypassing a 8,5′-cyclo-2′-deoxyadenosine lesion by human DNA polymerase η at atomic resolution. *Proc. Natl. Acad. Sci. USA* **2018**, *115*, 10660–10665. [CrossRef] [PubMed]
82. Beard, W.A.; Wilson, S.H. Structure and mechanism of DNA polymerase β. *Chem. Rev.* **2006**, *106*, 361–382. [CrossRef] [PubMed]

83. Beard, W.A.; Wilson, S.H. Structure and mechanism of DNA polymerase β. *Biochemistry* **2014**, *53*, 2768–2780. [CrossRef]
84. Jiang, Z.; Xu, M.; Lai, Y.; Laverde, E.E.; Terzidis, M.A.; Masi, A.; Chatgilialoglu, C.; Liu, Y. Bypass of a 5′,8-cyclopurine-2′-deoxynucleoside by DNA polymerase beta during DNA replication and base excision repair leads to nucleotide misinsertions and DNA strand breaks. *DNA Repair* **2015**, *33*, 24–34. [CrossRef]
85. Xu, M.; Gabison, J.; Liu, Y. Trinucleotide repeat deletion via a unique hairpin bypass by DNA polymerase beta and alternate flap cleavage by flap endonuclease 1. *Nucl. Acids Res.* **2013**, *41*, 1684–1697. [CrossRef]
86. Lai, Y.; Budworth, H.; Beaver, J.M.; Chan, N.L.; Zhang, Z.; McMurray, C.T.; Liu, Y. Crosstalk between MSH2-MSH3 and polβ promotes trinucleotide repeat expansion during base excision repair. *Nat. Commun.* **2016**, *7*, 12465. [CrossRef]
87. Gonzalez-Hunt, C.P.; Wadhwa, M.; Sanders, L.H. DNA damage by oxidative stress: Measurement strategies for two genomes. *Curr. Opin. Toxicol.* **2018**, *7*, 87–94. [CrossRef]
88. Cooke, M.S.; Evans, M.D.; Dizdaroglu, M.; Lunec, J. Oxidative DNA damage: Mechanisms, mutation, and disease. *FASEB J.* **2003**, *17*, 1195–1214. [CrossRef] [PubMed]
89. Yu, Y.; Cui, Y.; Niedernhofer, L.J.; Wang, Y. Occurrence, biological consequences, and human health relevance of oxidative stress-induced DNA damage. *Chem. Res. Toxicol.* **2016**, *29*, 2008–2039. [CrossRef] [PubMed]
90. Randerath, K.; Zhou, G.D.; Somers, R.L.; Robbins, J.H.; Brooks, P.J. A ^{32}P-postlabeling assay for the oxidative DNA lesion 8,5′-cyclo-2′-deoxyadenosine in mammalian tissues: Evidence that four type II I-compounds are dinucleotides containing the lesion in the 3′ nucleotide. *J. Biol. Chem.* **2001**, *276*, 36051–36057. [CrossRef] [PubMed]
91. Zhou, G.D.; Moorthy, B. Detection of bulky endogenous oxidative DNA lesions derived from 8,5′-cyclo-2′-deoxyadenosine by ^{32}P-postlabeling assay. *Curr. Protoc. Toxicol.* **2015**, *64*, 17.17.1–17.17.14. [PubMed]
92. Reddy, M.V.; Randerath, K. Nuclease P1-mediated enhancement of sensitivity of ^{32}P-postlabeling test for structurally diverse DNA adducts. *Carcinogenesis* **1986**, *7*, 1543–1551. [CrossRef] [PubMed]
93. Gupta, R.C.; Arif, J.M. An improved (32)P-postlabeling assay for the sensitive detection of 8-oxodeoxyguanosine in tissue DNA. *Chem. Res. Toxicol.* **2001**, *14*, 951–957. [CrossRef] [PubMed]
94. Zhou, G.D.; Randerath, K.; Donnelly, K.C.; Jaiswal, A.K. Effects of NQO1 deficiency on levels of cyclopurines and other oxidative DNA lesions in liver and kidney of young mice. *Int. J. Cancer* **2004**, *112*, 877–883. [CrossRef] [PubMed]
95. Longo, J.A.; Nevaldine, B.; Longo, S.L.; Winfield, J.A.; Hahn, P.J. An assay for quantifying DNA double-strand break repair that is suitable for small numbers of unlabeled cells. *Radiat. Res.* **1997**, *147*, 35–40. [CrossRef]
96. Okahashi, Y.T.; Iwamoto, N.; Suzuki, S.; Shibutani, S.; Sugiura, S.; Itoh, T.; Nishiwaki, S.; Ueno, T.; Mori, T. Quantitative detection of 4-hydroxyequilenin-DNA adducts in mammalian cells using an immunoassay with a novel monoclonal antibody. *Nucl. Acids Res.* **2010**, *38*, e133. [CrossRef] [PubMed]
97. Cooke, M.S.; Olinski, R.; Loft, S. European standards committee on urinary (DNA) lesion analysis. *Cancer Epidemiol. Biomarkers Prev.* **2008**, *17*, 3–14. [CrossRef]
98. Iwamoto, T.; Brooks, P.J.; Nishiwaki, T.; Nishimura, K.; Kobayashi, N.; Sugiura, S.; Mori, T. Quantitative and in situ detection of oxidatively generated DNA damage 8,5′-cyclo-2′-deoxyadenosine using an immunoassay with a novel monoclonal antibody. *Photochem. Photobiol.* **2014**, *90*, 829–836. [PubMed]
99. Mori, T.; Nakane, H.; Iwamoto, T.; Krokidis, M.G.; Chatgilialoglu, C.; Tanaka, K.; Kaidoh, T.; Hasegawa, M.; Sugiura, S. High levels of oxidatively generated DNA damage 8,5′-cyclo-2′-deoxyadenosine accumulate in the brain tissues of xeroderma pigmentosum group A gene knockout mice. *DNA Repair* **2019**, in press. [CrossRef]
100. Guo, C.; Li, X.; Wang, R.; Yu, J.; Ye, M.; Mao, L.; Zhang, S.; Zheng, S. Association between oxidative DNA damage and risk of colorectal cancer: Sensitive determination of urinary 8-hydroxy-2′-deoxyguanosine by UPLC–MS/MS analysis. *Sci. Rep.* **2016**, *6*, 32581. [CrossRef] [PubMed]
101. Liu, S.; Wang, Y. Mass spectrometry for the assessment of the occurrence and biological consequences of DNA adducts. *Chem. Soc. Rev.* **2015**, *44*, 7829–7854. [CrossRef] [PubMed]
102. Jaruga, P.; Theruvathu, J.; Dizdaroglu, M.; Brooks, P.J. Complete release of (5′*S*)-8,5′-cyclo-2′-deoxyadenosine from dinucleotides, oligodeoxynucleotides and DNA, and direct comparison of its levels in cellular DNA with other oxidatively induced DNA lesions. *Nucl. Acids Res.* **2004**, *32*, e87. [CrossRef] [PubMed]
103. Dizdaroglu, M.; Coskun, E.; Jaruga, P. Measurement of oxidatively induced DNA damage and its repair, by mass spectrometric techniques. *Free Radic. Res.* **2015**, *49*, 525–548. [CrossRef] [PubMed]

104. Terzidis, M.A.; Ferreri, C.; Chatgilialoglu, C. Radiation-induced formation of purine lesions in single and double stranded DNA: Revised quantification. *Front. Chem.* **2015**, *3*, 18. [CrossRef]
105. Belmadoui, N.; Boussicault, F.; Guerra, M.; Ravanat, J.-L.; Chatgilialoglu, C.; Cadet, J. Radiation-induced formation of purine 5′,8-cyclonucleosides in isolated and cellular DNA: High stereospecificity and modulating effect of oxygen. *Org. Biomol. Chem.* **2010**, *8*, 3211–3219. [CrossRef]
106. Jaruga, P.; Xiao, Y.; Nelson, B.C.; Dizdaroglu, M. Measurement of (5′R)- and (5′S)-8,5′-cyclo-2′-deoxyadenosines in DNA in vivo by liquid chromatography/isotope-dilution tandem mass spectrometry. *Biochem. Biophys. Res. Commun.* **2009**, *386*, 656–660. [CrossRef]
107. Chatgilialoglu, C.; Krokidis, M.G.; Papadopoulos, K.; Terzidis, M.A. Purine 5′,8-cyclo-2′-deoxynucleoside lesions in irradiated DNA. *Radiat. Phys. Chem.* **2016**, *128*, 75–81. [CrossRef]
108. Cleaver, J.E.; Lam, E.T.; Revet, I. Disorders of nucleotide excision repair: The genetic and molecular basis of heterogeneity. *Nat. Rev. Genet.* **2009**, *10*, 756–768. [CrossRef] [PubMed]
109. Di Giovanna, J.J.; Kraemer, K.H. Shining a light on xeroderma pigmentosum. *J. Invest. Dermatol.* **2012**, *132*, 785–796. [CrossRef]
110. Brooks, P.J. The Case for 8,5′-Cyclopurine-2′-Deoxynucleosides as Endogenous DNA Lesions That Cause Neurodegeneration in Xeroderma Pigmentosum. *Neuroscience* **2007**, *145*, 1407–1417. [CrossRef] [PubMed]
111. Hirai, Y.; Kodama, Y.; Moriwaki, S.; Noda, A.; Cullings, H.M.; Macphee, D.G.; Kodama, K.; Mabuchi, K.; Kraemer, K.H.; Land, C.E.; et al. Heterozygous individuals bearing a founder mutation in the XPA DNA repair gene comprise nearly 1% of the Japanese population. *Mutat. Res.* **2006**, *601*, 171–178. [CrossRef] [PubMed]
112. Satokata, I.; Tanaka, K.; Miura, N.; Miyamoto, I.; Satoh, Y.; Kondo, S.; Okada, Y. Characterization of a splicing mutation in group A xeroderma pigmentosum. *Proc. Natl. Acad. Sci. USA* **1990**, *87*, 9908–9912. [CrossRef] [PubMed]
113. Robbins, J.H.; Brumback, R.A.; Mendiones, M.; Barrett, S.F.; Carl, J.R.; Cho, S.; Denckla, M.B.; Ganges, M.B.; Gerber, L.H.; Guthrie, R.A.; et al. Neurological disease in xeroderma pigmentosum. Documentation of a late onset type of the juvenile onset form. *Brain* **1991**, *114*, 1335–1361. [CrossRef] [PubMed]
114. Andrews, A.D.; Barrett, S.F.; Robbins, J.H. Xeroderma pigmentosum neurological abnormalities correlate with colony-forming ability after ultraviolet radiation. *Proc. Natl. Acad. Sci. USA* **1978**, *75*, 1984–1988. [CrossRef]
115. Robbins, J.H. A childhood neurodegeneration due to defective DNA repair: A novel concept of disease based on studies xeroderma pigmentosum. *J. Child Neurol.* **1989**, *4*, 143–146. [CrossRef]
116. Theruvathu, J.A.; Jaruga, P.; Dizdaroglu, M.; Brooks, P.J. The oxidatively induced DNA lesions 8,5′-cyclo-2′-deoxyadenosine and 8-hydroxy-2′-deoxyadenosine are strongly resistant to acid-induced hydrolysis of the glycosidic bond. *Mech. Ageing Dev.* **2007**, *128*, 494–502. [CrossRef]
117. Das, R.S.; Samaraweera, M.; Morton, M.; Gascón, J.A.; Basu, A.K. Stability of N-glycosidic bond of (5′S)-8,5′-cyclo-2′-deoxyguanosine. *Chem. Res. Toxicol.* **2012**, *25*, 2451–2461. [CrossRef]
118. Wang, J.; Clauson, C.L.; Robbins, P.D.; Niedernhofer, L.J.; Wang, Y. The oxidative DNA lesions 8,5′-cyclopurines accumulate with aging in a tissue-specific manner. *Aging Cell* **2012**, *11*, 714–716. [CrossRef]
119. Brooks, P.J. The cyclopurine deoxynucleosides: DNA repair, biological effects, mechanistic insights, and unanswered questions. *Free Radic. Biol. Med.* **2017**, *107*, 90–100. [CrossRef] [PubMed]
120. You, C.; Dai, X.; Yuan, B.; Wang, J.; Wang, J.; Brooks, P.J.; Niedernhofer, L.J.; Wang, Y. A quantitative assay for assessing the effects of DNA lesions on transcription. *Nat. Chem. Biol.* **2012**, *8*, 817–822. [CrossRef] [PubMed]
121. Marietta, C.; Brooks, P.J. Transcriptional bypass of bulky DNA lesions causes new mutant RNA transcripts in human cells. *EMBO Rep.* **2007**, *8*, 388–393. [CrossRef] [PubMed]
122. Abraham, J.; Brooks, J.P. Divergent effects of oxidatively induced modification to the C8 of 2′-deoxyadenosine on transcription factor binding: 8,5′(s)-cyclo-2′-deoxyadenosine inhibits the binding of multiple sequence specific transcription factors, while 8-oxo-2′-deoxyadenosine increases binding of CREB and NF-kappa B to DNA. *Environ. Mol. Mutagen.* **2011**, *52*, 287–295. [PubMed]
123. Arczewska, K.D.; Tomazella, G.G.; Lindvall, J.M.; Kassahun, H.; Maglioni, S.; Torgovnick, A.; Henriksson, J.; Matilainen, O.; Marquis, B.J.; Nelson, B.C.; et al. Active transcriptomic and proteomic reprogramming in the C. elegans nucleotide excision repair mutant xpa-1. *Nucl. Acids Res.* **2013**, *41*, 5368–5381. [CrossRef] [PubMed]

124. D'Errico, M.; Pascucci, B.; Iorio, E.; Van Houten, B.; Dogliotti, E. The role of CSA and CSB protein in the oxidative stress response. *Mech. Ageing Dev.* **2013**, *134*, 261–269. [CrossRef]
125. Feichtinger, R.G.; Sperl, W.; Bauer, J.W.; Kofler, B. Mitochondrial dysfunction: A neglected component of skin diseases. *Exp. Dermatol.* **2014**, *23*, 607–614. [CrossRef]
126. Fang, E.F.; Scheibye-Knudsen, M.; Brace, L.E.; Kassahun, H.; SenGupta, T.; Nilsen, H.; Mitchell, J.R.; Croteau, D.L.; Bohr, V.A. Defective mitophagy in XPA via PARP-1 hyperactivation and NAD(+)/SIRT1 reduction. *Cell* **2014**, *157*, 882–896. [CrossRef]
127. Itoh, M.; Hayashi, M.; Shioda, K.; Minagawa, M.; Isa, F.; Tamagawa, K.; Morimatsu, Y.; Oda, M. Neurodegeneration in hereditary nucleotide repair disorders. *Brain Dev.* **1999**, *21*, 326–333. [CrossRef]
128. D'Errico, M.; Parlanti, E.; Teson, M.; Degan, P.; Lemma, T.; Calcagnile, A.; Iavarone, I.; Jaruga, P.; Ropolo, M.; Pedrini, A.M.; et al. The role of CSA in the response to oxidative DNA damage in human cells. *Oncogene* **2007**, *26*, 4336–4343. [CrossRef] [PubMed]
129. Brooks, P.J. Blinded by the UV light: How the focus on transcription-coupled NER has distracted from understanding the mechanisms of Cockayne syndrome neurologic disease. *DNA Repair* **2013**, *12*, 656–671. [CrossRef] [PubMed]
130. Dizdaroglu, M. Oxidatively induced DNA damage and its repair in cancer. *Mutat. Res.* **2015**, *763*, 212–245. [CrossRef] [PubMed]
131. Dedon, P.C.; Tannenbaum, S.R. Reactive nitrogen species in the chemical biology of inflammation. *Arch. Biochem. Biophys.* **2004**, *423*, 12–22. [CrossRef] [PubMed]
132. Helleday, T.; Eshtad, S.; Nik-Zainal, S. Mechanisms underlying mutational signatures in human cancers. *Nat. Rev. Genet.* **2014**, *15*, 585–598. [CrossRef] [PubMed]
133. Krokidis, M.G.; Terzidis, M.A.; Efthimiadou, E.; Zervou, S.K.; Kordas, G.; Papadopoulos, K.; Hiskia, A.; Kletsas, D.; Chatgilialoglu, C. Purine 5',8-cyclo-2'-deoxynucleoside lesions: Formation by radical stress and repair in human breast epithelial cancer cells. *Free Radic. Res.* **2017**, *51*, 470–482. [CrossRef]
134. Matta, J.; Morales, L.; Ortiz, C.; Adams, D.; Vargas, W.; Casbas, P.; Dutil, J.; Echenique, M.; Suárez, E. Estrogen receptor expression is associated with DNA repair capacity in breast cancer. *PLoS ONE* **2016**, *11*, e0152422. [CrossRef] [PubMed]
135. Acu, I.D.; Liu, T.; Suino-Powell, K.; Mooney, S.M.; D'Assoro, A.B.; Rowland, N.; Muotri, A.R.; Correa, R.G.; Niu, Y.; Kumar, R.; et al. Coordination of centrosome homeostasis and DNA repair is intact in MCF-7 and disrupted in MDA-MB 231 breast cancer cells. *Cancer Res.* **2010**, *70*, 3320–3328. [CrossRef]
136. Nyaga, S.G.; Jaruga, P.; Lohani, A.; Dizdaroglu, M.; Evans, M.K. Accumulation of oxidatively induced DNA damage in human breast cancer cell lines following treatment with hydrogen peroxide. *Cell Cycle* **2007**, *6*, 1472–1478. [CrossRef]
137. Rodriguez, H.; Jaruga, P.; Leber, D.; Nyaga, S.G.; Evans, M.K.; Dizdaroglu, M. Lymphoblasts of women with BRCA1 mutations are deficient in cellular repair of 8,5'- Cyclopurine-2'-deoxynucleosides and 8-hydroxy-2'-deoxyguanosine. *Biochemistry* **2007**, *46*, 2488–2496. [CrossRef]
138. Egler, R.A.; Fernandes, E.; Rothermund, K.; Sereika, S.; de Souza-Pinto, N.; Jaruga, P.; Dizdaroglu, M.; Prochownik, E.V. Regulation of reactive oxygen species, DNA damage, and c-Myc function by peroxiredoxin 1. *Oncogene* **2005**, *24*, 8038–8050. [CrossRef] [PubMed]
139. Mitra, D.; Luo, X.; Morgan, A.; Wang, J.; Hoang, M.P.; Lo, J.; Guerrero, C.R.; Lennerz, J.K.; Mihm, M.C.; Wargo, J.A.; et al. An ultraviolet-radiation-independent pathway to melanoma carcinogenesis in the red hair/fair skin background. *Nature* **2012**, *491*, 449–453. [CrossRef]
140. Krokidis, M.; Louka, M.; Efthimiadou, E.; Zervou, S.-K.; Papadopoulos, K.; Hiskia, A.; Ferreri, C.; Chatgilialoglu, C. Membrane lipidome reorganization and accumulation of tissue DNA lesions in tumor-bearing mice: An exploratory study. *Cancers* **2019**, *11*, 480. [CrossRef] [PubMed]
141. De Magahaes, J.P. How ageing processes influence cancer. *Nat. Rev. Cancer* **2013**, *13*, 357–365. [CrossRef] [PubMed]
142. Robinson, A.R.; Yousefzadeh, M.J.; Rozgaja, T.A.; Wang, J.; Li, X.; Tilstra, J.S.; Feldman, C.H.; Gregg, S.Q.; Johnson, C.H.; Skoda, E.M.; et al. Spontaneous DNA damage to the nuclear genome promotes senescence, redox imbalance and aging. *Redox Biol.* **2018**, *17*, 259–273. [CrossRef] [PubMed]
143. Yu, Y.; Guerrero, C.R.; Liu, S.; Amato, N.J.; Sharma, Y.; Gupta, S.; Wang, Y. Comprehensive assessment of oxidatively induced modifications of DNA in a rat model of human Wilson's disease. *Mol. Cell. Proteomics* **2016**, *15*, 810–817. [CrossRef] [PubMed]

144. Jaruga, P.; Xiao, Y.; Vartanian, V.; Lloyd, R.S.; Dizdaroglu, M. Evidence for the involvement of DNA repair enzyme NEIL1 in nucleotide excision repair of (5′R)- and (5′S)-8, 5′-cyclo-2′-deoxyadenosines. *Biochemistry* **2010**, *49*, 1053–1055. [CrossRef]
145. Kirkali, G.; de Souza-Pinto, N.C.; Jaruga, P.; Bohr, V.A.; Dizdaroglu, M. Accumulation of (5′S)-8,5′-cyclo-2′-deoxyadenosine in organs of Cockayne syndrome complementation group B gene knockout mice. *DNA Repair* **2009**, *8*, 274–278. [CrossRef]
146. Kirkali, G.; Tunca, M.; Genc, S.; Jaruga, P.; Dizdaroglu, M. Oxidative DNA damage in polymorphonuclear leukocytes of patients with familial Mediterranean fever. *Free Radic. Biol. Med.* **2008**, *44*, 386–393. [CrossRef]
147. Kant, M.; Akış, M.; Çalan, M.; Arkan, T.; Bayraktar, F.; Dizdaroglu, M.; İşlekel, H. Elevated urinary levels of 8-oxo-2′-deoxyguanosine, (5′R)- and (5′S)-8,5′-cyclo-2′-deoxyadenosines, and 8-iso-prostaglandin F2α as potential biomarkers of oxidative stress in patients with prediabetes. *DNA Repair* **2016**, *48*, 1–7. [CrossRef]
148. Cadet, J.; Di Mascio, P.; Wagner, J.R. Radiation-induced (5′R)- and (5′S)-purine 5′,8-cyclo-2′-deoxyribon ucleosides in human cells: A revisited analysis of HPLC–MS/MS measurements. *Free Radic. Res.* **2019**. [CrossRef] [PubMed]

8

VAS2870 Inhibits Histamine-Induced Calcium Signaling and vWF Secretion in Human Umbilical Vein Endothelial Cells

Pavel V. Avdonin [1,*], Elena Yu. Rybakova [1], Piotr P. Avdonin [1], Sergei K. Trufanov [1], Galina Yu. Mironova [1], Alexandra A. Tsitrina [1] and Nikolay V. Goncharov [2,3]

1 Koltsov Institute of Developmental Biology, Moscow 119334, Russia; alenka3107@mail.ru (E.Y.R.); ppavdonin@gmail.com (P.P.A.); gad.91@inbox.ru (S.K.T.); wereshelen@gmail.com (G.Y.M.); sashulka.s@gmail.com (A.A.T.)

2 Sechenov Institute of Evolutional Institute of Evolutionary Physiology and Biochemistry, Saint Petersburg 194223, Russia; ngoncharov@gmail.com

3 Research Institute of Hygiene, Occupational Pathology and Human Ecology, Leningrad Region 188663, Russia

* Correspondence: pvavdonin@yandex.ru

Abstract: In this study, we investigated the effects of NAD(P)H oxidase (NOX) inhibitor VAS2870 (3-benzyl-7-(2-benzoxazolyl)thio-1,2,3-triazolo[4,5-d]pyrimidine) on the histamine-induced elevation of free cytoplasmic calcium concentration ($[Ca^{2+}]_i$) and the secretion of von Willebrand factor (vWF) in human umbilical vein endothelial cells (HUVECs) and on relaxation of rat aorta in response to histamine. At 10 μM concentration, VAS2870 suppressed the $[Ca^{2+}]_i$ rise induced by histamine. Inhibition was not competitive, with IC50 3.64 and 3.22 μM at 1 and 100 μM concentrations of histamine, respectively. There was no inhibition of $[Ca^{2+}]_i$ elevation by VAS2870 in HUVECs in response to the agonist of type 1 protease-activated receptor SFLLRN. VAS2870 attenuated histamine-induced secretion of vWF and did not inhibit basal secretion. VAS2870 did not change the degree of histamine-induced relaxation of rat aortic rings constricted by norepinephrine. We suggest that NOX inhibitors might be used as a tool for preventing thrombosis induced by histamine release from mast cells without affecting vasorelaxation.

Keywords: histamine; calcium; endothelial cells; NADPH-oxidase; VAS2870; von Willebrand factor; aorta; relaxation

---

1. Introduction

Histamine plays an important role as chemical mediator in multiple physiological and pathophysiological processes in central and peripheral tissues. Mast cells and basophils are important sources of histamine, which is released from granule stores in response to several stimuli. The pleiotropic effects of histamine are mediated via four G protein-coupled receptor (GPCR) subtypes (H_1R–H_4R), which differ in their distribution, ligand binding, signaling pathways, and functions [1]. Histamine acts as a full agonist on the receptors, with subtype-specific differences in affinity. H_1R is ubiquitously expressed and mediates its effects by $G_{q/11}$ activation via phospholipase C (PLC) and to a minor degree by PLA_2 and PLD. Activation of H_1R leads to an increase of inositol-1,4,5-triphosphate (IP_3) and 1,2-diacylglycerol (DAG), associated with an increase of the intracellular Ca^{2+} concentration, followed by activation of protein kinase C (PKC). Other signaling pathways include the stimulation of adenylyl cyclase via H_2R with the formation of cAMP, inducing the generation of NO; the NF-κB cascade leads to a release or increase in the expression of (pro)inflammatory mediators (P-selectin, ICAM-1, VCAM-1, iNOS, IL-1beta, IL-6, TNF-alpha, etc.) [2]. Brain vascular endothelial cells (ECs) express histamine H1 and H2 receptors, which regulate brain capillary permeability. Also, histamine receptor genes Hrh3 and Hrh4, for corresponding H3 and H4 receptors, are expressed in rat brain ECs, which are potentially important for the regulation of blood–brain barrier permeability, including trafficking of immunocompetent cells [3]. H4 receptors play a dominant role in histamine-induced eosinophil adhesion to endothelium [4].

Reactive oxygen species (ROS) in large amounts clearly have detrimental effects on cell physiology, whereas low concentrations of ROS are permanently produced in cells and play a role as signaling molecules. An imbalance in ROS production and defense mechanisms can lead to pathological vascular remodeling. Among the possible sources of ROS within endothelial and/or neighboring cells are mitochondria, NADPH-oxidases, xanthine oxidase, peroxidases, NO-synthases, cytochrome P450, cyclooxygenases, lipoxygenases, monoamine oxidases, and the hemoglobin of red blood cells [5–7]. Among these possible sources of ROS, nicotinamide adenine dinucleotide phosphate (NADPH) oxidases (NOXs) play a central role. NOXs are present in neutrophils, where they mostly determine the non-specific immune response [8]. In the ischemic-reperfusion condition, NOXs substantially determine tissue injury by generating ROS in ECs and smooth muscle cells (SMCs) of blood vessels [9].

In ECs, four NOX isoforms include the superoxide-generating enzymes NOX1, NOX2, and NOX5 and the hydrogen-peroxide-generating enzyme NOX4 [10]. NOX2 is the most studied isoform of NOX, which generates superoxide anion in professional phagocytes and many other cells. In ECs, it is the predominant form, and all four subunits of NOX2 are present [11,12]. It is activated by angiotensin II [13], oxidized low-density lipoproteins [14], cytokines, and growth factors [15]. Superoxide anion generated by NOX2 causes the inactivation of nitric oxide and promotes the development of endothelial dysfunction, hypertension, and inflammation [16]. NOX4 predominantly generates hydrogen peroxide (H_2O_2). H_2O_2 produced by endothelial NOX4 potentiates vasorelaxation induced by acetylcholine and histamine [17]. Activators of other NOXs do not affect the activity of NOX4, which is constitutive, since the generation of H_2O_2 in cells is dependent on the level of expression of NOX4 [18]. NADPH oxidases of the DUOX group (DUOX1 and DUOX2) also generate H_2O_2 [19,20], though their existence and/or functional significance in ECs and SMCs has not been clearly shown.

Calcium-dependent NOX5 has been implicated in the oxidative damage of ECs in atherosclerosis and hypertension [21]. It is noteworthy that NOX5 is expressed in primates but does not occur naturally in rodents. In transgenic mice expressing human NOX5 in a vascular SMC-specific manner, agonist-induced vasoconstriction was exaggerated, and endothelium-dependent vasorelaxation was attenuated [22].

The influence of ROS on Ca^{2+} signaling in EC has been demonstrated in several works [23–30]. ROS can activate various calcium channels of the endoplasmic reticulum and plasma membrane, such as InsP3- and ryanodine-sensitive channels, and some cation channels of the TRP superfamily [25,31,32]. NOX-derived ROS are critical for the generation of Ca^{2+} oscillations in response to histamine in human aortic endothelial cells [33]. Recently, we demonstrated an involvement of two-pore channels in an H_2O_2-induced increase in the level of calcium ions in the cytoplasm of human umbilical vein endothelial cells (HUVECs) [34]. This could be coupled with exocytosis of the von Willebrand factor (vWF) in these cells in response to H_2O_2 [35]. The vWF release from EC is induced by superoxide anion [30]. In the present work, we studied the role of NOX in histamine-induced Ca^{2+} rise and vWF secretion in HUVECs and NOX involvement in rat aorta relaxation in response to histamine. To solve these questions, we used VAS2870 as a tool inhibitor of NOXs [36]. VAS2870 belongs to the triazolopyrimidines, which are regarded as the most specific inhibitors of NOXs [37].

2. Methods

2.1. Reagents

VAS2870, histamine, and SFLLRN were from SigmaAldrich (St. Louis, MO, USA); CalciumGreen/AM and dihydroethidium were from Thermo Fischer Scientific (Waltham, MA, USA). VAS2870 was dissolved at a 10 mM concentration in DMSO. Before being added to the cells, it was dissolved to the required concentrations in physiological salt solution (PSS). DMSO at appropriate concentrations was used as a vehicle control. PSS was added as a vehicle control for histamine and SFLLRN.

2.2. Cell Culture

Human umbilical vein endothelial cells (HUVECs) were isolated according to [38]. The cells were grown in plastic dishes pre-coated with gelatin, using M199 medium with Earl's salts and 20 mM HEPES containing 20% fetal calf serum (SigmaAldrich), 300 μg/mL endothelial growth supplement, isolated from rabbit brain, 100 μg/mL heparin and gentamicin. We used the cells on early passages (1–4). Accutase® was applied for passaging the cells (SigmaAldrich).

2.3. Measurement of Free Cytoplasmic Calcium Concentration in HUVECs

HUVECs grown in 96-well plates were loaded with 1 μM CalciumGreen/AM dissolved with 0.02% Pluronic F-127 in M199 during 1 h at 37 °C in a CO_2 incubator (New Brunswick Scientific, Edison, NJ, USA). Measurement of $[Ca^{2+}]_i$ was performed in physiological salt solution (PSS) containing NaCl (145 mM), KCl (5 mM), $MgCl_2$ (1 mM), $CaCl_2$ (1 mM), HEPES (5 mM), and D-glucose (10 mM), at pH 7.4. Fluorescence was registered at 485 nm (excitation) and 530 nm (emission) at 25 °C using a Synergy 4 Microplate Reader (BioTek, Winooski, VT, USA). The changes in $[Ca^{2+}]_i$ in HUVECs are presented as the ratio of the increment in F_{530} and initial F_{530}. Each curve in in the graph is a superposition of three curves recorded simultaneously from three wells in a plate.

2.4. Registration of ROS Generation in HUVECs

Kinetics of ROS generation in HUVECs was registered with the fluorescent indicator dihydroethidium. The fluorescence was measured by a Synergy 4 Microplate Reader (BioTek) with excitation filter 485 nm (bandwidth 20 nm) and emission 620 nm (bandwidth 40 nm). The bandwidth of the emission filter is enough wide to cover a large part of the emission of the oxidation products of DHE [39]. The cells grown in 96-well plates were incubated with different concentrations of VAS2870 or vehicle control (dimethylsulfoxide at final concentrations from 0.006% to 0.2%) during 5 min and then DHE at final concentration 2.5 μM was added. The slight inhibitory effect of DMSO on DHE oxidation at a concentration of 0.2% was taken into account when analyzing the data. The oxidation kinetics of DHE determined by the increase in fluorescence was linear during the 60 min of incubation both in the absence and in presence of VAS2870. The increase in fluorescence in the absence of VAS2870 was taken as 100%. The results are presented as the mean values obtained in three experiments with different cell preparations. In each experiment, the experimental point was a mean value of the fluorescence from six wells in a 96-well plate.

2.5. Measurement of vWF Secretion

HUVECs grown in 48-well plates were incubated in PSS at 30 °C during 5 min with or without 10 μM VAS2870. Then, histamine at final concentration or vehicle were added and the cells were incubated for 30 min. The extracellular fluid from each well was collected in a separate tube and frozen. Later, it was used for the measurement of vWF concentration with a TECHNOZYM vWF:Ag ELISA kit (Technoclone, Vienna, Austria). The values of the secreted vWF are the means of six measurements.

2.6. *Registration of Aorta Contraction*

Aorta were isolated from male Wistar rats weighing 250–300 g. The rats were anesthetized with 25% urethane (4 mL/kg) and decapitated. All manipulations with the animals were performed in accordance with the guide for the care and use of Laboratory animals of the Bioethics committee of the Koltsov Institute of Developmental Biology and European Convention for the Protection of Vertebrate Animals used for Experimental and Other Scientific Purposes. Aorta were cleaned from connective tissue and cut into rings with a width of around 2 mm. Experiments were performed on a four-channel wire myograph (ADInstruments, Bella Vista, New South Wales, Australia) using LabChart 7.3.7 program for data acquisition and analysis. The rings were mounted on the holders in chambers filled with Krebs–Henseleit solution (37°C) perfused with 95% O_2/5% CO_2 and extended with a force of 1 g. Contractility of the vessel rings and intactness of its endothelium were tested by adding 10^{-7} M norepinephrine and then 10^{-5} M carbachol. After washing, norepinephrine and carbachol were added again. Aortic rings with relaxation of at least 50% were used for the experiments.

2.7. *Statistics*

Data are presented as mean ± SEM. The number of measurements is presented in the legend to the figures. In each case at least three independent experiments were performed with different cell or vessel preparations. Statistical significance was calculated using Excel 2003 and MedCalc 18.9.1 statistical software according the Student-Neuman-Keuls test. The IC50 values were determined with GraphPad Prism 8.0.

3. Results and Discussion

We studied the effect of VAS2870 on $[Ca^{2+}]_i$ elevation induced by histamine. At a concentration of 10 μM, VAS2870 almost completely suppressed the calcium signal of 1 mM histamine, while there was no significant inhibition of the action of PAR1 agonist SFLLRN (Figure 1). The effect of VAS2870 was not competitive toward histamine, since wat is observed at a low concentration of VAS2870, which is two orders of magnitude lower than the concentration of histamine. We determined the concentration-dependence curves for the inhibition by VAS2870 of $[Ca^{2+}]_i$ elevation at different histamine concentrations (Figure 2c). The IC50 for VAS2870 calculated from these data was 3.64 and 3.22 μM at 1 and 100 μM concentrations of histamine, respectively. These results suggest the existence of a specific mechanism of histamine-induced $[Ca^{2+}]_i$ mobilization in HUVECs. The following data provide evidence in favor of this conclusion. It has been shown that two-pore channels activated by NAADP are involved in histamine-induced $[Ca^{2+}]_i$ mobilization in HUVEC via H1 receptors, while calcium signaling of thrombin is independent from NAADP [40,41].

VAS2870 in micromolar concentrations is widely used to suppress NOX activity in different types of cells [42–45]. The IC50 for VAS2870's effect on NOX activity in a cell-free system with membranes and cytosol from human neutrophils was 10.6 μM, and in intact HL-60 cells the IC50 was 2 μM [43]. However, according to Gatto et al. [46], in neutrophils, ROS generation is inhibited by VAS2870 at submicromolar concentrations with an IC50 of 77 nM. We determined which concentration of VAS2870 suppresses ROS production in HUVECs. For this purpose, dihydroethidium (DHE) was used. VAS2870 reduced the rate of DHE oxidation in HUVECs with an IC50 of 2.47 μM (Figure 2d). At a 20 μM concentration, VAS2870 suppressed nearly half of ROS formation measured with DHE, and its effect reached a plateau. The reason for this might be that oxidation of DHE in cells is caused only partially by superoxide anion [39]. The values of IC50 for the inhibition of ROS generation and suppression of histamine-induced $[Ca^{2+}]_i$ elevation in HUVECs were very close, which suggests a link between these processes. VAS2870 is an inhibitor of NADPH-oxidase isoforms NOX1/2/4 [47,48] and does not suppress NOX5 [22]. It has been demonstrated that VAS2870 reverses oxidative stress, which is caused by NOX1/2 activation [49]. We propose that the suppression of histamine-induced $[Ca^{2+}]_i$ increase in HUVECs might occur due to the inhibition of NOX2 because this form of NOX is under the control of several intracellular regulatory pathways [9], while the activity of NOX4 is mostly regulated by the level of expression. In addition, it has been shown that increased expression of NOX4 in endothelial cells enhances endothelium-dependent relaxation [17].

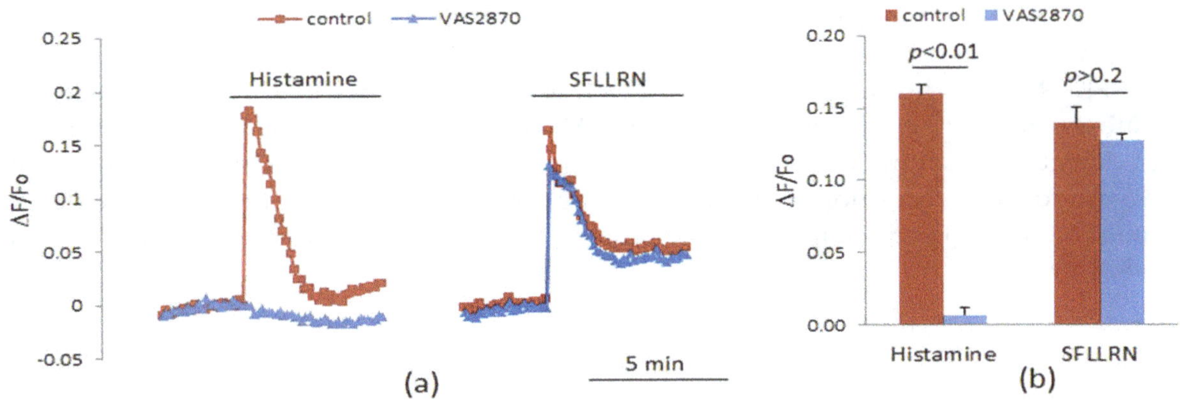

Figure 1. The influence of VAS2870 (10 μM) on calcium signaling in response to histamine (10 μM) and agonist of PAR1 SFLLRN (2 μg/mL). (**a**) The curves represent the data of one of three experiments with similar results; (**b**) The bars represent the mean values of $[Ca^{2+}]_i$ increase in three independent experiments. $[Ca^{2+}]_i$ was measured as a relative increase in CalciumGreen fluorescence. In control experiments, the vehicle, dimethylsulfoxide (DMSO) at a final concentration 0.1%, was added instead of VAS2870.

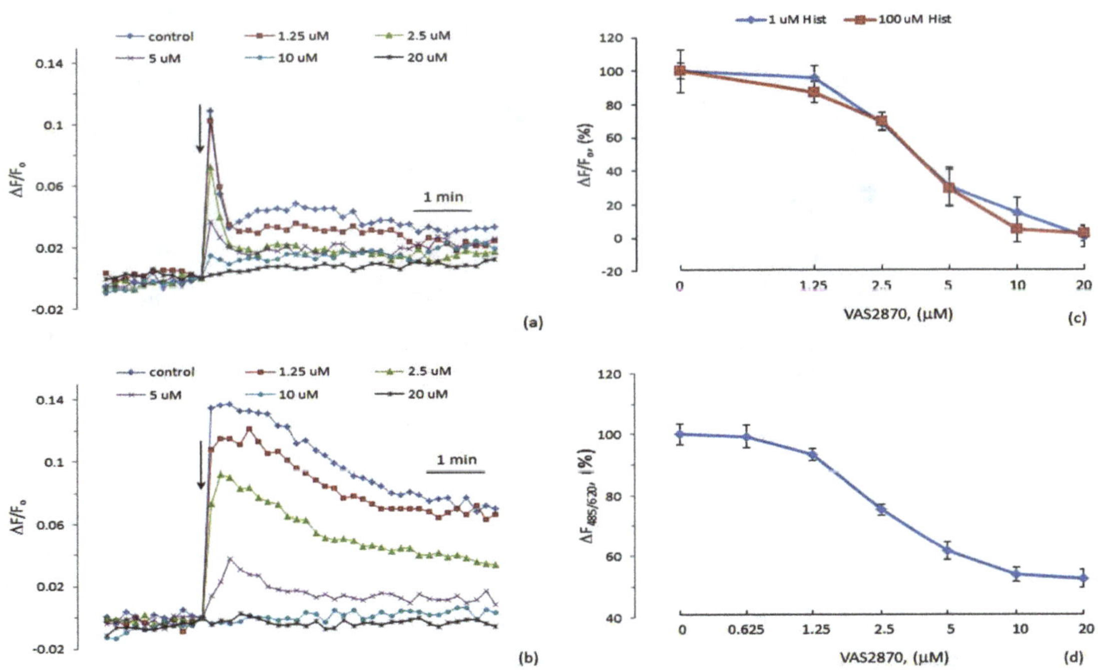

Figure 2. The influence of different concentrations of VAS2870 on histamine-induced $[Ca^{2+}]_i$ increases and on the generation of reactive oxygen species (ROS) in human umbilical vein endothelial cells (HUVECs). (**a**,**b**) The kinetics of $[Ca^{2+}]_i$ elevation in HUVECs in response to 1 and 100 μM histamine, respectively. (**c**) Concentration-dependent curves of the inhibition of $[Ca^{2+}]_i$ increases. (**d**) Concentration-dependence curve of the inhibition of dihydroethidium (DHE) oxidation by VAS2870. Each point in (**c**,**d**) is a mean of three values from three independent experiments on different cell preparations. In control experiments, it was demonstrated that DMSO at concentrations on which it was applied to the cells along with VAS2870 did not suppress histamine-induced $[Ca^{2+}]_i$ elevation and oxidation of DHE.

Evidence in favor of NOX involvement in histamine-induced calcium signaling was published quite a long time ago by Hu and co-workers [33], wherein they demonstrated that NOX-derived ROS were critical for generating the Ca^{2+} oscillations in response to histamine in human aortic ECs. Recently, it was demonstrated that in microglial cells, histamine stimulates ROS formation due to the activation of NOX1 via H1 receptors [50]. We expected that histamine might affect ROS generation in HUVECs. However, there was no increase in the rate of DHE oxidation in the presence of histamine (Figure 3). SFLLRN also did not produce any effect. It might be suggested that either histamine does not stimulate the generation of superoxide anion in HUVECs, or the method with DHE is not sensitive enough to detect the increment in ROS formation due to histamine. It should be mentioned that in [33] 2′,7′-dihydrodichlorofluorescin diacetate was used as a fluorescent probe. Elucidation of the mechanism of NOX involvement in signal transduction from histamine receptors in HUVECs requires further research.

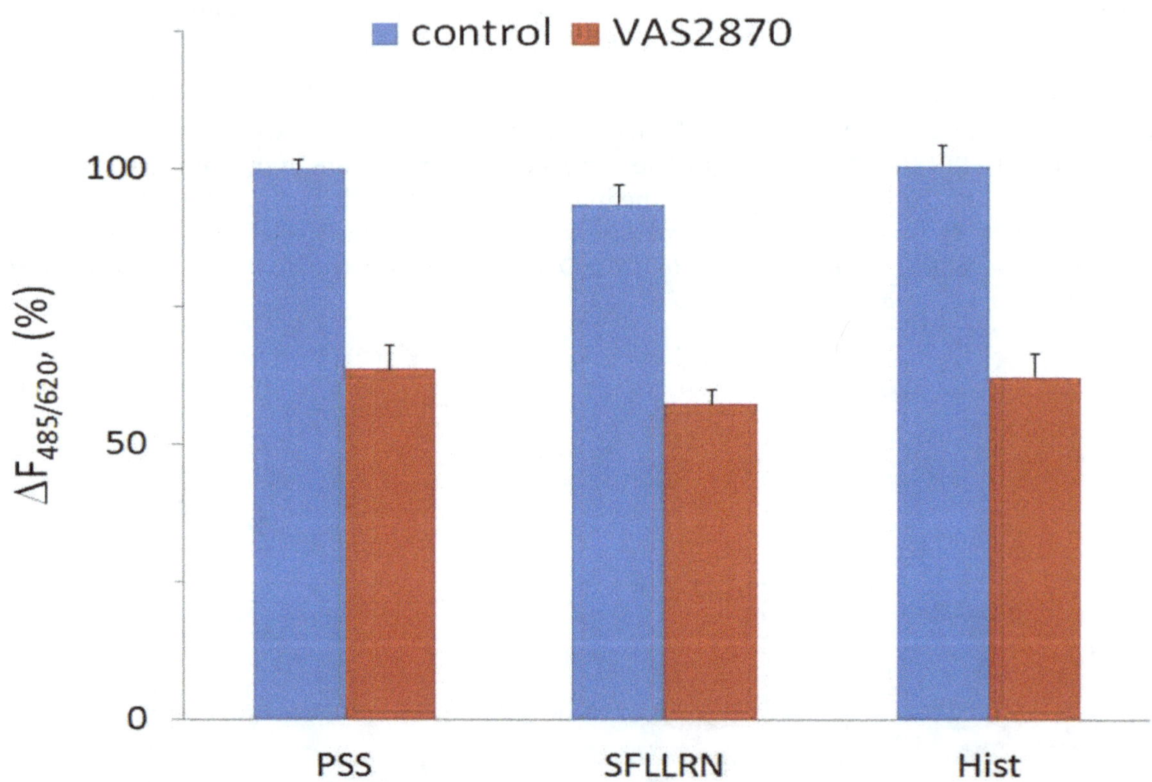

Figure 3. The influence of SFLLRN and histamine on DHE oxidation in the absence and presence of VAS2870. Physiological salt solution (PSS) was used as a vehicle control for the agonists. Each value is a mean of three values from three independent experiments with different cell preparations. DMSO at final concentration 0.1% was used as a vehicle control for VAS2870. The difference between the values in the absence and presence of VAS2870 was statistically significant ($p < 0.01$).

The next task was to study how VAS2870 can affect physiological reactions of ECs activated by histamine, which causes secretion of vWF in HUVECs due to the activation of H1 receptors. This secretion is mediated by the elevation of cytoplasmic calcium ion concentration [40,51]. We showed that, in the presence of VAS2870, the effect of histamine on vWF secretion was attenuated, though VAS2870 did not inhibit the basal secretion of vWF (Figure 4). This result correlates with our data on the inhibition of histamine's effect on $[Ca^{2+}]_i$ by VAS2870.

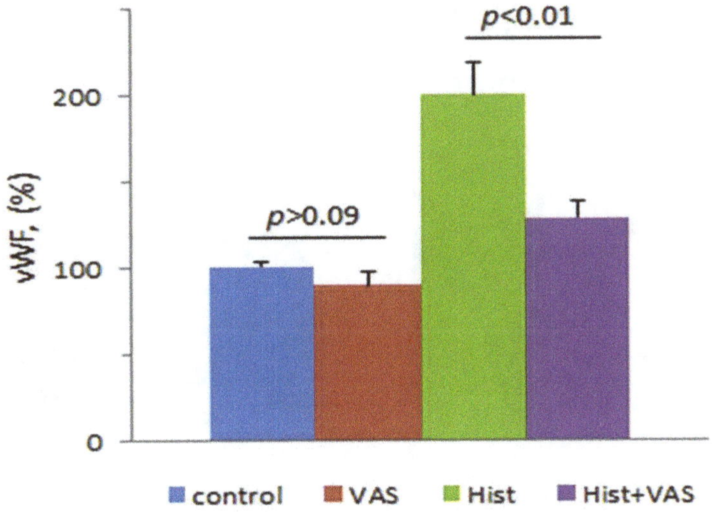

Figure 4. The influence of histamine on the secretion of von Willebrand factor (vWF) in the absence or presence of VAS2870. HUVECs were incubated at 30 °C during 5 min with 10 μM VAS2870 or vehicle (DMSO at final concentration 0.1% in the wells of a 48-well plate. Then, 100 μM histamine, or PSS as a vehicle control, were added to the cells, and they were additionally incubated for 30 min. Each value is a mean of the data from three independent experiments.

Among the physiological effects of histamine, relaxation of blood vessels is one of the most important, so the next task was to evaluate the influence of VAS2870 on this function. We determined the histamine-induced decrease of the contraction force of rat aortic rings preconstricted with norepinephrine. The rings were preincubated for 30 or 60 min with VAS2870 or vehicle before adding norepinephrine, followed by the addition of histamine. At concentrations of 10 and 100 μM, histamine induced the relaxation of the rings. Surprisingly, there was no attenuation of histamine-induced relaxation after preincubation of the rings with VAS2870 (Figure 5). It can be assumed that in rat aortic ECs VAS2870 does not produce as strong an inhibition of histamine-induced $[Ca^{2+}]_i$ rise as in HUVECs. It is known that histamine elevates $[Ca^{2+}]$i in HUVEC due to its mobilization from endoplasmic reticulum via the channels activated by inositol 1,4,5-trisphosphate and from endolysosomal vesicles via two-pore channels [40,41]. The relative roles of these two calcium signaling mechanisms might be different in ECs from rat aorta and human umbilical vein. Another possible explanation is that the role of calcium ions entering the cytoplasm from different sources is different in the case of vasorelaxation and vWF secretion. It has been demonstrated that the suppression of Ca^{2+} release from endolysosomal vesicles inhibits vWF secretion induced by histamine [40]. On the other hand, endothelium-dependent vasorelaxation depends on store-operated Ca^{2+} entry [52].

In [49], it was demonstrated that VAS2870 improves endothelium-dependent relaxation of spontaneously hypertensive rat (SHR) aortas induced by acetylcholine. The reason for the difference of these data and our results about the lack of VAS2870's effect on endothelium-dependent vasorelaxation is the increased level of ROS in the aorta of SHR due to the elevated expression of NOX1 and NOX2. The inhibition of their activity by VAS2870 improved impaired the acetylcholine-induced relaxation of spontaneously hypertensive rat aortas [49]. In control normotensive Wistar-Kioto rats, VAS2870 does not affect endothelium-dependent relaxation. There is evidence that VAS2870 can normalize arteriolar flow-induced dilation caused by oxidative stress at hyperinsulinemia [53]. In experiments with murine aorta rings it was demonstrated that their incubation in conditions of hyperglycemia-induced oxidative stress causes impairment of vasodilation mediated by PAR2 agonists, and VAS2870 improved this function [54]. From all these data it can be concluded that VAS2870 is able to improve endothelial-dependent vasorelaxation at pathological state associated with oxidative stress when NOX1 and NOX2 activities are increased, while in normal vessels it does not affect relaxation. This is also supported by our data.

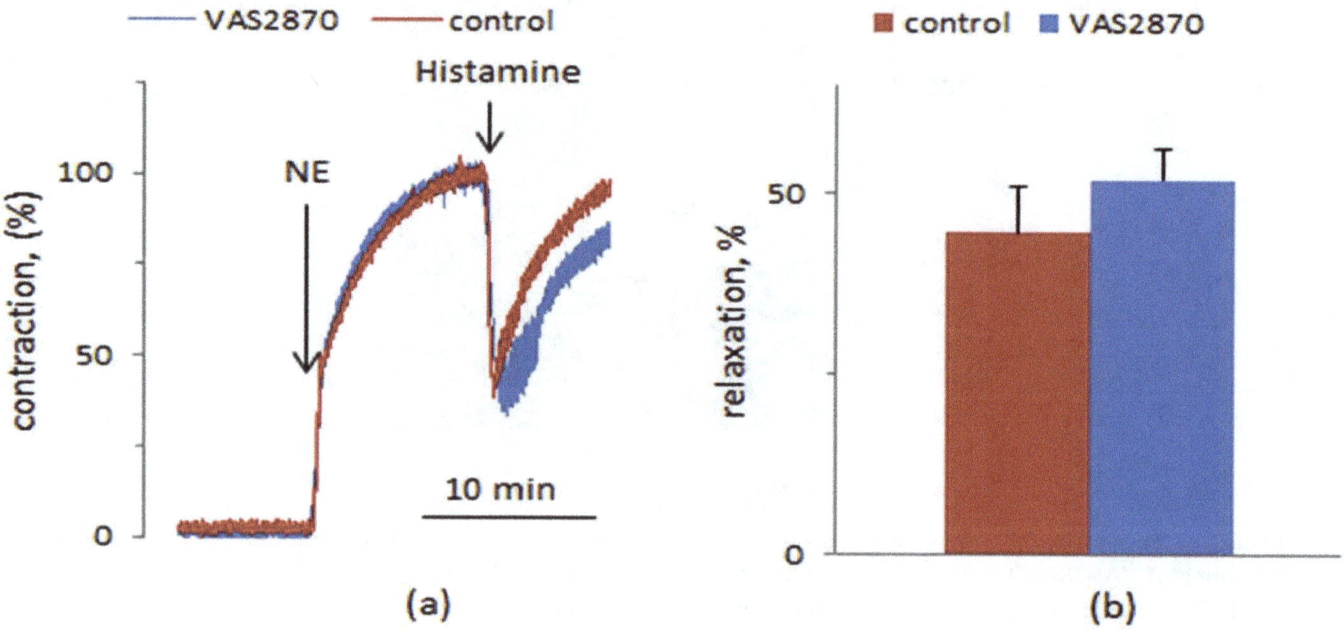

Figure 5. Relaxation of rat aortic rings preconstricted by norepinephrine (NE) in response to 100 μM histamine in the absence and presence of VAS2870. VAS2870 at a concentration of 10 μM or vehicle was added to the rings 30 min or 1 h before NE. (**a**) Typical curves of the contraction and relaxation. (**b**) The degrees of relaxation in the absence and presence of VAS2870 ($n = 10$ for control and for VAS2870 from five independent experiments). In each experiment there were measurements from two control and two experimental aortic rings.

Inhibition of vWF secretion by VAS2870 indicates that ROS produced by NOXs are involved in this reaction. Direct activation of vWF secretion was demonstrated for superoxide anion [30]. We have recently shown that H_2O_2 induces the exocytosis of vWF in HUVECs [35]. The physiological significance of the stimulation of vWF exocytosis and secretion by ROS is not clear. However, this effect could take place in pathophysiological conditions when hyperactivation of vWF secretion occurs. It was demonstrated that histamine released from mast cells is one of the factors initiating deep vein thrombosis due to excessive secretion of vWF [55]. Thus, it could be suggested that NOX inhibitors might be used as a tool for preventing deep vein thrombosis induced by histamine release from mast cells without affecting vasorelaxation. In fact, the number of pathological states which could be targeted by VAS2870 or other NOX inhibitors together with or alternatively to other pharmaceuticals has been increasing over the past several years [56].

Endothelial activation initiates multiple processes, including hemostasis and inflammation. The molecules that contribute to these processes are co-stored in secretory granules, the Weibel-Palade bodies (WPBs) being the most known and important in ECs. It was previously shown that Ca^{2+}-elevating agonists can deplete the cell of almost all WPBs, whereas cAMP-elevating agonists selectively release a pool of mature WPBs that contain vWF but little or no P-selectin [57]. NOX inhibitors can control the release of granule content from ECs to allow differentiated responses, and the development of selected agonists for tuning the endothelial response is currently an urgent problem [58].

Author Contributions: P.V.A. proposed the project, designed experiments, wrote the paper. E.Y.R., P.P.A., S.K.T., G.Y.M., A.A.T. performed experiments, analyzed data; N.V.G. wrote the paper.

References

1. Panula, P.; Chazot, P.L.; Cowart, M.; Gutzmer, R.; Leurs, R.; Liu, W.L.; Stark, H.; Thurmond, R.L.; Haas, H.L. International Union of Basic and Clinical Pharmacology. XCVIII. Histamine Receptors. *Pharmacol. Rev.* **2015**, *67*, 601–655. [PubMed]
2. Bakker, R.A.; Timmerman, H.; Leurs, R. Histamine receptors: Specific ligands, receptor biochemistry, and signal transduction. *Clin. Allergy Immunol.* **2002**, *17*, 27–64. [PubMed]
3. Karlstedt, K.; Jin, C.; Panula, P. Expression of histamine receptor genes Hrh3 and Hrh4 in rat brain endothelial cells. *Br. J. Pharmacol.* **2013**, *170*, 58–66. [CrossRef] [PubMed]
4. Grosicki, M.; Wojcik, T.; Chlopicki, S.; Kiec-Kononowicz, K. In vitro study of histamine and histamine receptor ligands influence on the adhesion of purified human eosinophils to endothelium. *Eur. J. Pharmacol.* **2016**, *777*, 49–59. [CrossRef] [PubMed]
5. Mailloux, R.J. Teaching the fundamentals of electron transfer reactions in mitochondria and the production and detection of reactive oxygen species. *Redox Biol.* **2015**, *4*, 381–398. [CrossRef] [PubMed]
6. Sturza, A.; Duicu, O.M.; Vaduva, A.; Danila, M.D.; Noveanu, L.; Varro, A.; Muntean, D.M. Monoamine oxidases are novel sources of cardiovascular oxidative stress in experimental diabetes. *Can. J. Physiol. Pharmacol.* **2015**, *93*, 555–561. [CrossRef] [PubMed]
7. Goncharov, N.V.; Avdonin, P.V.; Nadeev, A.D.; Zharkikh, I.L.; Jenkins, R.O. Reactive oxygen species in pathogenesis of atherosclerosis. *Curr. Pharm. Des.* **2015**, *21*, 1134–1146. [CrossRef] [PubMed]
8. Segal, A.W. How neutrophils kill microbes. *Annu. Rev. Immunol.* **2005**, *23*, 197–223. [CrossRef] [PubMed]
9. Lassegue, B.; San Martin, A.; Griendling, K.K. Biochemistry, physiology, and pathophysiology of NADPH oxidases in the cardiovascular system. *Circ. Res.* **2012**, *110*, 1364–1390. [CrossRef] [PubMed]
10. Drummond, G.R.; Sobey, C.G. Endothelial NADPH oxidases: Which NOX to target in vascular disease? *Trends Endocrinol. Metab.* **2014**, *25*, 452–463. [CrossRef] [PubMed]
11. Jones, S.A.; O'Donnell, V.B.; Wood, J.D.; Broughton, J.P.; Hughes, E.J.; Jones, O.T. Expression of phagocyte NADPH oxidase components in human endothelial cells. *Am. J. Physiol* **1996**, *271*, H1626–H1634. [CrossRef] [PubMed]
12. Meyer, J.W.; Holland, J.A.; Ziegler, L.M.; Chang, M.M.; Beebe, G.; Schmitt, M.E. Identification of a functional leukocyte-type NADPH oxidase in human endothelial cells: A potential atherogenic source of reactive oxygen species. *Endothelium* **1999**, *7*, 11–22. [CrossRef] [PubMed]
13. Thakur, S.; Du, J.; Hourani, S.; Ledent, C.; Li, J.M. Inactivation of adenosine A2A receptor attenuates basal and angiotensin II-induced ROS production by Nox2 in endothelial cells. *J. Biol. Chem.* **2010**, *285*, 40104–40113. [CrossRef] [PubMed]
14. Heinloth, A.; Heermeier, K.; Raff, U.; Wanner, C.; Galle, J. Stimulation of NADPH oxidase by oxidized low-density lipoprotein induces proliferation of human vascular endothelial cells. *J. Am. Soc. Nephrol.* **2000**, *11*, 1819–1825. [PubMed]
15. Schroder, K. Isoform specific functions of Nox protein-derived reactive oxygen species in the vasculature. *Curr Opin Pharmacol* **2010**, *10*, 122–126. [CrossRef] [PubMed]
16. Konior, A.; Schramm, A.; Czesnikiewicz-Guzik, M.; Guzik, T.J. NADPH oxidases in vascular pathology. *Antioxid Redox Signal.* **2014**, *20*, 2794–2814. [CrossRef] [PubMed]
17. Ray, R.; Murdoch, C.E.; Wang, M.; Santos, C.X.; Zhang, M.; Alom-Ruiz, S.; Anilkumar, N.; Ouattara, A.; Cave, A.C.; Walker, S.J.; et al. Endothelial Nox4 NADPH oxidase enhances vasodilatation and reduces blood pressure in vivo. *Arterioscler. Thromb. Vasc. Biol.* **2011**, *31*, 1368–1376. [CrossRef] [PubMed]
18. Martyn, K.D.; Frederick, L.M.; von Loehneysen, K.; Dinauer, M.C.; Knaus, U.G. Functional analysis of Nox4 reveals unique characteristics compared to other NADPH oxidases. *Cell Signal.* **2006**, *18*, 69–82. [CrossRef] [PubMed]
19. Geiszt, M.; Witta, J.; Baffi, J.; Lekstrom, K.; Leto, T.L. Dual oxidases represent novel hydrogen peroxide sources supporting mucosal surface host defense. *FASEB J.* **2003**, *17*, 1502–1504. [CrossRef] [PubMed]
20. Ris-Stalpers, C. Physiology and pathophysiology of the DUOXes. *Antioxid Redox Signal.* **2006**, *8*, 1563–1572. [CrossRef] [PubMed]
21. Touyz, R.M.; Anagnostopoulou, A.; Camargo, L.L.; Rios, F.J.; Montezano, A.C. Vascular Biology of Superoxide-Generating NADPH Oxidase 5-Implications in Hypertension and Cardiovascular Disease. *Antioxid Redox Signal.* **2018**. [CrossRef] [PubMed]

22. Montezano, A.C.; De Lucca Camargo, L.; Persson, P.; Rios, F.J.; Harvey, A.P.; Anagnostopoulou, A.; Palacios, R.; Gandara, A.C.P.; Alves-Lopes, R.; Neves, K.B.; et al. NADPH Oxidase 5 Is a Pro-Contractile Nox Isoform and a Point of Cross-Talk for Calcium and Redox Signaling-Implications in Vascular Function. *J. Am. Heart Assoc.* **2018**, *7*. [CrossRef] [PubMed]
23. Doan, T.N.; Gentry, D.L.; Taylor, A.A.; Elliott, S.J. Hydrogen peroxide activates agonist-sensitive Ca(2+)-flux pathways in canine venous endothelial cells. *Biochem J.* **1994**, *297*, 209–215. [CrossRef] [PubMed]
24. Dreher, D.; Junod, A.F. Differential effects of superoxide, hydrogen peroxide, and hydroxyl radical on intracellular calcium in human endothelial cells. *J. Cell Physiol.* **1995**, *162*, 147–153. [CrossRef] [PubMed]
25. Hecquet, C.M.; Ahmmed, G.U.; Vogel, S.M.; Malik, A.B. Role of TRPM2 channel in mediating H2O2-induced Ca^{2+} entry and endothelial hyperpermeability. *Circ. Res.* **2008**, *102*, 347–355. [CrossRef] [PubMed]
26. Siflinger-Birnboim, A.; Lum, H.; Del Vecchio, P.J.; Malik, A.B. Involvement of Ca^{2+} in the H2O2-induced increase in endothelial permeability. *Am. J. Physiol.* **1996**, *270*, L973–L978. [CrossRef] [PubMed]
27. Suresh, K.; Servinsky, L.; Reyes, J.; Baksh, S.; Undem, C.; Caterina, M.; Pearse, D.B.; Shimoda, L.A. Hydrogen peroxide-induced calcium influx in lung microvascular endothelial cells involves TRPV4. *Am. J. Physiol. Lung Cell Mol. Physiol.* **2015**, *309*, L1467–L1477. [CrossRef] [PubMed]
28. Volk, T.; Hensel, M.; Kox, W.J. Transient Ca2+ changes in endothelial cells induced by low doses of reactive oxygen species: Role of hydrogen peroxide. *Mol. Cell Biochem.* **1997**, *171*, 11–21. [CrossRef] [PubMed]
29. Wesson, D.E.; Elliott, S.J. The H2O2-generating enzyme, xanthine oxidase, decreases luminal Ca2+ content of the IP3-sensitive Ca2+ store in vascular endothelial cells. *Microcirculation* **1995**, *2*, 195–203. [CrossRef] [PubMed]
30. Vischer, U.M.; Jornot, L.; Wollheim, C.B.; Theler, J.M. Reactive oxygen intermediates induce regulated secretion of von Willebrand factor from cultured human vascular endothelial cells. *Blood* **1995**, *85*, 3164–3172. [PubMed]
31. Poteser, M.; Graziani, A.; Rosker, C.; Eder, P.; Derler, I.; Kahr, H.; Zhu, M.X.; Romanin, C.; Groschner, K. TRPC3 and TRPC4 associate to form a redox-sensitive cation channel. Evidence for expression of native TRPC3-TRPC4 heteromeric channels in endothelial cells. *J. Biol. Chem.* **2006**, *281*, 13588–13595. [CrossRef] [PubMed]
32. Hu, Q.; Zheng, G.; Zweier, J.L.; Deshpande, S.; Irani, K.; Ziegelstein, R.C. NADPH oxidase activation increases the sensitivity of intracellular Ca2+ stores to inositol 1,4,5-trisphosphate in human endothelial cells. *J. Biol. Chem.* **2000**, *275*, 15749–15757. [CrossRef] [PubMed]
33. Hu, Q.; Yu, Z.X.; Ferrans, V.J.; Takeda, K.; Irani, K.; Ziegelstein, R.C. Critical role of NADPH oxidase-derived reactive oxygen species in generating Ca2+ oscillations in human aortic endothelial cells stimulated by histamine. *J. Biol. Chem.* **2002**, *277*, 32546–32551. [CrossRef] [PubMed]
34. Avdonin, P.V.; Nadeev, A.D.; Tsitrin, E.B.; Tsitrina, A.A.; Avdonin, P.P.; Mironova, G.Y.; Zharkikh, I.L.; Goncharov, N.V. Involvement of two-pore channels in hydrogen peroxide-induced increase in the level of calcium ions in the cytoplasm of human umbilical vein endothelial cells. *Dokl. Biochem. Biophys.* **2017**, *474*, 209–212. [CrossRef] [PubMed]
35. Avdonin, P.V.; Tsitrina, A.A.; Mironova, G.Y.; Avdonin, P.P.; Zharkikh, I.L.; Nadeev, A.D.; Goncharov, N.V. Hydrogen Peroxide Stimulates Exocytosis of Von Willebrand Factor in Human Umbilical Vein Endothelial Cells. *Biol. Bulletin* **2017**, *44*, 531–537. [CrossRef]
36. Stielow, C.; Catar, R.A.; Muller, G.; Wingler, K.; Scheurer, P.; Schmidt, H.H.; Morawietz, H. Novel Nox inhibitor of oxLDL-induced reactive oxygen species formation in human endothelial cells. *Biochem. Biophys. Res. Commun* **2006**, *344*, 200–205. [CrossRef] [PubMed]
37. Wind, S.; Beuerlein, K.; Eucker, T.; Muller, H.; Scheurer, P.; Armitage, M.E.; Ho, H.; Schmidt, H.H.; Wingler, K. Comparative pharmacology of chemically distinct NADPH oxidase inhibitors. *Br. J. Pharmacol.* **2010**, *161*, 885–898. [CrossRef] [PubMed]
38. Goncharov, N.V.; Sakharov, I.; Danilov, S.M.; Sakandelidze, O.G. Use of collagenase from the hepatopancreas of the Kamchatka crab for isolating and culturing endothelial cells of the large vessels in man. *Biull. Eksp. Biol. Med.* **1987**, *104*, 376–378. [CrossRef] [PubMed]
39. Kalyanaraman, B.; Hardy, M.; Podsiadly, R.; Cheng, G.; Zielonka, J. Recent developments in detection of superoxide radical anion and hydrogen peroxide: Opportunities, challenges, and implications in redox signaling. *Arch. Biochem Biophys* **2017**, *617*, 38–47. [CrossRef] [PubMed]

40. Esposito, B.; Gambara, G.; Lewis, A.M.; Palombi, F.; D'Alessio, A.; Taylor, L.X.; Genazzani, A.A.; Ziparo, E.; Galione, A.; Churchill, G.C.; et al. NAADP links histamine H1 receptors to secretion of von Willebrand factor in human endothelial cells. *Blood* **2011**, *117*, 4968–4977. [CrossRef] [PubMed]
41. Zharkich, I.L.; Nadeev, A.D.; Tsitrin, E.B.; Goncharov, N.V.; Avdonin, P.V. Suppression of Histamine Induced Relaxation of Rat Aorta and Calcium Signaling in Endothelial Cells by Two Pore Channel Blocker trans-NED19 and Hydrogen Peroxide. *Biol. Bulletin* **2016**, *43*, 365–372. [CrossRef]
42. Yang, J.; Li, J.; Wang, Q.; Xing, Y.; Tan, Z.; Kang, Q. Novel NADPH oxidase inhibitor VAS2870 suppresses TGFbetadependent epithelialtomesenchymal transition in retinal pigment epithelial cells. *Int. J. Mol. Med.* **2018**, *42*, 123–130. [PubMed]
43. Ten Freyhaus, H.; Huntgeburth, M.; Wingler, K.; Schnitker, J.; Baumer, A.T.; Vantler, M.; Bekhite, M.M.; Wartenberg, M.; Sauer, H.; Rosenkranz, S. Novel Nox inhibitor VAS2870 attenuates PDGF-dependent smooth muscle cell chemotaxis, but not proliferation. *Cardiovasc. Res.* **2006**, *71*, 331–341. [CrossRef] [PubMed]
44. Hosseini, E.; Ghasemzadeh, M.; Atashibarg, M.; Haghshenas, M. ROS scavenger, N-acetyl-l-cysteine and NOX specific inhibitor, VAS2870 reduce platelets apoptosis while enhancing their viability during storage. *Transfusion* **2019**. [CrossRef] [PubMed]
45. Munhoz, A.C.; Riva, P.; Simoes, D.; Curi, R.; Carpinelli, A.R. Control of Insulin Secretion by Production of Reactive Oxygen Species: Study Performed in Pancreatic Islets from Fed and 48-Hour Fasted Wistar Rats. *PLoS One* **2016**, *11*, e0158166. [CrossRef] [PubMed]
46. Gatto, G.J., Jr.; Ao, Z.; Kearse, M.G.; Zhou, M.; Morales, C.R.; Daniels, E.; Bradley, B.T.; Goserud, M.T.; Goodman, K.B.; Douglas, S.A.; et al. NADPH oxidase-dependent and -independent mechanisms of reported inhibitors of reactive oxygen generation. *J. Enzyme Inhib. Med. Chem.* **2013**, *28*, 95–104. [CrossRef] [PubMed]
47. Sancho, P.; Fabregat, I. The NADPH oxidase inhibitor VAS2870 impairs cell growth and enhances TGF-beta-induced apoptosis of liver tumor cells. *Biochem. Pharmacol.* **2011**, *81*, 917–924. [CrossRef] [PubMed]
48. Kleinschnitz, C.; Grund, H.; Wingler, K.; Armitage, M.E.; Jones, E.; Mittal, M.; Barit, D.; Schwarz, T.; Geis, C.; Kraft, P.; et al. Post-stroke inhibition of induced NADPH oxidase type 4 prevents oxidative stress and neurodegeneration. *PLoS Biol.* **2010**, *8*. [CrossRef] [PubMed]
49. Wind, S.; Beuerlein, K.; Armitage, M.E.; Taye, A.; Kumar, A.H.; Janowitz, D.; Neff, C.; Shah, A.M.; Wingler, K.; Schmidt, H.H. Oxidative stress and endothelial dysfunction in aortas of aged spontaneously hypertensive rats by NOX1/2 is reversed by NADPH oxidase inhibition. *Hypertension* **2010**, *56*, 490–497. [CrossRef] [PubMed]
50. Rocha, S.M.; Saraiva, T.; Cristovao, A.C.; Ferreira, R.; Santos, T.; Esteves, M.; Saraiva, C.; Je, G.; Cortes, L.; Valero, J.; et al. Histamine induces microglia activation and dopaminergic neuronal toxicity via H1 receptor activation. *J. Neuroinflamm.* **2016**, *13*, 137. [CrossRef] [PubMed]
51. Lorenzi, O.; Frieden, M.; Villemin, P.; Fournier, M.; Foti, M.; Vischer, U.M. Protein kinase C-delta mediates von Willebrand factor secretion from endothelial cells in response to vascular endothelial growth factor (VEGF) but not histamine. *J. Thromb. Haemost* **2008**, *6*, 1962–1969. [CrossRef] [PubMed]
52. Kassan, M.; Zhang, W.; Aissa, K.A.; Stolwijk, J.; Trebak, M.; Matrougui, K. Differential role for stromal interacting molecule 1 in the regulation of vascular function. *Pflugers Arch.* **2015**, *467*, 1195–1202. [CrossRef] [PubMed]
53. Mahmoud, A.M.; Ali, M.M.; Miranda, E.R.; Mey, J.T.; Blackburn, B.K.; Haus, J.M.; Phillips, S.A. Nox2 contributes to hyperinsulinemia-induced redox imbalance and impaired vascular function. *Redox Biol.* **2017**, *13*, 288–300. [CrossRef] [PubMed]
54. El-Daly, M.; Pulakazhi Venu, V.K.; Saifeddine, M.; Mihara, K.; Kang, S.; Fedak, P.W.M.; Alston, L.A.; Hirota, S.A.; Ding, H.; Triggle, C.R.; et al. Hyperglycaemic impairment of PAR2-mediated vasodilation: Prevention by inhibition of aortic endothelial sodium-glucose-co-Transporter-2 and minimizing oxidative stress. *Vascul Pharmacol.* **2018**, *109*, 56–71. [CrossRef] [PubMed]
55. Ponomaryov, T.; Payne, H.; Fabritz, L.; Wagner, D.D.; Brill, A. Mast Cells Granular Contents Are Crucial for Deep Vein Thrombosis in Mice. *Circ. Res.* **2017**, *121*, 941–950. [CrossRef] [PubMed]
56. Ma, M.W.; Wang, J.; Dhandapani, K.M.; Wang, R.; Brann, D.W. NADPH oxidases in traumatic brain injury - Promising therapeutic targets? *Redox Biol.* **2018**, *16*, 285–293. [CrossRef] [PubMed]
57. Cleator, J.H.; Zhu, W.Q.; Vaughan, D.E.; Hamm, H.E. Differential regulation of endothelial exocytosis of P-selectin and von Willebrand factor by protease-activated receptors and cAMP. *Blood* **2006**, *107*, 2736–2744. [CrossRef] [PubMed]

9

Increasing the TRPM2 Channel Expression in Human Neuroblastoma SH-SY5Y Cells Augments the Susceptibility to ROS-Induced Cell Death

Xinfang An [1], Zixing Fu [1], Chendi Mai [1], Weiming Wang [1], Linyu Wei [1], Dongliang Li [1], Chaokun Li [1,*] and Lin-Hua Jiang [1,2,*]

1 Sino-UK Joint Laboratory for Brain Function and Injury and Department of Physiology and Neurobiology, Xinxiang Medical University, Xinxiang 453003, China; anxinfang2018@126.com (X.A.); fuzixing1991@126.com (Z.F.); mcdhsm@163.com (C.M.); wangweiming2015@126.com (W.W.); wly98124@126.com (L.W.); xyldl8@126.com (D.L.)

2 School of Biomedical Sciences, Faculty of Biological Sciences, University of Leeds, Leeds LS2 JT, UK

* Correspondence: lichaokun@hotmail.com (C.L.); l.h.jiang@leeds.ac.uk (L.-H.J.)

Abstract: Human neuroblastoma SH-SY5Y cells are a widely-used human neuronal cell model in the study of neurodegeneration. A recent study shows that, 1-methyl-4-phenylpyridine ion (MPP), which selectively causes dopaminergic neuronal death leading to Parkinson's disease-like symptoms, can reduce SH-SY5Y cell viability by inducing H_2O_2 generation and subsequent TRPM2 channel activation. MPP-induced cell death is enhanced by increasing the TRPM2 expression. By contrast, increasing the TRPM2 expression has also been reported to support SH-SY5Y cell survival after exposure to H_2O_2, leading to the suggestion of a protective role for the TRPM2 channel. To clarify the role of reactive oxygen species (ROS)-induced TRPM2 channel activation in SH-SY5Y cells, we generated a stable SH-SY5Y cell line overexpressing the human TRPM2 channel and examined cell death and cell viability after exposure to H_2O_2 in the wild-type and TRPM2-overexpressing SH-SY5Y cells. Exposure to H_2O_2 resulted in concentration-dependent cell death and reduction in cell viability in both cell types. TRPM2 overexpression remarkably augmented H_2O_2-induced cell death and reduction in cell viability. Furthermore, H_2O_2-induced cell death in both the wild-type and TRPM2-overexpressing cells was prevented by 2-APB, a TRPM2 inhibitor, and also by PJ34 and DPQ, poly(ADP-ribose) polymerase (PARP) inhibitors. Collectively, our results show that increasing the TRPM2 expression renders SH-SY5Y cells to be more susceptible to ROS-induced cell death and reinforce the notion that the TRPM2 channel plays a critical role in conferring ROS-induced cell death. It is anticipated that SH-SY5Y cells can be useful for better understanding the molecular and signaling mechanisms for ROS-induced TRPM2-mediated neurodegeneration in the pathogenesis of neurodegenerative diseases.

Keywords: human neuroblastoma SH-SY5Y cells; TRPM2 channel; ROS; neuronal cell death

---

1. Introduction

Mammalian cells express a large family of transient receptor potential (TRP) cationic channels that are activated by poly-modality and play a role in a diversity of physiological and pathological processes [1–5]. These channels are often grouped based on sequence relatedness into canonical (TRPC), vanilloid (TRPV), melastatin (TRPM), mucolipins (TRPML), polycystins (TRPP) and ankyrin (TRPA) subfamilies. The TRPM2 channel is gated by intracellular ADP-ribose (ADPR) [6] and has been recognised as a key molecular mechanism that confers cells with a prominent sensitivity to reactive oxygen species (ROS), thanks to the well-documented ability of ROS to induce poly(ADPR) polymerase (PARP)-dependent generation of ADPR [7–10]. The TRPM2 channel is expressed in both excitable and non-excitable cells [7–10]. More than a decade worth of studies have shown the TRPM2 channel as a common mechanism mediating cell death after exposure to high levels of ROS or diverse pathological stimuli or factors that are known to induce ROS generation. For example, increasing evidence shows that the TRPM2 channel plays an important role in mediating neuronal cell death in the brain, as a result of exposure to ROS, amyloid-β peptide, 1-methyl-4-phenylpyridine ion (MPP) and ischemia-reperfusion [11–24]. Therefore, ROS-induced TRPM2 channel activation has been proposed to contribute to the pathogenesis of ischemia stroke and neurodegenerative conditions such as Alzheimer's disease (AD) and Parkinson's diseases (PD) [25–30].

Human neuroblastoma SH-SY5Y cells have been widely used as a human neuronal cell model in the study of the molecular and signaling mechanisms for neurodegeneration, particularly those related to PD because of the catecholaminergic, albeit strictly speaking not dopaminergic, neuronal properties [31]. Studies from other groups and us have consistently documented functional expression of the TRPM2 channel in SH-SY5Y cells [21,23,24,32–34], but the findings regarding the role of the TRPM2 channel in SH-SY5Y cells are intriguing. A recent study using the 3-(4,5-dimethylthiazol-2-yl)-2,5-diphenyltetrazolium (MTT) assay has shown that concentration-dependent reduction in cell viability after exposure to 10 to 400 μM H_2O_2 or 125 to 500 μM MPP via inducing H_2O_2 generation [23]. The cytotoxicity induced by both H_2O_2 and MPP was significantly suppressed by inhibition of the TRPM2 channel with flufenamic acid (FFA), a TRPM2 channel inhibitor [7]. In addition, MPP-induced detrimental effect on the cell viability was attenuated by small inference RNA (siRNA)-mediated knockdown of the TRPM2 expression and, conversely, enhanced by increasing the TRPM2 expression [23]. These results strongly support that the TRPM2 channel has a critical role in mediating H_2O_2/MPP-induced SH-SY5Y cell death [23]. However, an earlier study using the MTT assay showed that increasing the TRPM2 expression in SH-SY5Y cells resulted in higher cell viability after exposure to 50 to 100 μM H_2O_2 for 6 or 24 h [32], leading to the recent proposal of a protective role for the TRPM2 channel in cell survival [33,34].

To clarify the role of ROS-induced TRPM2 channel activation in SH-SY5Y cells is critical for the question of whether or not this human cell model can be used to gain insights into the molecular and signaling mechanisms for TRPM2-dependent neurodegeneration. For example, in addition to the revelation of a role for the TRPM2 channel in mediating MPP-induced SH-SY5Y cell death, the recent study by Sun et al. [23] has found strong up-regulation of the TRPM2 expression in MPP-treated SH-SY5Y cells as observed in the substantia nigra pars compacta (SNpc) of the brains from PD patients and PD mice induced by injection with 1-methyl-4-phenyl-tetrahydropyridine to selectively destroy dopaminergic neurons in the SNpc region. These observations raise an interesting question with respect to the role of the TRPM2 channel in mediating dopaminergic neuronal death and, therefore, the pathogenesis of PD. In the present study, we aimed to test the hypothesis that increasing the TRPM2 expression results in greater susceptibility of SH-SY5Y cells to ROS-induced cell death. We generated a stable SH-SY5Y cell line overexpressing the human TRPM2 channel and used the propidium iodide (PI) staining assay to directly examine H_2O_2-induced cell death in the wild-type (WT) and TRPM2-overexpressing cells. To make our results more comparable to those reported by previous studies, we also determined cell viability after exposure to H_2O_2 in the WT and TRPM2-overexpressing cells. Our results from measuring cell death and cell viability provide consistent evidence to show that exposure to H_2O_2 induces SH-SY5Y cell death, regardless of the TRPM2 expression level, and that increasing the TRPM2 expression augments the susceptibility to ROS-induced cell death.

2. Materials and Methods

2.1. General Chemicals and Culture Medium

General chemicals used in this study were purchased from Sigma, except those indicated specifically. Dulbecco's modified Eagle's medium (DMEM) and penicillin/streptomycin were from

GIBCO. Foetal bovine serum (FBS) and trypsin-ethylenediaminetetraacetic acid (EDTA) were from Beyotime Biotechnology (Nantong, China).

2.2. Cell Culturing

SH-SY5Y cells were kindly provided by Dr. J.A. Sim (University of Manchester, Manchester, UK) and maintained in standard cell culture medium (DMEM supplemented with 10% FBS, 100 units/mL of penicillin and 100 μg/mL of streptomycin) in a tissue incubator (Thermofisher, Waltham, MA, USA) at 37 °C in the presence of 5% CO_2. Cells were passaged every 3 to 4 days or when they became 70% to 80% confluent.

2.3. Propidium Iodide (PI) Staining Cell Death Assay

Cell death was examined using the PI staining assay as described in our previous studies [24,35]. In brief, cells were seeded in 24-well plates at 2×10^4 cells per well in 500 μL of standard culture medium and incubated overnight. After cells were treated with H_2O_2, PI, and Hoechst 33342 were added into the culture medium with the final concentrations of 2 μg/mL and 1 μg/mL, respectively. Cells were incubated at 37 °C for 30 min. In experiments examining the effects of inhibitors on H_2O_2-induced cell death, cells were treated with the inhibitor at indicated concentrations 30 min before and during exposure to H_2O_2. Images were captured using an AxioVert A1 fluorescence microscope (Zeiss). For each condition in every independent experiment, three wells of cells were used, one randomly selected field in each well was imaged, and at least 500 cells from three images/wells were analyzed. Cell death was presented as a percentage of PI-positive cells in all cells identified by Hoechst 33342 staining in the same fields.

2.4. Generation of SH-SY5Y Cells Overexpressing the Human TRPM2 Channel

The internal ribosome entry site (IRES) bicistronic construct that enables expression of separate TRPM2 protein and enhanced green fluorescent protein (GFP) under the control of the same promoter, pTRPM2-IRES-GFP, was generated by ligation of the full-length human TRPM2 cDNA [6] and pGFP-IRES fragment that were both amplified by polymerase chain reaction (PCR). Primers for PCR are available upon request. PCR was performed using Phusion polymerase (New England Biolabs, Beijing, China) and the following protocols: 120 s at 98 °C, followed by 25 cycles for TRPM2 or 20 cycles for pGFP-IRES fragment, and a final step of 10 min at 72 °C. Each cycle comprised 30 s at 98 °C, 30 s at 58 °C, and 5 min at 72 °C for TRPM2 or 30 s at 98 °C, 30 s at 58 °C, and 3 min at 72 °C for pGFP-IRES. Ligated PCR products were transformed into competent DH5α *Escherichia coli* (Takara, Beijing, China). Positive colonies were identified by PCR and sequencing the PCR products. Plasmids were purified, and the construct was further confirmed by DNA sequencing.

For transfection, SH-SY5Y cells were cultured in standard culture medium in six-well plates and transfected with the pTRPM2-IRES-GFP construct using Xfect transfection reagent (Clonetech, Beijing, China) according to the manufacturer's instructions. Transfected cells were identified by examining GFP expression using a fluorescence microscope 48 h post transfection and also using flow cytometry 72 h post transfection. GFP-positive SH-SY5Y cells were incubated in standard culture medium containing 400 μg/mL G418 for another 2 weeks. Individual cells were inoculated into a 96-well plate and, after being cultured in G418-containing standard culture medium for 1 month, cells showing strong GFP expression were selected and expanded. The TRPM2 protein expression in the stable cell line used in this study was further verified by Western blotting.

2.5. Flow Cytometry

Cells were cultured for 3 days before they were harvested for analysis using flow cytometry. Approximately 10,000 cells were analyzed for positive expression of GFP, using a flow cytometer (BD Biosciences, Beijing, China) and 488 nm/512 nm filters.

2.6. Western Blotting

The TRPM2 protein expression was examined using standard sodium dodecyl sulfate polyacrylamide gel electrophoresis (SDS-PAGE) and Western blotting. In brief, cell lysates were prepared in the radioimmuno-precipitation assay buffer (Beyotime Biotechnology, Nantong, China) containing 1 mM phenylmethane sulfonyl fluoride. The protein concentrations in cell lysates were determined using a bicinchoninic acid assay kit (Beyotime Biotechnology). Twenty microliters of cell lysate containing 80 μg proteins alongside protein markers (Beyotime Biotechnology) were separated by electrophoresis on 15% SDS-PAGE gels and transferred to nitrocellular membranes (Millipore, Burlington, MA, USA). The membranes were blocked by 5% non-fat milk in Tris-buffered saline containing 0.05% Tween 20 (TBST) and then incubated with the primary anti-TRPM2 antibody at a dilution of 1:200 (ab87050, Abcam, Shanghai, China) or anti-glyceraldehyde 3-phosphate dehydrogenase (GAPDH) at 1:1000 (Hangzhou Goodhere Biotechnology Co, Hangzhou, China) at 4 °C overnight. After extensive washing in TBST, the membranes were incubated with the secondary horseradish peroxidase-conjugated goat anti-rabbit IgG antibody at 1:800 (Affinity Biosciences, Cincinnati, OH, USA) at room temperature for 1 h. After extensive washing in TBST, the proteins were visualized using an enhanced chemiluminescence kit (Beyotime Biotechnology), and the images were captured using an Amersham Imager 600 system (GE Healthcare, Chicago, IL, USA).

2.7. Cell Counting Kit-8 (CCK-8) Cell Viability Assay

The cell viability was examined using Cell Counting Kit-8 (CCK-8) assay kits (Dojindo Molecular Technologies, Shanghai, China) according to the manufacturer's instructions. Cells were seeded in 96-cell plates at 1×10^4 cells per well in 100 μL of standard culture medium and incubated overnight. After cells were treated with H_2O_2 at indicated concentrations, 10 μL of CCK-8 reagent was added to each well and incubated in a tissue culture incubator at 37 °C for 1 h. The absorbance at 450 nm was determined, using a Bio-Tek800 microplate reader (BioTek Instruments, Winooski, VT, USA). Three wells of cells were used for each condition for every independent experiment. The cell viability in H_2O_2-treated cells was presented as a percentage of that in control or untreated cells in parallel experiments.

2.8. Data Presentation and Statistical Analysis

The results are presented as mean ± standard error mean (SEM), where appropriate, with "n" representing the number of independent experiments. Comparisons were performed using GraphPad Prism software, with $p < 0.05$ considered to be statistically significant. One-way analysis of variance (ANOVA) and post hoc Tukey's test were used for multiple groups to compare cell death or cell viability for the same type of cells, that is, WT or TRPM2-overexpressing cells. Two-way ANOVA and post hoc Tukey's test were used to compare cell death or cell viability between the WT and TRPM2-overexpressing cells that were treated with the same concentration of H_2O_2.

3. Results

3.1. H_2O_2 Induces Cell Death via PARP-Dependent TRPM2 Channel Activation

We started with using the PI staining assay to detect cell death following exposure to 100 to 500 μM H_2O_2 for 24 h in the WT SH-SY5Y cells. Representative images are shown in Figure 1A, and the results from three independent experiments are summarized in Figure 1B. Exposure to H_2O_2 resulted in concentration-dependent cell death. The cell death was negligible (3.8 ± 1.0%, n = 3) under control conditions or in untreated cells, but there was a noticeable increase in cell death after exposure to H_2O_2, with 11.4 ± 2.0%, 27.6 ± 6.0%, 51.9 ± 6.2%, and 95.0 ± 1.4% at 100, 300, 400, and 500 μM H_2O_2, respectively, although the increase reached a statistically significant level ($p < 0.05$) only for 400 and 500 μM H_2O_2.

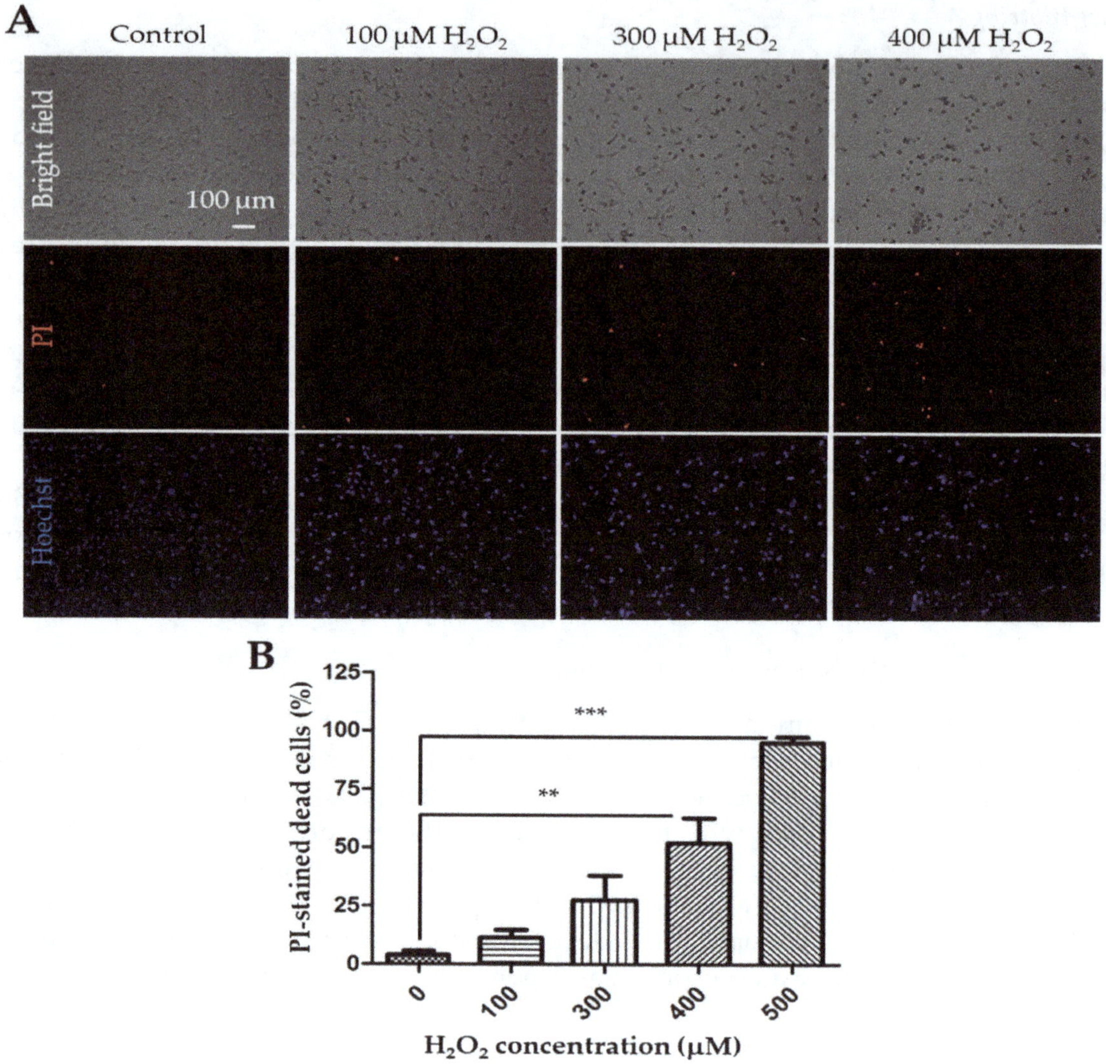

Figure 1. Exposure to H_2O_2 induces concentration-dependent cell death in SH-SY5Y cells. (**A**) Representative bright field and fluorescence images showing propidium iodide (PI) and Hoechst staining in control cells and cells after exposure to 100, 300 or 400 μM H_2O_2 for 24 h. (**B**) Mean ± standard error mean (SEM) percentage of cell death after exposure to indicated concentrations of H_2O_2 from 3 independent experiments. **, $p < 0.01$ and ***, $p < 0.001$, compared to control cells without exposure to H_2O_2.

As has been shown in many cell types, ROS can activate the TRPM2 channel via a PARP-dependent mechanism [7]. To confirm that H_2O_2-induced cell death resulted from PARP-dependent TRPM2 channel activation in SH-SY5Y cells, we moved on to examine the effects of pharmacological inhibitors of PARP. Cells were treated 30 min before and during 24 h exposure to 400 μM H_2O_2 with 30 μM PJ34 and 10 μM 3,4-dihydro-5[4-(1-piperindinyl)butoxy]-1(2H)-isoquinoline (DPQ), two structurally distinct PARP inhibitors that have been shown in our recent study to be effective in inhibiting H_2O_2-induced TRPM2-mediated microglial cell death [34]. Indeed, cell death was reduced from 69.1 ± 3.0% in cells treated with H_2O_2 alone to 24.3 ± 1.0% in cells that were also treated with PJ34, indicating a strong inhibition (Figure 2A,B; n = 3; $p < 0.001$). Similarly, H_2O_2-induced cell death was reduced to 15.4 ± 1.1% in DPQ-treated cells from 62.3 ± 5.9% in cells treated with dimethyl sulfoxide (DMSO), the solvent used to prepare DPQ stock solution (Figure 2A,B; n = 3; $p < 0.001$). We next examined the effect of treatment with 100 μM 2-APB, a TRPM2 channel blocker that can inhibit H_2O_2-induced TRPM2-mediated cell death [34]. Treatment with 2-APB also strongly suppressed H_2O_2-induced cell death from 62.3 ± 5.9% in DMSO-treated cells to 12.4 ± 2.4% in 2-APB-treated cells (Figure 2A,B; n = 3; $p < 0.001$). Treatment with DMSO or the inhibitor alone resulted in no or minimal increase in cell death

(Figure 2B). Collectively, these results provide evidence to support the notion that PARP-dependent TRPM2 channel activation mediates H_2O_2-induced cell death in SH-SY5Y cells, which is consistent with a recent study that reports a role for the TRPM2 channel in H_2O_2-induced cell death in SH-SY5Y cells based on the measurement of cell viability [23].

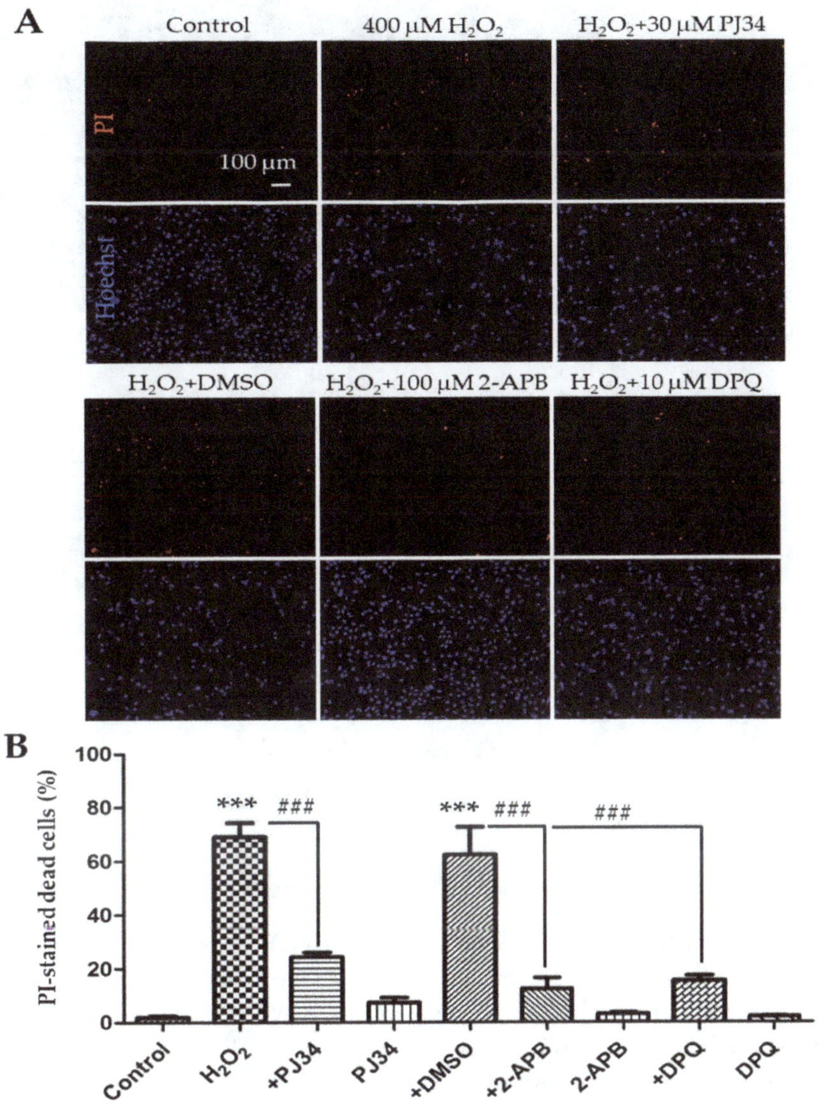

Figure 2. Effects of treatment with polymerase (PARP) or melastatin 2 (TRPM2) channel inhibitor on H_2O_2-induced cell death in SH-SY5Y cells. (**A**) Representative fluorescence images showing staining with PI and Hoechst of control cells and cells treated with indicated conditions. Treatment with the inhibitors or dimethyl sulfoxide (DMSO) started 30 min before and during exposure to 400 µM H_2O_2 for 24 h. (**B**) Mean ± SEM percentage of PI-positive cells under indicated conditions, as shown in panel A, from three independent experiments. ***, $p < 0.001$ compared with untreated cells, and ###, $p < 0.001$ compared to cells treated with H_2O_2 alone or treated with H_2O_2 and DMSO.

3.2. TRPM2 Overexpression Augments the Susceptibility to H_2O_2-Induced Cell Death

To examine the effect of increasing the TRPM2 expression in SH-SY5Y cells on H_2O_2-induced cell death, we made a mammalian expression construct using the IRES technology to generate a stable SH-SY5Y cell line expressing the human TRPM2 protein and GFP under the control of the same promoter (see Materials and Methods). Positively transfected SH-SY5Y cells were initially identified by the positive GFP expression as shown using fluorescent microscopic imaging (Figure 3A) and flow cytometry analysis (Figure 3B). As shown by Western blotting, SH-SY5Y cells showed endogenous expression of the TRPM2 protein with the expected molecular weight (171 kDa), as reported in previous studies [23,24] and, as anticipated, the TRPM2 protein expression was noticeably increased in the GFP-positive cells (Figure 3C).

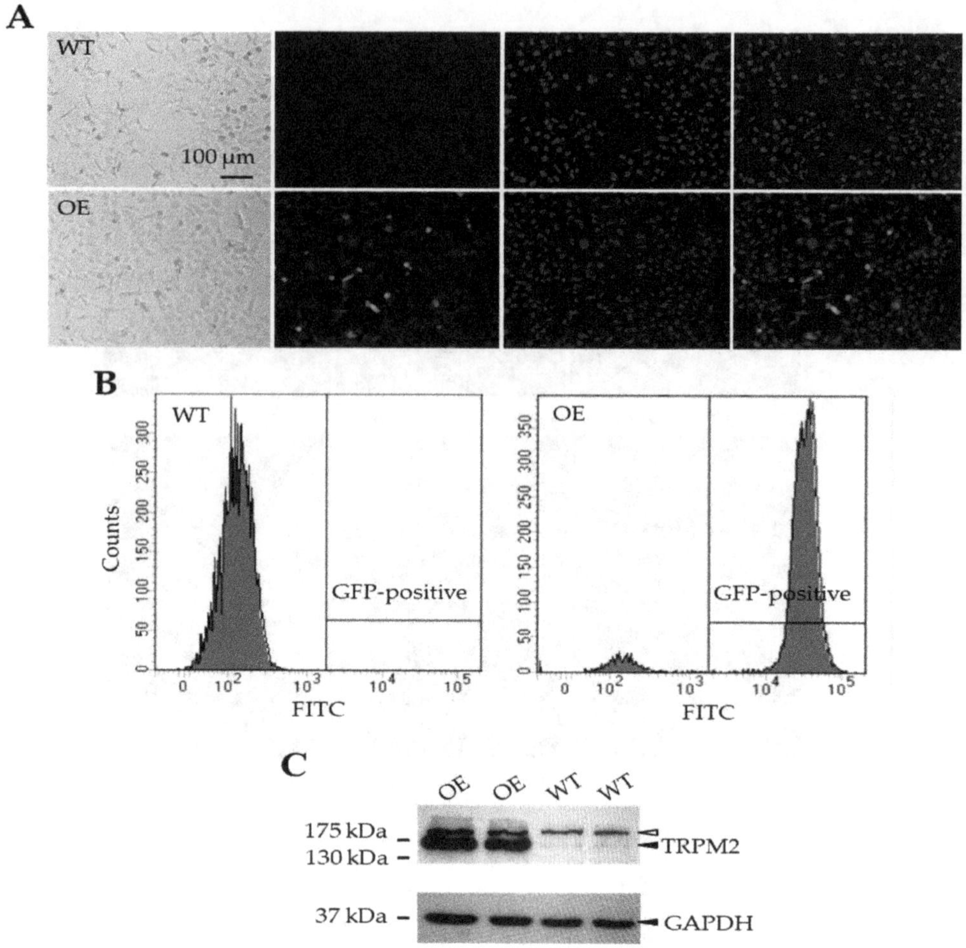

Figure 3. Generation of TRPM2-overexpressing SH-SY5Y stable cells. (**A**) Represent bright field and fluorescence images of wild-type (WT) and TRPM2-overexpressing (OE) cells. (**B**) Flow cytometry analysis of green fluorescent protein (GFP) expression in the WT (left) and OE (right) cells. (**C**) Western blots showing expression of TRPM2 and glyceraldehyde 3-phosphate dehydrogenase (GAPDH) in the WT and OE cells. The solid arrowheads in the top and bottom panels indicate TRPM2 and GAPDH, respectively. The empty arrowhead in the top panel indicates a non-specific and high molecular weight protein, which was present in both the WT and OE cells with a similar level. There was a strong increase in the TRPM2 protein expression in the OE cells (lanes 1–2) as compared to the endogenous TRPM2 protein expression in the WT cells (lanes 3–4), where the GAPDH expression level was similar in both cell types.

We next performed parallel experiments using the PI staining assay to compare H_2O_2-induced cell death in the WT and TRPM2-overexpressing SH-SY5Y cells after exposure to 30, 100, and 300 µM H_2O_2 for 24 h. The results from three independent experiments are summarized in Figure 4A. After exposure to 30 µM H_2O_2, there was a slight but statistically insignificant increase in cell death in the TRPM2-overexpressing cells as compared to that in the WT cells. Cell death upon exposure to 100 µM H_2O_2 was increased from 11.3 $\pm$ 1.9% in the WT cells to 93.3 $\pm$ 1.0% in the TRPM2-overexpressing cells (Figure 4A; n = 3; $p < 0.001$). Similarly, cell death induced by exposure to 300 µM H_2O_2 was increased from 25.5 $\pm$ 4.8% in the WT cells to 96.2 $\pm$ 0.4% in the TRPM2-overexpressing cells (Figure 4A; n = 3; $p < 0.001$). These results clearly indicate that increasing the TRPM2 expression in SH-SY5Y cells remarkably augments the susceptibility to H_2O_2-induced cell death. To verify that such increased H_2O_2-induced cell death was mediated by PARP-dependent activation of the TRPM2 channel as shown above in the WT cells, we examined the effects of PJ34, DPQ or 2-APB on H_2O_2-induced cell death in the TRPM2-overexpressing cells. We used the same experimental conditions for the TRPM2-overexpressing cell as for the WT cells described above, that is, exposure of cells to 400 μM H_2O_2 for 24 h and treatment of cells with 30 μM PJ34, 10 μM DPQ, or 100 μM 2-APB 30 min before and during exposure to H_2O_2. The results are shown in Figure 4B,C. Cell death in the TRPM2-overexpressing cells was reduced from 91.7 ± 0.5% in cells treated with H_2O_2 alone to 13.1 ± 0.8% (n = 3; $p < 0.001$) in cells that were also treated with PJ34. Similarly, H_2O_2-induced cell death of 82.0 ± 3.8% in DMSO-treated cells was reduced to 17.7 ± 1.5% (n = 3; $p < 0.001$) in DPQ-treated cells or 7.1 ± 1.0% in 2-APB-treated cells (n = 3; $p < 0.001$). Again, treatment with DMSO or the inhibitor alone caused no or minimal cell death in the TRPM2-overexpressing cells (Figure 4C). These pharmacological results are readily reconciled with the above-described observation that increasing the TRPM2 expression enhanced the susceptibility to ROS-induced cell death and, taken together, provide consistent evidence to support that ROS-induced PARP-dependent activation of the TRPM2 channel, regardless of its expression, results in cell death.

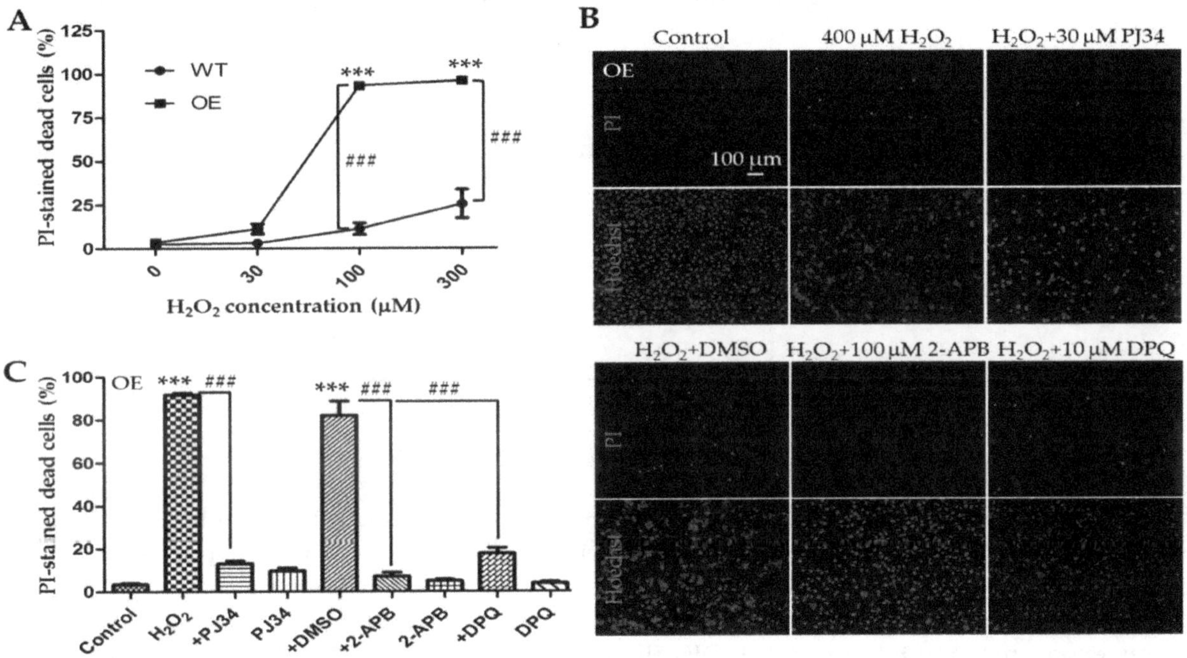

Figure 4. Increasing the TRPM2 expression in SH-SY5Y cells augments the susceptibility to H_2O_2-induced cell death. (**A**) Mean ± SEM percentage of PI positive cells in the WT and TRPM2-overexpressing (OE) cells after exposure to indicated concentrations of H_2O_2 from three independent experiments. ***, $p < 0.001$, compared to cells without exposure to H_2O_2 in the OE cells; ###, $p < 0.001$ compared between the WT and OE cells exposed to the same concentration of H_2O_2. (**B**) Representative fluorescence images showing staining with PI and Hoechst in the OE cells treated with indicated inhibitors, 30 min before and during exposure to 400 μM H_2O_2 for 24 h. (**C**) Mean ± SEM percentage of PI-positive cells under indicated conditions in the OE cells from three independent experiments. ***, $p < 0.001$ compared to control cells; ###, $p < 0.001$ compared to cells treated with H_2O_2 alone or treated with H_2O_2 and DMSO.

3.3. TRPM2 Overexpression Reduces the Cell Viability after Exposure to H_2O_2

As mentioned above, previous studies mainly used the MTT assay to examine the cell viability and the effects of increasing the TRPM2 expression on the cell viability after exposure to H_2O_2 or MPP [23,32]. The MTT assay is a metabolic assay that measures the live cells, whereas the PI assay is used to detect cell death, particularly necrotic cell death. To make more direct or meaningful comparisons, we finally examined the effects of exposure to 30 to 400 μM H_2O_2 for 24 h on the cell viability of the WT and TRPM2-overexpressing SH-SY5Y cells using the CCK-8 assay, which is based on the same principle as the MTT assay. Our results from three independent experiments are summarized in Figure 5. Exposure to H_2O_2 resulted in a concentration-dependent reduction in cell viability in both the WT and TRPM2-overexpressing cells. In the WT cells, the mean cell viability was 101.5 ± 5.0%, 101.5 ± 5.9%, 52.1 ± 0.6%, and 31.6 ± 4.7% after 24 h exposure to 30, 100, 300, and 400 μM H_2O_2, respectively. The cell viability was statistically significantly lower after exposure to 300 μM ($p < 0.05$) or 400 μM H_2O_2 ($p < 0.001$), but not 30 or 100 μM H_2O_2. These results are similar to those in a recent study reporting a significant reduction in cell viability after exposure to 200 or 400 μM, but not 10 or 100 μM H_2O_2 [23]. In the TRPM2-overexpressing cells, the mean cell viability was 92.6 ± 2.8%, 87.8 ± 3.8%, 5.6 ± 0.4%, and 3.2 ± 0.2% after exposure to 30, 100, 300, and 400 μM H_2O_2, respectively. The cell viability was significantly lower after exposure to 300 μM or 400 μM H_2O_2 ($p < 0.001$ in both cases), but not 30 or 100 μM H_2O_2 (Figure 5). The mean cell viability in the TRPM2-overexpressing cells was lower compared to that in the WT cells after exposure to the same concentration of H_2O_2, but the difference reached a statistically significant level only for 300 μM ($p < 0.001$) and 400 μM H_2O_2 ($p < 0.05$) (Figure 5). These results from measuring cell viability also show that TRPM2 overexpression increases the susceptibility of SH-SY5Y cells to ROS-induced cell death.

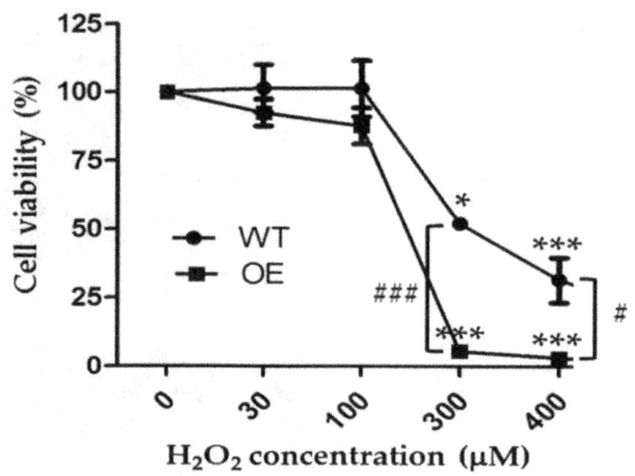

Figure 5. Increasing the TRPM2 expression in SH-SY5Y cell enhances H_2O_2-induced reduction in cell viability. Mean ± SEM cell viability, as a percentage of that under control conditions, in the WT and TRPM2-overexpressing (OE) cells after exposure to indicated concentrations of H_2O_2 for 24 h. The results were from three independent experiments. *, $p < 0.05$ and ***, $p < 0.001$, compared to cells without exposure to H_2O_2 in the WT or OE cells, respectively; #, $p < 0.05$ and ###, $p < 0.001$ compared between the WT and OE cells treated with the same concentration of H_2O_2.

4. Discussion

The present study shows that exposure of SH-SY5Y cells to H_2O_2 induced concentration-dependent cell death as well as confirming the finding that H_2O_2 induced a concentration-dependent reduction in cell viability. Furthermore, we show that increasing the TRPM2 expression enhanced H_2O_2-induced cell death and reduction in cell viability, consistently indicating that increasing the TRPM2 expression in SH-SY5Y cells augments the susceptibility to H_2O_2-induced cell death.

The TRPM2 channel has been shown to mediate ROS-induced cell death in many cell types [7]. In SH-SY5Y cells, previous studies using the MTT assay showed a concentration-dependent reduction in cell viability after exposure to 50 to 100 μM H_2O_2 [32] or 10 to 400 μM H_2O_2 or 125 to 500 μM MPP via inducing H_2O_2 generation [23]. The present study also demonstrated concentration-dependent cell death after exposure to 30 to 500 μM H_2O_2, using the PI staining assay (Figures 1 and 4A), and also a concentration-dependent reduction in cell viability after exposure to 30 to 400 μM H_2O_2, using the CCK-8 assay (Figure 5). However, regarding the effects of H_2O_2 on cell viability, there were noticeable differences between the present and previous studies. The earlier study reported that exposure to 50 or 100 μM H_2O_2 for 6 or 24 h reduced the cell viability [32]. The present study found a significant reduction in cell viability only after 24 h exposure to higher concentrations of H_2O_2 (300 and 400 μM) but not to lower concentrations including 100 μM H_2O_2, the condition used in the aforementioned previous study [32]. In accordance with our study, Sun et al. have recently shown that the cell viability was significantly reduced after exposure to 200 and 400 μM H_2O_2, but not 10 or 100 μM H_2O_2 [23]. They also reported reduced cell viability following exposure to MPP at high concentrations (250 and 500 μM) but not at a low concentration (125 μM) [23]. The reasons for these discrepancies remain unclear but the cells used, particularly the proliferative properties of the cells, are likely to contribute, considering that cell proliferation can influence the number of live cells which is measured using the MTT and CCK-8 assays.

In the present study, we showed significant inhibition of H_2O_2-induced cell death by 2-APB (Figure 2). Similarly, Sun et al. have reported suppression by FFA of H_2O_2-induced reduction in cell viability [23]. Both 2-APB and FFA are known to inhibit the TRPM2 channel, but neither of them is specific to the TRPM2 channel [7]. Nonetheless, the inhibition of H_2O_2-induced cell death by 2-APB in this study and by FFA in the study by Sun et al. [23] are in support of the notion that the TRPM2 channel activation is critical for H_2O_2-induced cell death or reduction in cell viability. We also showed that H_2O_2-induced cell death was strongly inhibited by PJ34 and DPQ (Figure 2), confirming an important role for PARP in ROS-induced TRPM2 channel activation and ensuing cell death [7]. We showed here that increasing the TRPM2 expression augmented H_2O_2-induced reduction in cell viability as well as H_2O_2-induced cell death (Figures 4A and 5). Sun et al. have reported that MPP-induced reduction in cell viability was strongly attenuated by FFA and also by siRNA-mediated knockdown of the TRPM2 expression and by contrast exacerbated by increasing the TRPM2 expression [23]. The observations that H_2O_2-induced cell death in the TRPM2-overexpressing cells was largely prevented by 2-APB, PJ34, and DPQ (Figure 4B,C) strongly disfavor, albeit not completely ruling out, the possibility that alterations in TRPM2-independent mechanism(s) give rise to the increased susceptibility to H_2O_2-induced cell death. Overall, our results suggest that ROS-induced PARP-dependent activation of the TRPM2 channel, regardless of its expression level, leads to SH-SY5Y cell death. These findings reinforce the notion that the TRPM2 channel is a key and common meditator for ROS-induced cell death in SH-SY5Y cells [23,24] as reported in neurons [11–13,21] and other types of cells [7,35–38].

However, studies examining the effect of increasing the TRPM2 expression in SH-SY5Y have led to the proposal of a different role for the TRPM2 channel [32–34]. An earlier study reported that the reduction in cell viability after exposure to 50 and 100 μM H_2O_2 for 6 or 24 h was attenuated by increasing the TRPM2 expression, suggesting a role for the TRPM2 channel in supporting cell survival and proliferation [32]. As we showed here, increasing the TRPM2 expression resulted in no significant protection against the reduction in cell viability after exposure to 30 and 100 μM H_2O_2 for 24 h, but a further reduction in cell viability after exposure to 300 and 400 μM H_2O_2 for 24 h (Figure 5). The study by Sun et al., while not examining the effect on H_2O_2-induced reduction in cell viability, has shown that increasing the TRPM2 expression enhanced the reduction in cell viability after exposure to a high concentration of MPP [23], as we observed here after exposure to high concentrations of H_2O_2. As discussed above, it is highly likely that the discrepancy in part arises from the cells used. It is worth mentioning that increasing the TRPM2 expression has been shown to induce cell death or reduce cell viability after exposure to H_2O_2 in other cell types. For example, in human embryonic kidney 293 cells that are largely void of endogenous TRPM2 expression, overexpression of the recombinant TRPM2 channel strongly reduced cell viability following exposure to H_2O_2, determined using the trypan blue exclusion assay, or increased cell death after exposure to H_2O_2, shown by staining with PI and Alexa Fluor 488-annexin V [36]. Likewise, in human monocytic U937 cells which endogenously express the TRPM2 channel, increasing the TRPM2 expression enhanced H_2O_2-induced reduction in cell viability based on the trypan blue exclusion assay and apoptotic cell death determined by staining with Alexa Fluor 594-annexin V [37].

As already mentioned above, cell death has been assessed directly by PI staining for necrotic cell death or annexin V staining for apoptotic cell death and also indirectly using the MTT and CCK-8 assays that measure the number of live cells. In this study, we assessed H_2O_2-induce cell death using the PI staining and CCK-8 assays. As discussed above, the results from these two measurements overall provide consistent evidence to show that exposure to H_2O_2 resulted in concentration-dependent cell death and that increasing the TRPM2 expression enhanced the susceptibility to H_2O_2-induced cell death (Figure 1, Figure 4A, and Figure 5). Nonetheless, a close examination of the data reveals a noticeable difference. Increasing the TRPM2 expression in SH-SY5Y cells conferred a significant increase in cell death (Figure 4A) but resulted in a modest reduction in cell viability after exposure to 100 μM H_2O_2 (Figure 5). Such a difference may result from the experimental conditions used (e.g., a high cell seeding density used for the CCK8 assay versus a low cell seeding density for the PI staining assay) as well as the aforementioned proliferative properties of the cells. Therefore, cautions needs to be exercised in conducting quantitative comparisons and interpretation of the results obtained using different experimental conditions and/or methods.

Recent studies, using TRPM2-knockout mice in combination with disease models, support a critical role for the TRPM2 channel in mediating ROS-induced neuronal death and in contributing to the pathogenesis of ischemia-reperfusion brain damage and neurodegenerative diseases, such as AD [27–30]. SH-SY5Y cells have been useful as a human neuronal cell model in the study of neurodegeneration [31]. In a recent study using SH-SY5Y cells, we have recently revealed TRPM2 channel activation as a critical step in a positive feedback signaling mechanism that causes lysosomal and mitochondrial dysfunction to drive delayed neuronal cell death [24]. Such information is helpful in gaining mechanistic insights into TRPM2-dependent delayed neuronal death responsible for ischemia-reperfusion brain damage [15,18]. As already introduced above, in addition to the revelation of the importance of the TRPM2 channel in mediating MPP-induced SH-SY5Y cell death, Sun et al. have shown strong up-regulation of the TRPM2 expression in MPP-treated SH-SY5Y cells as observed in the SNpc region of the brains from PD patients and also PD mice [23]. These new findings point to the TRPM2 channel for its potential role in mediating loss of dopaminergic neurons, the key event in the pathogenesis of PD [39]. It is anticipated that SH-SY5Y cells should be useful for researchers to gain a better understanding of TRPM2-dependent signaling mechanisms for neurodegeneration that are relevant to the pathogenesis of PD.

5. Conclusions

The present study provides evidence to show that PARP-dependent activation of the TRPM2 channel in SH-SY5Y cell mediates ROS-induced cell death and increasing the TRPM2 expression augments the susceptibility to ROS-induced death. As shown by recent studies, using SH-SY5Y cells as a human neuronal cell model should help in interrogating TRPM2-dependent signaling mechanisms in neuronal cell death and related neurodegenerative diseases.

Author Contributions: L.-H.J. and C.L. conceived the research and designed the experiments. X.A., Z.F., C.M. and W.W. performed the experiments. X.A. and L.-H.J. analyzed data and prepared figures. L.W. and D.L. provided technical supports and intellectual inputs. L.-H.J. wrote and revised the manuscript.

References

1. Clapham, D.E. TRP channels as cellular sensors. *Nature* **2003**, *426*, 517–524. [CrossRef] [PubMed]
2. Montell, C.; Birnbaumer, L.; Flockerzi, V. The TRP channels, a remarkably functional family. *Cell* **2002**, *108*, 595–598. [CrossRef]
3. Venkatachalam, K.; Montell, C. TRP channels. *Annu. Rev. Biochem.* **2007**, *76*, 387–417. [CrossRef] [PubMed]

4. Fleig, A.; Penner, R. The TRPM ion channel subfamily: Molecular, biophysical and functional features. *Trends Pharmacol. Sci.* **2004**, *25*, 633–639. [CrossRef]
5. Nilius, B.; Owsianik, G.; Voets, T.; Peters, J.A. Transient receptor potential cation channels in disease. *Physiol. Rev.* **2007**, *87*, 165–217. [CrossRef] [PubMed]
6. Perraud, A.L.; Fleig, A.; Dunn, C.A.; Bagley, L.A.; Launay, P.; Schmitz, C.; Stokes, A.J.; Zhu, Q.; Bessman, M.J.; Penner, R.; et al. ADP-ribose gating of the calcium-permeable LTRPC2 channel revealed by Nudix motif homology. *Nature* **2001**, *411*, 595–599. [CrossRef] [PubMed]
7. Jiang, L.H.; Yang, W.; Zou, J.; Beech, D.J. TRPM2 channel properties, functions and therapeutic potentials. *Expert Opin. Ther. Targets* **2010**, *14*, 973–988. [CrossRef]
8. Sumoza-Toledo, A.; Penner, R. TRPM2: A multifunctional ion channel for calcium signaling. *J. Physiol.* **2011**, *589*, 1515–1525. [CrossRef]
9. Knowles, H.; Li, Y.; Perraud, A.L. The TRPM2 ion channel, an oxidative stress and metabolic sensor regulating innate immunity and inflammation. *Immunol. Res.* **2013**, *55*, 241–248. [CrossRef]
10. Ru, X.; Yao, X. TRPM2: A multifunctional ion channel for oxidative stress sensing. *Sheng Li Xue Bao* **2014**, *66*, 7–15.
11. Fonfria, E.; Marshall, I.C.; Boyfield, I.; Skaper, S.D.; Hughes, J.P.; Owen, D.E.; Zhang, W.; Miller, B.A.; Benham, C.D.; McNulty, S. Amyloid beta-peptide(1-42) and hydrogen peroxide-induced toxicity are mediated by TRPM2 in rat primary striatal cultures. *J. Neurochem.* **2005**, *95*, 715–723. [CrossRef] [PubMed]
12. Kaneko, S.; Kawakami, S.; Hara, Y.; Wakamori, M.; Itoh, E.; Minami, T.; Takada, Y.; Kume, T.; Katsuki, H.; Mori, Y.; et al. A critical role of TRPM2 in neuronal cell death by hydrogen peroxide. *J. Pharmacol. Sci.* **2006**, *101*, 66–76. [CrossRef] [PubMed]
13. Bai, J.Z.; Lipski, J. Differential expression of TRPM2 and TRPV4 channels and their potential role in oxidative stress-induced cell death in organotypic hippocampal culture. *Neurotoxicology* **2010**, *31*, 204–214. [CrossRef] [PubMed]
14. Jia, J.; Verma, S.; Nakayama, S.; Quillinan, N.; Grafe, M.R.; Hurn, P.D.; Herson, P.S. Sex differences in neuroprotection provided by inhibition of TRPM2 channels following experimental stroke. *J. Cereb. Blood Flow Metab.* **2011**, *31*, 2160–2168. [CrossRef]
15. Verma, S.; Quillinan, N.; Yang, Y.F.; Nakayama, S.; Cheng, J.; Kelley, M.H.; Herson, P.S. TRPM2 channel activation following in vitro ischemia contributes to male hippocampal cell death. *Neurosci. Lett.* **2012**, *530*, 41–46. [CrossRef] [PubMed]
16. Nakayama, S.; Vest, R.; Traystman, R.J.; Herson, P.S. Sexually dimorphic response of TRPM2 inhibition following cardiac arrest-induced global cerebral ischemia in mice. *J. Mol. Neurosci.* **2013**, *51*, 92–98. [CrossRef]
17. Alim, I.; Teves, L.; Li, R.W.; Mori, Y.; Tymianski, M. Modulation of NMDAR subunit expression by TRPM2 channels regulates neuronal vulnerability to ischemic cell death. *J. Neurosci.* **2013**, *33*, 17264–71727. [CrossRef] [PubMed]
18. Ye, M.; Yang, W.; Ainscough, J.F.; Hu, X.P.; Li, X.; Sedo, A.; Zhang, X.H.; Zhang, X.; Chen, Z.; Li, X.M.; et al. TRPM2 channel deficiency prevents delayed cytosolic Zn^{2+} accumulation and CA1 pyramidal neuronal death after transient global ischemia. *Cell. Death Dis.* **2014**, *5*, e1541. [CrossRef]
19. Shimizu, K.; Quillinan, N.; Orfila, J.E.; Herson, P.S. Sirtuin-2 mediates male specific neuronal injury following experimental cardiac arrest through activation of TRPM2 ion channels. *Exp. Neurol.* **2016**, *275*, 78–83. [CrossRef]
20. Ostapchenko, V.G.; Chen, M.; Guzman, M.S.; Xie, Y.F.; Lavine, N.; Fan, J.; Beraldo, F.H.; Martyn, A.C.; Belrose, J.C.; Mori, Y.; et al. The transient receptor potential melastatin 2 (TRPM2) channel contributes to beta-amyloid oligomer-related neurotoxicity and memory impairment. *J. Neurosci.* **2015**, *35*, 15157–15169. [CrossRef]
21. Li, X.; Yang, W.; Jiang, L.-H. Alteration in intracellular Zn^{2+} homeostasis as a result of TRPM2 channel activation contributes to ROS-induced hippocampal neuronal death. *Front. Mol. Neurosci.* **2017**, *10*, 414. [CrossRef] [PubMed]
22. Li, X.; Jiang, L.-H. Multiple molecular mechanisms form a positive feedback loop driving amyloid β42 peptide-induced neurotoxicity via activation of the TRPM2 channel in hippocampal neurons. *Cell. Death Dis.* **2018**, *9*, 195. [CrossRef] [PubMed]
23. Sun, Y.; Sukumaran, P.; Selvaraj, S.; Cilz, N.I.; Schaar, A.; Lei, S.; Singh, B.B. TRPM2 promotes neurotoxin MPP^+/MPTP-induced cell death. *Mol. Neurobiol.* **2018**, *55*, 409–420. [CrossRef] [PubMed]

24. Li, X.; Jiang, L.-H. A critical role of the transient receptor potential melastatin 2 channel in a positive feedback mechanism for reactive oxygen species-induced delayed cell death. *J. Cell. Physiol.* **2019**, *234*, 3647–3660. [CrossRef] [PubMed]
25. Yamamoto, S.; Wajima, T.; Hara, Y.; Nishida, M.; Mori, Y. Transient receptor potential channels in Alzheimer's disease. *Biochim. Biophys. Acta* **2007**, *1772*, 958–967. [CrossRef] [PubMed]
26. Nazıroğlu, M. TRPM2 cation channels, oxidative stress and neurological diseases: Where are we now? *Neurochem. Res.* **2011**, *36*, 355–366. [CrossRef]
27. Li, C.; Meng, L.; Li, X.; Li, D.; Jiang, L.-H. Non-NMDAR neuronal Ca^{2+}-permeable channels in delayed neuronal death and as potential therapeutic targets for ischemic brain damage. *Expert Opin. Ther. Targets* **2015**, *19*, 879–892. [CrossRef]
28. Belrose, J.C.; Jackson, M.F. TRPM2: A candidate therapeutic target for treating neurological diseases. *Acta Pharmacol. Sin.* **2018**, *39*, 722–732. [CrossRef]
29. Turlova, E.; Feng, Z.P.; Sun, H.S. The role of TRPM2 channels in neurons, glial cells and the blood-brain barrier in cerebral ischemia and hypoxia. *Acta Pharmacol. Sin.* **2018**, *39*, 713–721. [CrossRef]
30. Jiang, L.-H.; Li, X.; Syed Mortadza, S.A.; Lovatt, M.; Yang, W. The TRPM2 channel nexus from oxidative damage to Alzheimer's pathologies: An emerging novel intervention target for age-related dementia. *Ageing Res. Rev.* **2018**, *47*, 67–79. [CrossRef]
31. Xicoy, H.; Wieringa, B.; Martens, G.J. The SH-SY5Y cell line in Parkinson's disease research: A systematic review. *Mol. Neurodegener.* **2017**, *12*, 10. [CrossRef]
32. Chen, S.J.; Zhang, W.; Tong, Q.; Conrad, K.; Hirschler-Laszkiewicz, I.; Bayerl, M.; Kim, J.K.; Cheung, J.Y.; Miller, B.A. Role of TRPM2 in cell proliferation and susceptibility to oxidative stress. *Am. J. Physiol. Cell Physiol.* **2013**, *304*, C548–C560. [CrossRef] [PubMed]
33. Bao, L.; Chen, S.J.; Conrad, K.; Keefer, K.; Abraham, T.; Lee, J.P.; Wang, J.; Zhang, X.Q.; Hirschler-Laszkiewicz, I.; Wang, H.G.; et al. Depletion of the human ion channel TRPM2 in neuroblastoma demonstrates its key role in cell survival through modulation of mitochondrial reactive oxygen species and bioenergetics. *J. Bio. Chem.* **2016**, *291*, 24449–24464. [CrossRef]
34. Hirschler-Laszkiewicz, I.; Chen, S.J.; Bao, L.; Wang, J.; Zhang, X.Q.; Shanmughapriya, S.; Keefer, K.; Madesh, M.; Cheung, J.Y.; Miller, B.A. The human ion channel TRPM2 modulates neuroblastoma cell survival and mitochondrial function through Pyk2, CREB, and MCU activation. *Am. J. Physiol. Cell. Physiol.* **2018**, *315*, C571–C586. [CrossRef] [PubMed]
35. Syed Mortadza, S.A.; Sim, J.A.; Stacey, M.; Jiang, L.-H. Signalling mechanisms mediating Zn^{2+}-induced TRPM2 channel activation and cell death in microglial cells. *Sci. Rep.* **2017**, *7*, 45032. [CrossRef] [PubMed]
36. Zhang, W.; Chu, X.; Tong, Q.; Cheung, J.Y.; Conrad, K.; Masker, K.; Miller, B.A. A novel TRPM2 isoform inhibits calcium influx and susceptibility to cell death. *J. Biol. Chem.* **2003**, *27*, 16222–16229. [CrossRef]
37. Zhang, W.; Hirschler-Laszkiewicz, I.; Tong, Q.; Conrad, K.; Sun, S.C.; Penn, L.; Barber, D.L.; Stahl, R.; Carey, D.J.; Cheung, J.Y.; et al. TRPM2 is an ion channel that modulates hematopoietic cell death through activation of caspases and PARP cleavage. *Am. J. Physiol. Cell Physiol.* **2006**, *290*, C1146–C1159. [CrossRef]
38. Jiang, L.-H.; Syed Mortadza, S.A. Transient receptor potential cation channel subfamily M member 2. In *Encyclopedia of Signaling Molecules*; Choi, S., Ed.; Springer International Publishing: Cham, Switzerland, 2018. [CrossRef]
39. Maiti, P.; Manna, J.; Dunbar, G.L. Current understanding of the molecular mechanisms in Parkinson's disease: Targets for potential treatments. *Transl. Neurodegener.* **2017**, *6*, 28. [CrossRef]

10

Beneficial Effects of Vitamins K and D3 on Redox Balance of Human Osteoblasts Cultured with Hydroxyapatite-Based Biomaterials

Ewa Ambrożewicz [1], Marta Muszyńska [1], Grażyna Tokajuk [2], Grzegorz Grynkiewicz [3], Neven Žarković [4] and Elżbieta Skrzydlewska [1,*]

1 Department of Analytical Chemistry, Medical University of Bialystok, 15-222 Bialystok, Poland; ewa.ambrozewicz@umb.edu.pl (E.A.); marta.muszynska@umb.edu.pl (M.M.)

2 Department of Integrated Dentistry, Medical University of Bialystok, 15-230 Bialystok, Poland; grazyna.t1@gmail.com

3 Pharmaceutical Research Institute, 01-793 Warsaw, Poland; g.grynkiewicz@ifarm.eu

4 Laboratory for Oxidative Stress, Rudjer Boskovic Institute, 10000 Zagreb, Croatia; zarkovic@irb.hr

* Correspondence: elzbieta.skrzydlewska@umb.edu.pl

Abstract: Hydroxyapatite-based biomaterials are commonly used in surgery to repair bone damage. However, the introduction of biomaterials into the body can cause metabolic alterations, including redox imbalance. Because vitamins D3 and K (K1, MK-4, MK-7) have pronounced osteoinductive, anti-inflammatory, and antioxidant properties, it is suggested that they may reduce the adverse effects of biomaterials. The aim of this study was to investigate the effects of vitamins D3 and K, used alone and in combination, on the redox metabolism of human osteoblasts (hFOB 1.19 cell line) cultured in the presence of hydroxyapatite-based biomaterials (Maxgraft, Cerabone, Apatos, and Gen-Os). Culturing of the osteoblasts in the presence of hydroxyapatite-based biomaterials resulted in oxidative stress manifested by increased production of reactive oxygen species and decrease of glutathione level and glutathione peroxidase activity. Such redox imbalance leads to lipid peroxidation manifested by an increase of 4-hydroxynonenal level, which is known to influence the growth of bone cells. Vitamins D3 and K were shown to help maintain redox balance and prevent lipid peroxidation in osteoblasts cultured with hydroxyapatite-based biomaterials. The strongest effect was observed for the combination of vitamin D3 and MK-7. Moreover, vitamins promoted growth of the osteoblasts, manifested by increased DNA biosynthesis. Therefore, it is suggested that the use of vitamins D3 and K may protect redox balance and support the growth of osteoblasts affected by hydroxyapatite-based biomaterials.

Keywords: hydroxyapatite-based biomaterials; osteoblast growth; redox balance; vitamins; lipid peroxidation; 4-hydroxynonenal; oxidative stress

---

1. Introduction

Hydroxyapatite-based biomaterials possess osteoconductive properties and can also act as three-dimensional scaffolds to support bone regeneration [1]. However, introducing biomaterials into the body can lead to metabolic alterations. Namely, interactions between osteoblasts and biomaterials may interfere with cellular metabolism, including redox balance leading to oxidative stress [2]. The ingredients released by biomaterials may also take part in this activity [3,4]. Compounds released from bone substitutes can have adverse effects on the viability and function of osteoblasts, even in the absence of physical contact [2,5,6]. There is also evidence that the production of pro-inflammatory cytokines by osteoblasts is increased in the presence of hydroxyapatite [7]. The lack of cytocompatibility of bone substitutes can hinder bone regeneration and prolong wound healing [8]. To reduce the cytotoxicity of biomaterials and enhance their desirable bioactivities, they are often modified by the inclusion of various compounds, including growth factors [9] and antioxidants, e.g., N-acetylcysteine [5], or combined with other components, such as matrix-derived proteins [10].

The physiology of bone metabolism is dependent on the synergistic interplay between vitamins D3 and K [11]. The active form of vitamin D3 is 1,25-dihydroxycholecalciferol (1,25(OH)2D3; calcitriol). Calcitriol forms a complex with a specific nuclear receptor, commonly known as the vitamin D receptor (VDR), to control the expression of key bone-related proteins in osteoblasts, including osteocalcin (OC), and to regulate calcium and phosphate homeostasis [12,13]. The antioxidant and anti-inflammatory properties of vitamin D3 have been confirmed both in vitro and in vivo. Under pathological conditions, vitamin D3 has been shown to enhance the level and activity of antioxidant proteins which protect cellular proteins from oxidative modification. Vitamin D3 activates the Nrf2-Keap1 antioxidant pathway leading to transcription of antioxidant proteins and has been shown to reduce levels of advanced glycation end-products in the aortic wall of diabetic rats [14,15]. Administering vitamin D3 to rats increased the activity of an antioxidant enzyme superoxide dismutase (SOD) in the liver. Furthermore, a decrease in vitamin D3 levels in the blood serum in obese children was correlated with a decrease in SOD levels [16–19]. Changes in antioxidant capacity correspond to an increase in lipid peroxidation [20]. Vitamin D3 also inhibits oxidative stress and endoplasmic reticulum stress in endothelial cells thereby reducing the risk of cardiovascular disease in humans [21] and preventing endothelial cell death [22]. In addition, vitamin D3 plays a role in T cell-mediated immunity [23].

The vitamin K family is a group of fat-soluble vitamins that regulate metabolic processes [24]. The two main forms of vitamin K are phylloquinone (K1) and menaquinone (K2), including the menaquinone-4 (MK-4) and menaquinone-7 (MK-7) homologues. Vitamin K1 is transported to the liver where it regulates the production of coagulation factors, whereas vitamin K2 is found in many tissues, including bone, where it regulates the activity of vitamin K-dependent proteins, such as matrix carboxyglutamic acid (Gla)-protein (MGP) and OC (GLA protein) [25]. Bone metabolism depends on interaction between vitamins D3 and K2. Vitamin D promotes vitamin K-dependent protein production while vitamin K activates proteins involved in bone metabolism. Vitamin K acts as a cofactor in the carboxylation of glutamic acid (Glu) to Gla and the metabolically active form of OC, which can bind and deposit calcium in the extracellular matrix [11,26]. In vitro, vitamin K2 was shown to promote osteoclast apoptosis in a dose-dependent manner and to suppress osteoblast apoptosis [27,28]. Moreover, it was found that vitamin K2 increases the number and activity of osteoclasts [28] and the MK-4 homologue promotes osteoblast differentiation and proliferation of osteoclasts [29].

Vitamin K participates in blood coagulation, signal transduction processes, and cell proliferation [30], and has the ability to alter redox balance in cells [31]. In its reduced form, vitamin K hydroquinone (KH2) protects phospholipid membranes from peroxidation by direct reactive oxygen species (ROS) uptake [32]. During the vitamin K cycle, hydroquinone is reconstituted with the participation of *vitamin K epoxide reductase complex subunit 1* (VKORC1) [31]. Vitamin K inhibits the activation of 12-lipoxygenase (12-LOX) to prevent the formation of ROS [33]. The antioxidant actions of vitamins K1 and K2 (MK-4) protect oligodendrocytes and neurons from oxidative stress caused by glutathione (GSH) deficiency [34]. Nanomolar concentrations of vitamins K1 and K2 (MK-4) were also shown to prevent oxidative stress-induced neuronal death [33]. Since oxidative stress is directly associated with inflammation, vitamin K may reduce oxidative stress by lowering levels of pro-inflammatory factors [35,36].

There is a growing body of evidence to suggest vitamin K and vitamin D have synergistic effects on bone metabolism [11,37,38]. The aim of the present study was to investigate the effect of vitamin D3 and K on cellular metabolism and to determine whether vitamins D3, K1, MK-4, and MK-7 could be potentially used to support bone regeneration alongside hydroxyapatite-based biomaterials. Therefore, redox balance in osteoblasts cultured in presence of biomaterials was determined. The biomaterials characterized by origin—human (Maxgraft), bovine (Cerabone), porcine (Apatos and Gen-Os), and composition—hydroxyapatite (Maxgraft, Cerabone, Apatos) and hydroxyapatite enriched with collagen (Gen-Os) were examined.

2. Materials and Methods

2.1. Materials

To evaluate the in vitro effect of vitamin K (K1, MK4, and MK7) and vitamin D3 on osteoblasts' metabolism cultured with biomaterial cells were cultured on commercial biomaterials of different origin and chemical composition. Vitamin K1 (Sigma–Aldrich, St. Louis, MO, USA), vitamin K2 (Menaquinone 4, MK4 (Sigma–Aldrich, St. Louis, MO, USA) and Menaquinone 7, MK7 (was chemically synthesized by the Pharmaceutical Research Institute, Warsaw, Poland)), and vitamin D3 (1,25-dihydroxyvitamin D3 (1,25(OH)2D3) (Santa Cruz Biotechnology, USA)), was added to cells. Four different hydroxyapatites: human origin (Maxgraft, Botiss biomaterials GmgH, Germany), bovine origin (Cerabone, Botiss Biomaterials GmbH, Germany), porcine origin (Apatos, Tecnoss, Italy), and porcine hydroxyapatite with preserved collagen (Gen-Os, Tecnoss, Italy) were used for experiments.

2.2. Cell Cultures and Treatments Used

All experiments were performed using a human fetal osteoblast cell line (hFOB 1.19) obtained from ATCC (American Type Culture Collection, Menassas, VA, USA). The optimal time to assess metabolic changes during osteoblast differentiation is about 20 days [39]. During this period, three stages of osteoblast differentiation can be distinguished: proliferation (until day 4), synthesis of extracellular matrix (between 12–16 days), and mineralization of the extracellular matrix of the bone (about day 20) [39,40]. Cells were cultured for 20 days at 39 °C in 5% CO_2 in air in a 1:1 mixture of Ham's F12 medium and Dulbecco's modified Eagles's medium containing 2.5 mmol/L L-Glutamine without phenol red and supplemented with 0.3 mg/mL G418, 100 U/mL penicillin, 100 μg/mL streptomycin, and 10% fetal bovine serum (FBS). Culture medium was renewed every 2 days.

For all experiments, osteoblasts were seeded on each of the hydroxyapatite-based biomaterials (Maxgraft, Cerabone, Apatos, Gen-Os; 100 mg/well) at a density of 1×10^5 cells/well in 6-well plates. The influence of vitamin D3 and vitamin K on osteoblast growth was investigated by dividing the cells into several treatment groups: cells treated with vitamin D3, cells treated with vitamin K (K1, MK-4, or MK-7), and cells treated with combinations of vitamin D3 and K (D3+K1, D3+MK-4, or D3+MK-7). Controls without any hydroxyapatite were cultured in parallel. Cells were supplemented with vitamin K1 and K2 (MK-4 and MK-7) at a concentration of 10 μmol/L [29,41] and/or vitamin D3 at concentration 1 nM in the presence of 10 mmol/L sodium β-glycerophosphate to accelerate the mineralization process. After 4, 8, 12, 16, and 20 days, cells were collected. Vitamin D3 was used in its metabolically active form of 1,25(OH)2D3, which has previously been shown to regulate osteoblast differentiation [42,43]. The MTT *(3-(4,5-dimethylthiazol-2-yl)-2,5-diphenyltetrazolium bromide)* colorimetric assay was used to study the effect of vitamin D3 on cells and to determine the optimal concentration of vitamin D following a procedure previously described in the literature [44]. The concentration of vitamin K used for the MTT assay did not affect cell viability.

To examine the effects of vitamins D3, K1, K2 (MK-4 and MK-7), and hydroxyapatite on osteoblast proliferation and differentiation, DNA content as well as alkaline phosphatase (ALP) and OC levels were determined. Redox status was estimated for each group based on measured ROS, GSH, and 4-hydroxynonenal (4-HNE) levels.

2.3. Determination of ROS Level

Electron spin resonance (ESR) spectrometer e-scans (Noxygen GmbH/Bruker Biospin GmbH, Germany) were used to detect total ROS generation using spin probe CMH

(1-hydroxy-3-methoxycarbonyl-2,2,5,5-tetra-methylpyrrolidine, 200 μmol/L), which selectively interacts with ROS to form a stable nitroxide CM-radical (carbamoyl radical) with a half-life of 4 h [43]. After 4, 8, 12, 16, and 20 days in culture, the cells were suspended in Krebs HEPES buffer (2-[4-(2-hydroxyethyl)piperazin-1-yl]ethanesulfonic acid) (KHB) and the CMH spin probe was added. Samples were incubated for 30 min at 37 °C in a mixture of N2 and 02 of 96:4 (02:N2) and electron paramagnetic resonance (EPR) was performed on cell lysates using the following acquisition parameters: field center 1.99 g, microwave power 20 mW, amplitude of modulation 2 G, sweep time 10 s, number of scans 10, range of deviation 60 G.

2.4. Determination of GSH Level

Non-enzymatic antioxidant—glutathione (GSH)—was quantified using capillary electrophoresis (CE) method of Maeso et al. [45]. Cells were sonicated with a mixture containing acetonitrile/water (ACN/H2O 62.5:37.5, v/v) and centrifuged at 30,000× g for 10 min. The separation of supernatant components was performed on a capillary with 47 cm total length (40 cm effective length) and 50 μm i.d. and was operated at 27 kV with UV detection at 200 ± 10 nm.

2.5. Determination of Glutathione Peroxidase Activity

Glutathione peroxidase (GSH-Px, EC 1.11.1.6) activity was estimated by spectrophotometry using a microplate reader (Infinite M200, Tecan), according to the method of Paglia and Valentine [46]. Enzyme activity was measured using an indirect method involving two conjugated reactions: glutathione oxidation reaction (catalyzed by glutathione peroxidase) and glutathione reduction reaction with simultaneous oxidation of reduced nicotinamide adenine dinucleotide phosphate (NADPH) to NADP. The cell lysate was combined with a mixture containing 1.5 mM ethylenediaminetetraacetic acid (EDTA; in 0.2 mol/L Tris-HCl buffer; pH 7.6), 15 mmol/L sodium azide, and 0.72 mmol/L NADPH in a 96-well plate and incubated for 5 min at 20 °C. Then glutathione reductase and 0.1 mL of 10 mmol/L H_2O_2 in 0.02 mmol/L Tris-HCl buffer (pH 7.6) were added to each sample. Absorbance was measured at 340 nm for 1 min relative to Tris-HCl buffer. One unit of GSH-Px activity was defined as the amount of enzyme catalyzing the oxidation of 1 μmol NADPH min^{-1} at 25 °C and pH 7.4. Enzyme specific activity was expressed as micro-units per mg of protein.

2.6. Determination of Lipid Peroxidation

Lipid peroxidation was estimated based on the level of 4-hydroxynonenal (4-HNE) as a derivative of O-pentafluorobenzyl-oxime-trimethyl silyl ether (O-PFB-oxime-TMS) [47], assessed using gas chromatography-mass spectrometry (GC/MS) in selected ion monitoring mode (SIM), according to a previously described method [48]. Briefly, aldehydes were derivatized by the addition of O-(2,3,4,5,6-pentafluorobenzyl) hydroxylamine (PFB) in PIPES buffer (piperazine-N,N′-bis(2-ethanesulfonic acid) in the presence of benzaldehyde-D6 as an internal standard (IS). The O-PFB-oxime aldehyde derivative was extracted using hexane. The hexane phase was evaporated under a stream of argon followed by the addition of N,O-bis(trimethylsilyl)trifluoroacetamide in 1% trimethylchlorosilane to form TMS ether of the hydroxyaldehyde group. A 1 μl aliquot was injected into the GC-MS. Derivatized aldehyde was analyzed using a 7000 quadrupole MS/MS (7890A GC, Agilent Technologies, USA) equipped with a HP-5 ms capillary column (0.25-mm internal diameter, 0.25-μm film thickness, 30-m length). The column temperature was initially set to 50 °C for 1 min, increased at a rate of 10 °C/min to 200 °C, then at 3 °C/min to 220 °C, 20 °C/min to 310 °C, and finally, maintained at 310 °C for 5 min. The injector temperature was maintained at 250 °C, the transfer line was held constant at 280 °C, and the source temperature was set to 230 °C. Derivatized aldehyde was detected by selected ion-monitoring GC/MS using the following ions: m/z 333.0 and 181.0 for 4-HNE-PFB-TMS and m/z 307.0 for nternal standard (IS) derivative.

2.7. DNA Proliferation Assay

A DNA fluorometric assay for measuring osteoblasts proliferation was used [49]. Briefly, medium was discarded, and cells were washed with a washing solution (0.8% NaCl, 0.04% KCl, 0.1% ethylenediaminetetraacetic acid EDTA, and 1% NaN3). The plates were then allowed to dry at room temperature for 10 min, wrapped with parafilm, and frozen at −70 ° C until analysis. After removal from the freezer, 0.01% dodecyl sodium sulfate (SDS) was added, and samples were incubated at room temperature for 15 min. Cell lysates in the amount of 10 μL were put into a 96-well plate with H33258 at a concentration of 2 μg/mL. The DNA level was measured under excitation of 360 nm and emission at 460 nm and calculated from the calibration curve for Calf Thymus DNA solution. The DNA level was expressed as ng of DNA per mg of protein.

2.8. Determination of ALP Activity

Alkaline phosphatase (ALP) activity was determined fluorometrically using the p-nitrophenyl phosphate (p-NPP) according to Reference [50] at 405 nm. Briefly, a lysis buffer (10 mmol/L Tris-HCl, pH 7.5, containing 0.5 mmol/L MgCl2, 0.1% Triton X-100) was added to cells, and after 10 min incubation on ice, cells were collected in tubes and stored at −70 °C until analysis. After thawing, samples were sonicated and centrifuged at 10,000× *g* for 15 min at 4 °C. Next, glycine buffer (25 mmol/L, pH 10.4) containing 2 mmol/L MgCl2 and 10 mmol/L p-NPP was added to samples, and then incubated at 37 °C for 25 min. The enzymatic reaction was stopped by the addition of 3 mol/L NaOH. The activity of ALP was expressed as the amount of enzyme (nmol) that catalyzed the reaction with substrate per minute per mg of protein.

2.9. Determination of OC Level

The osteocalcin (Gla-OC) concentration was determined in the culture medium. The osteocalcin level was determined using a commercially available human-specific enzyme-linked immunosorbent assay (ELISA) kit (Gla-type Osteocalcin (Gla-OC) EIA Kit, TaKaRa, Bio Inc.). The tests were performed according to the manufacturer's protocol. The OC level was expressed as ng of osteocalcin per ml of culture media.

2.10. Statistical Analysis

Data were analyzed using standard statistical analyses, one-way/two-way analysis of variance (ANOVA) to determine significant differences between different groups. All analyses on cells were performed on the results obtained from three independent experiments. The results are expressed as the mean ± standard deviation (SD) for $n = 3$. p-values less than 0.05 were considered significant.

3. Results

3.1. Biochemical Studies

The results obtained show that the growth of osteoblasts was associated with a gradual increase of ROS levels, the largest increase being observed between the days 8 and 12 (Figure 1). While vitamin D3 did not affect such changes in ROS levels for cultured osteoblasts, the other vitamins attenuated continuous increases of ROS. The most prominent reduction of ROS production was observed in MK-7-treated osteoblasts at day 12 and day 16 (16% and 23%, respectively), whereas ROS levels observed in the K1-treated osteoblasts decreased by 11–17% and in MK-4-treated osteoblasts, by 14–15%.

Opposite to the vitamins, the hydroxyapatite-based biomaterials further enhanced production of ROS by cultured osteoblasts, which was also shown to be significantly reduced by vitamins, depending on the type of biomaterials and vitamins used (Figure 1b–e). Osteoblasts cultured on hydroxyapatite in medium containing vitamin D3 with either vitamin K1 or K2 exhibited significantly lower ROS production, particularly at day 16 and 20, compared to controls (10–20%) or cells grown on hydroxyapatite-based biomaterials alone (10–20%).

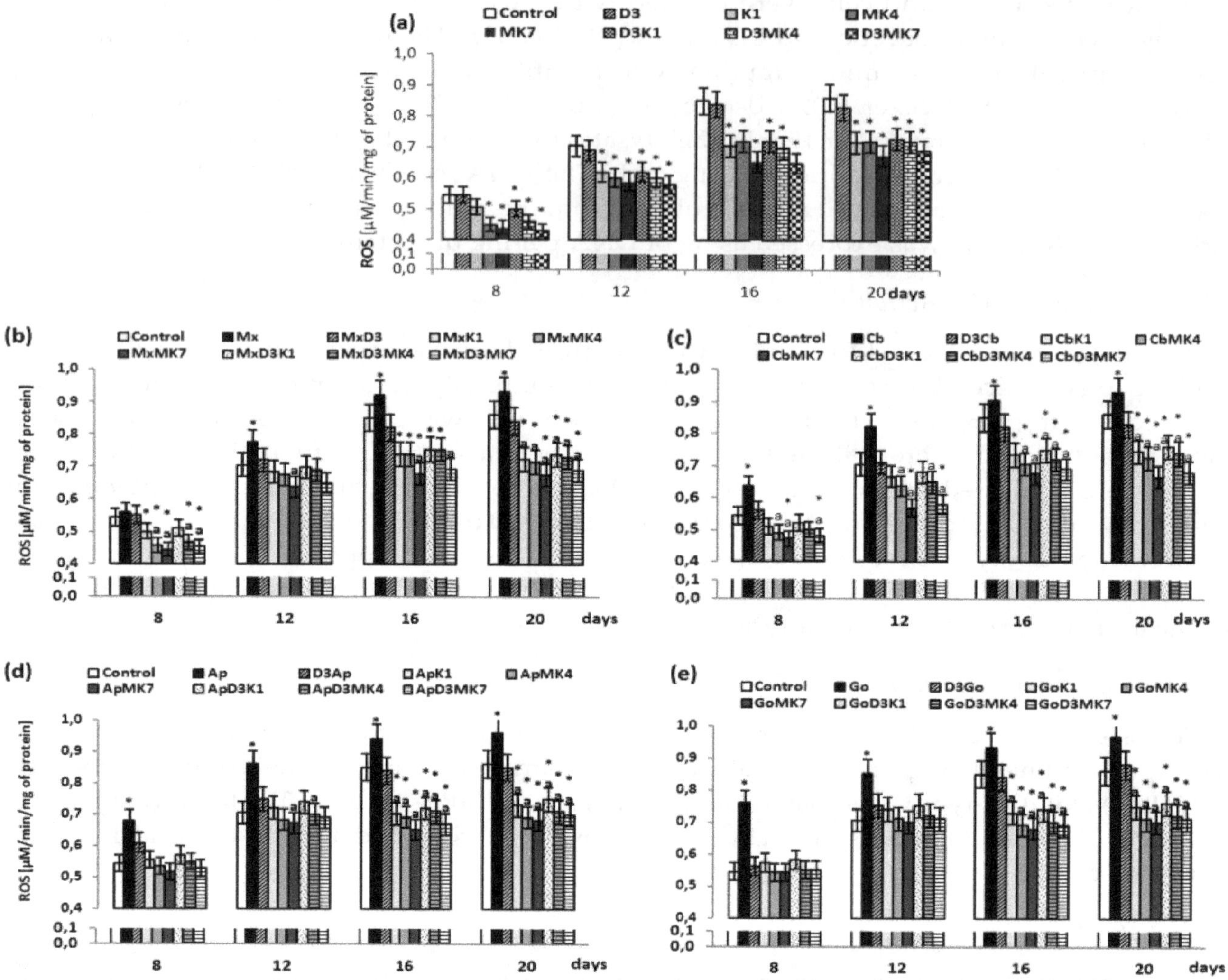

Figure 1. Reactive oxygen species (ROS) level in osteoblasts after incubation with hydroxyapatites and treated with vitamins D3, K1, MK4, and MK7 after 8, 12, 16, and 20 days. The results are expressed as the μM/min/mg of protein and are shown as the mean ± SD (n = 5). The values for the control cells and the treated cells were significantly different according to unpaired Student's t-test. * Statistically significant differences versus control, $p < 0.05$; (**a**) statistically significant differences versus group hydroxyapatites (Mx (Maxgraft), Cb (Cerabone), Ap (Apatos), Go (Gen-Os), respectively for graphs (**b**–**e**)); $p < 0.05$.

Levels of the main cytosolic antioxidant, GSH, gradually increased in osteoblast cultures until day 12; however, a decrease was observed at day 20 (Figure 2). The addition of vitamin D3 as well as K1 and K2 to culture medium significantly increased GSH concentrations compared to controls. Vitamin K1 and K2 had slightly less of an effect than vitamin D3. The largest increases in GSH concentration were observed between day 8 and 20. Furthermore, the combination of vitamin D3 and vitamin K1 or K2 resulted in enhanced GSH levels compared to cultures treated with only the vitamin K variants. Cells cultured with hydroxyapatite (Apatos and Gen-Os), exhibited lower GSH levels (Figure 2b–e); moreover, the addition of vitamin D3 and a mixture of vitamins D3 and K1 or K2 led to a significant increase in GSH levels compared to controls. The largest increase was observed between day 8 and day 20 in osteoblast cultures treated with vitamin D3 and K1 or K2 in the presence of Maxgraft (17–58%) or Cerabone (28–54%).

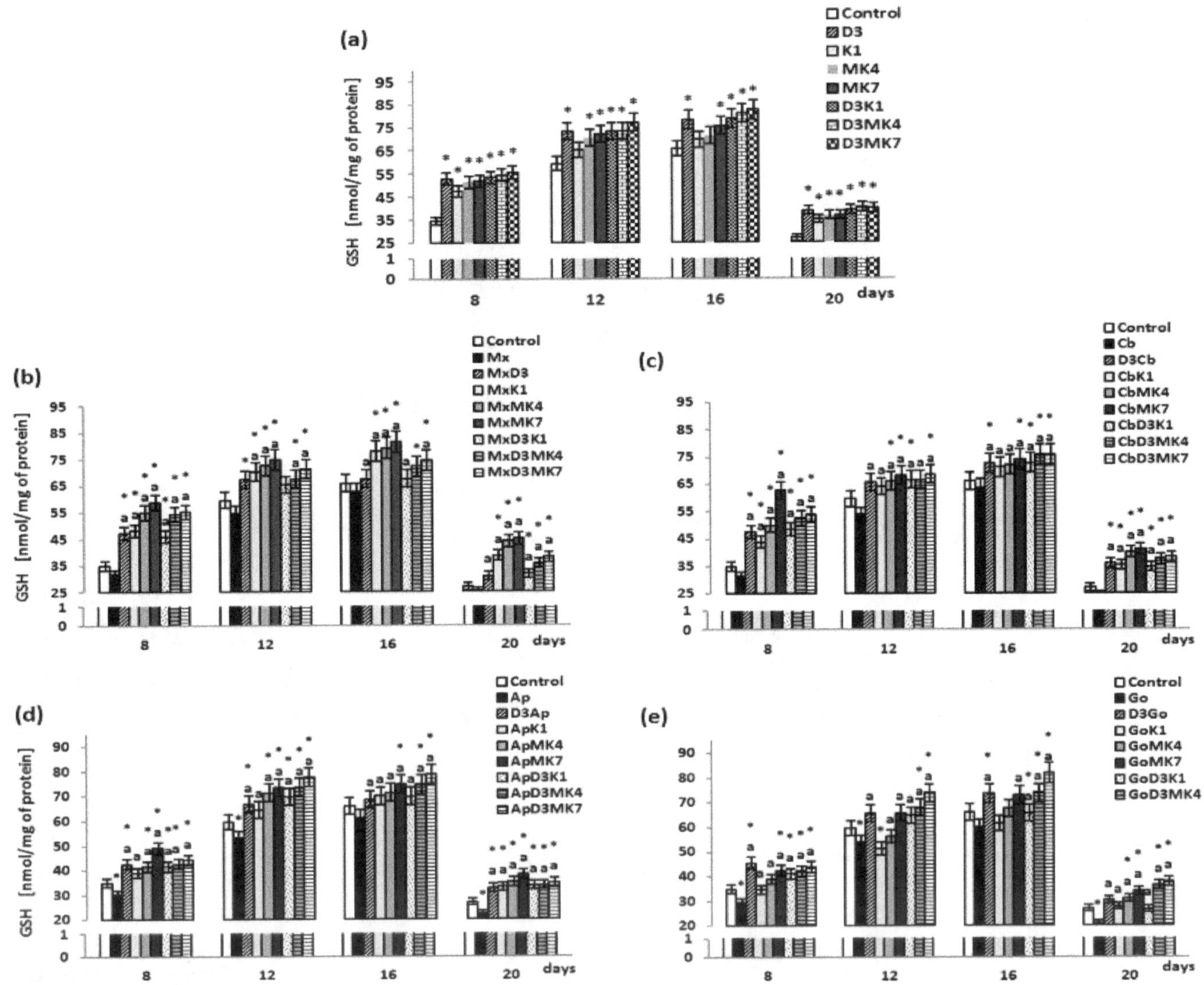

Figure 2. Glutathione (GSH) level in osteoblasts after incubation with hydroxyapatites and treated with vitamins D3, K1, MK4, and MK7 after 8, 12, 16, and 20 days. The results are expressed as the nmol/mg of protein and are shown as the mean ± SD ($n = 5$). The values for the control cells and the treated cells were significantly different according to unpaired Student's t-test. * Statistically significant differences versus control, $p < 0.05$; (**a**) statistically significant differences versus group hydroxyapatites (Mx (Maxgraft), Cb (Cerabone), Ap (Apatos), Go (Gen-Os), respectively for graphs (**b**–**e**)); $p < 0.05$.

Observed changes in GSH level were associated with changes in GSH-Px activity since GSH is its co-substrate. Vitamin D3 as well as MK-4 and MK-7 significantly enhanced the activity of GSH-Px by approximately 16–13% for vitamin D3 alone between 8 and 20 days and by approximately 14% for D3+MK-4 and D3+MK-7 on day 8 and day 20, compared to controls (Figure 3a). Therefore, the combination of D3+MK-4 or D3+MK-7 significantly enhanced GSH-Px activity. The most effective treatment was D3+MK-7 which resulted in a 28% increase after 16 days. The presence of hydroxyapatite-based biomaterials, in particular Apatos and Gen-Os, led to a decrease in GSH-Px activity of approximately 11% (Figure 3b–e). Osteoblasts cultured in the presence of hydroxyapatite but treated with vitamin D3 were characterized by significantly higher GSH-Px activity (9–12% and 11–19% for Apatos and Gen-Os, respectively) compared to cells grown only in the presence of hydroxyapatite. However, a combination of vitamins, D3+MK-4 or D3+MK-7, led to a significant increase in GSH-Px activity even in the presence of hydroxyapatite. For Apatos, GSH-Px activity was enhanced by approximately 11–17% and 20% with D3+MK-4 and D3+MK-7 treatment, respectively; for Cerabone, approximately 15–24% for D3+MK-4 and 18–24% for D3+MK-7, and for Gen-Os, 7–16% for D3+MK-4, and 9–19% for D3+MK-7.

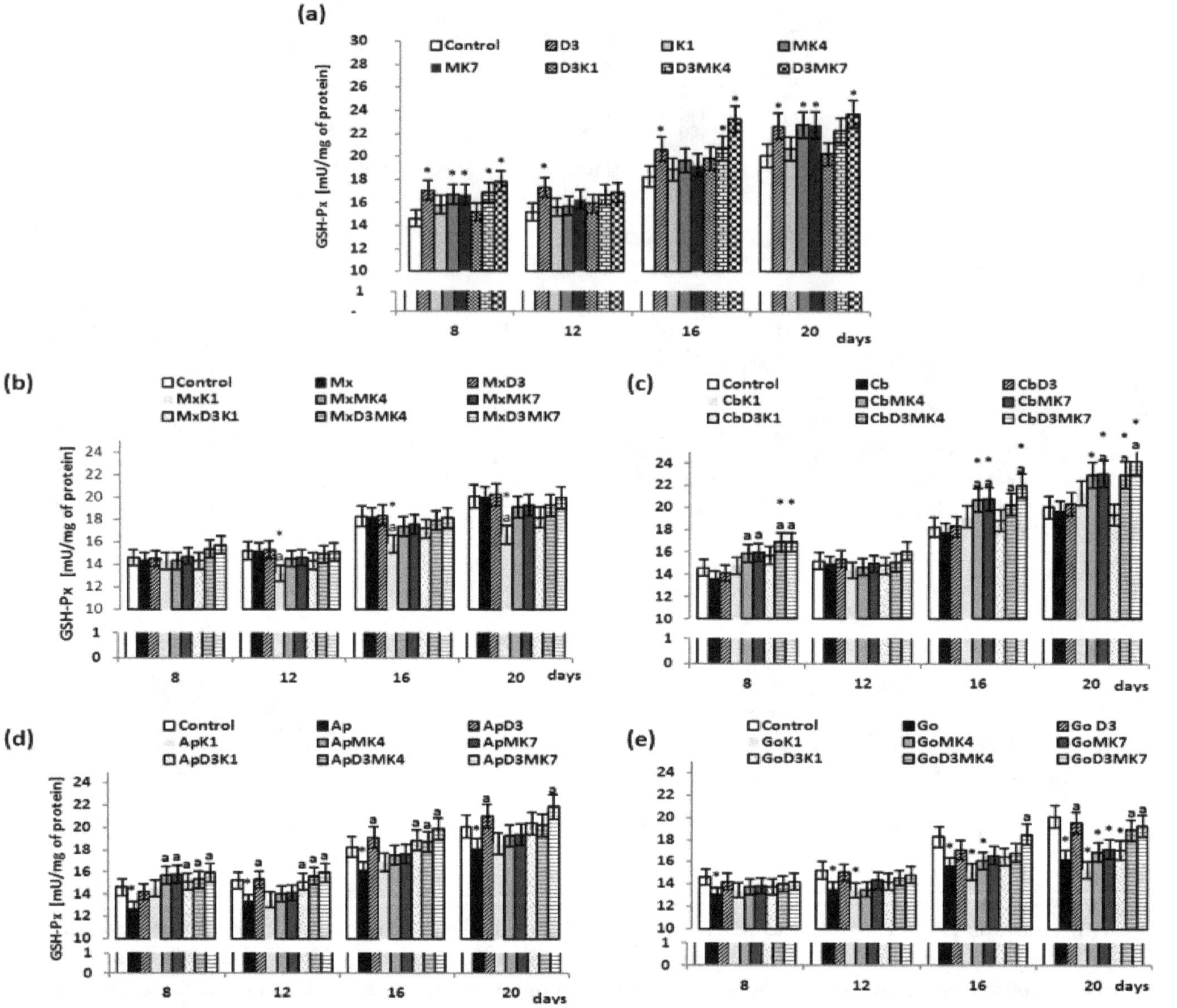

Figure 3. GSH-Px activity in osteoblasts after incubation with hydroxyapatites and treated with vitamins D3, K1, MK4, and MK7 after 8, 12, 16, and 20 days. The results are expressed as the mU/mg of protein and are shown as the mean ± SD ($n = 5$). The values for the control cells and the treated cells were significantly different according to unpaired Student's t-test. * Statistically significant differences versus control, $p < 0.05$; (**a**) statistically significant differences versus group hydroxyapatites (Mx (Maxgraft), Cb (Cerabone), Ap (Apatos), Go (Gen-Os), respectively for graphs (**b**–**e**)); $p < 0.05$.

Furthermore, biomaterials and vitamins affected redox balance in cultured osteoblasts by modifying lipid metabolism, in the manner similar as observed for ROS levels. Namely, the growth of osteoblasts was associated with continuous increase of 4-HNE levels (Figure 4). However, different from the lack of influence of vitamin D3 on the ROS change, the most significant effects on the 4-HNE production were observed in vitamin D3-treated cultures (Figure 4a), followed by MK-7, with a reduction of approximately 12–19%. As was noticed in the case of the change of ROS (Figure 1), the presence of hydroxyapatite biomaterials also caused a significant increase in lipid peroxidation, i.e., enhanced production of 4-HNE, while vitamins attenuated lipid peroxidation induced by the osteoblast growth also if cultured in the presence of hydroxyapatite.

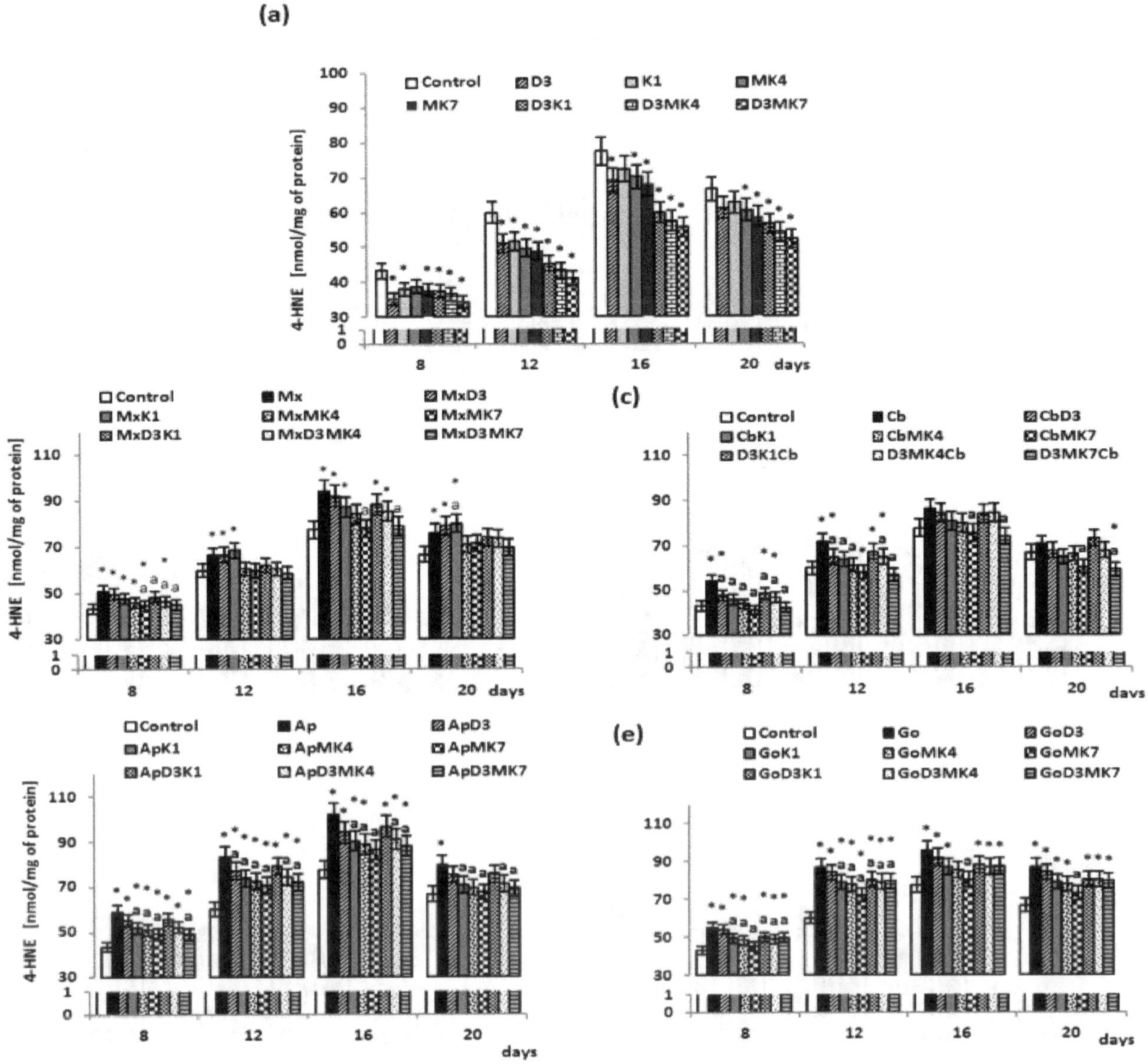

Figure 4. 4-HNE level in osteoblasts after incubation with hydroxyapatites and treated with vitamins D3, K1, MK4, and MK7 after 8, 12, 16, and 20 days. The results are expressed as the nmol/mg of protein and are shown as the mean ± SD ($n = 5$). The values for the control cells and the treated cells were significantly different according to unpaired Student's t-test. * Statistically significant differences versus control. $p < 0.05$; (**a**) statistically significant differences versus group hydroxyapatites (Mx (Maxgraft), Cb (Cerabone), Ap (Apatos), Go (Gen-Os), respectively for graphs (**b**–**e**)), $p < 0.05$.

3.2. Osteoblast Growth

Significant increase in the osteoblasts' DNA levels were observed on day 12 and thereafter, remaining constant until day 20 (Figure 5). Vitamin D3 did not affect DNA levels, whereas vitamin K1 accelerated cell proliferation by 12–23%, MK-4 by 15–27%, and MK-7 by 16–51% (Figure 5A). Addition of vitamin D3 to vitamin K1 and K2 did not change the rate of osteoblast proliferation. All hydroxyapatites, with the exception of Maxgraft, inhibited osteoblast proliferation (Figure 5b–e). The most notable changes were observed for Cerabone, Apatos, and Gen-Os on day 4 and 8 of culture, with a decline in DNA levels of 23%, 24%, and 37% (day 4), and 12%, 15%, and 36% (day 8), compared to controls. When vitamin D3 was present, osteoblast proliferation was not affected. In contrast, the combinations D3+K1, D3+MK-4, and D3+MK-7 increased DNA levels in both controls with no hydroxyapatite (10–34%) and osteoblasts cultured in the presence of hydroxyapatite (10–50%); however, D3+MK-7 was most effective.

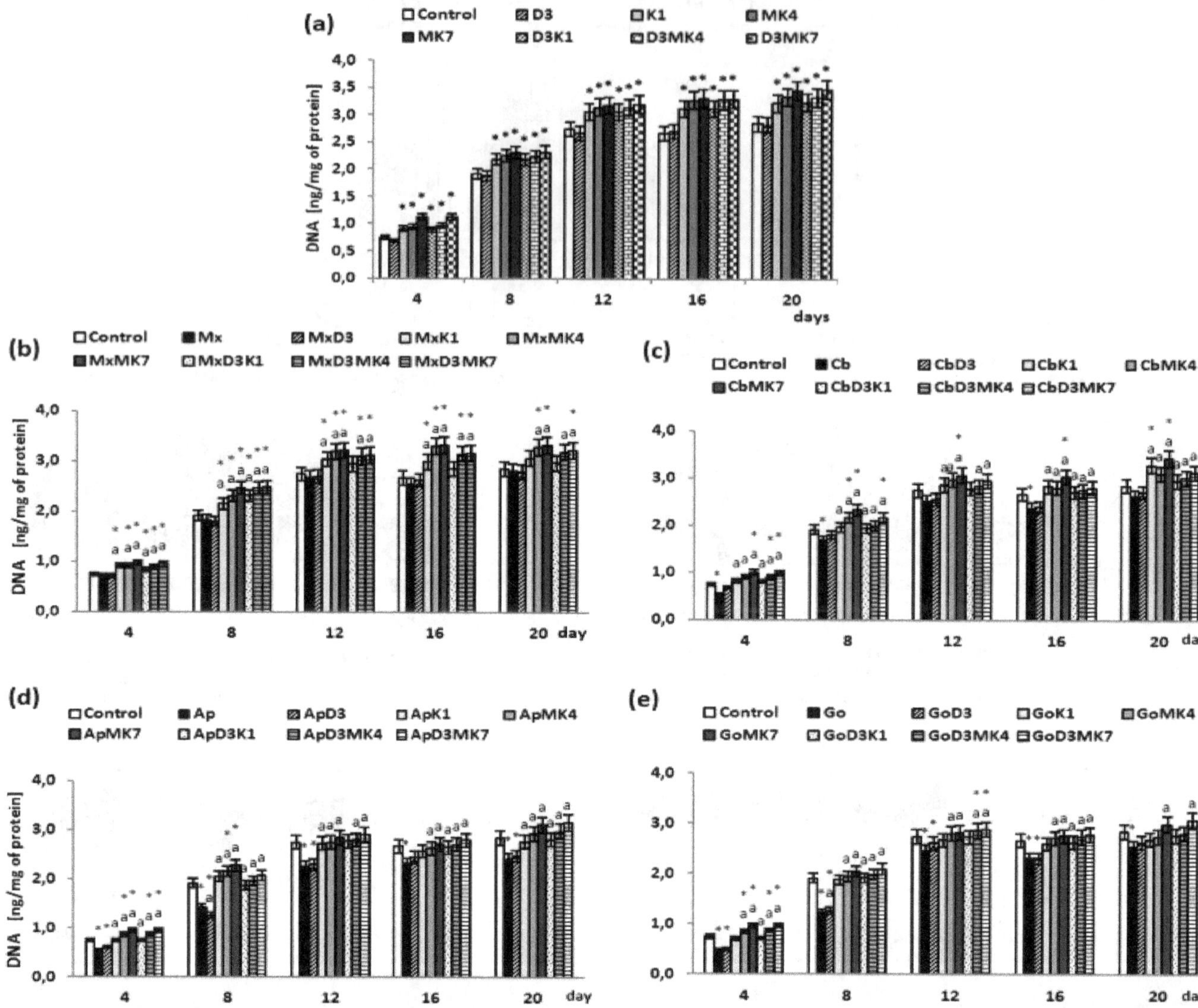

Figure 5. DNA levels in osteoblasts after incubation with hydroxyapatites and treated with vitamins D3, K1, MK4, and MK7 after 4, 8, 12, 16, and 20 days. The results are expressed as the ng/mg of protein and are shown as the mean ± SD (n = 5). The values for the control cells and the treated cells were significantly different according to unpaired Student's t-test. * Statistically significant differences versus control, $p < 0.05$; (**a**) statistically significant differences versus hydroxyapatites (Mx (Maxgraft), Cb (Cerabone), Ap (Apatos), Go (Gen-Os), respectively for graphs (**b**–**e**)), $p < 0.05$;

The osteoblast differentiation process, characterized by increasing ALP activity over time, reached a maximum after 16 days of culture, and thereafter, began to decrease (Figure 6). Vitamin D3 significantly enhanced ALP activity (40–150%) compared to the control (Figure 6a). Vitamin K1 and K2 did not affect ALP activity; however, the combination of vitamins D3+K1, D3+MK-4, and D3+MK-7 significantly increased ALP activity. At day 20, the largest increase was observed in MK-7-treated cultures, with a 2.5-fold increase. The presence of hydroxyapatite-based biomaterials did not significantly affect ALP activity, with the exception of an increase observed with Maxgraft (Figure 6b–e). Osteoblasts cultured with hydroxyapatite and treated with vitamin D3 exhibited significantly higher ALP activity compared to cells grown with only hydroxyapatite and control (no hydroxyapatite). Moreover, the addition of vitamin D3+K1, D3+MK-4 or D3+MK-7 led to a significant increase in ALP activity in the presence of hydroxyapatite. The combination of vitamin

D3 and MK-7 led to the largest increase in ALP activity, particularly evident on day 4 (a two-fold to four-fold increase).

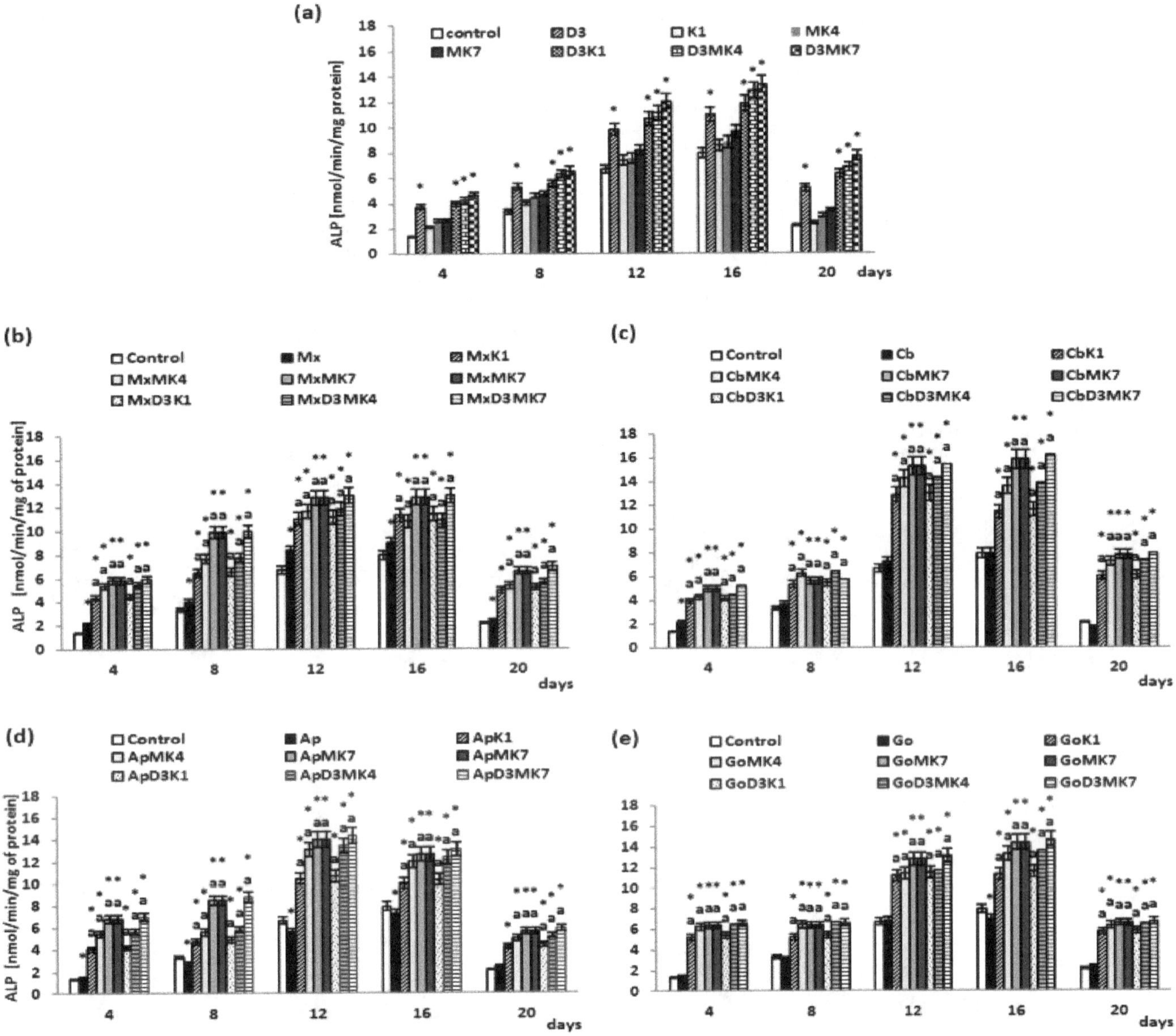

Figure 6. ALP activity in osteoblasts after incubation with hydroxyapatites and treated with vitamins D3, K1, MK4, and MK7 after 4, 8, 12, 16, and 20 days. The results are expressed as the nmol/min/mg of protein and are shown as the mean ± SD ($n = 5$). The values for the control cells and the treated cells were significantly different according to unpaired Student's t-test. * Statistically significant differences versus control, $p < 0.05$; (**a**) statistically significant differences versus hydroxyapatites (Mx (Maxgraft), Cb (Cerabone), Ap (Apatos), Go (Gen-Os), respectively for graphs (**b**–**e**)), $p < 0.05$.

Cell culture media was collected, and OC levels were estimated as a marker of mineralization. The OC concentration gradually increased from day 8 to day 20 (Figure 7). Furthermore, synthesis of OC was enhanced by vitamin D3 treatment compared to control (Figure 7a). Vitamin K1 did not affect OC levels, whereas K2 caused an increase in OC production of 16–18% with MK-4 and 22–23% with MK-7, at day 16 and 20, respectively. The level of OC also increased in the cell culture media of D3+K1 (41–66%), D3+MK-4 (47–82%), and D3+MK-7 (55–116%) groups compared to controls. A decrease in OC concentration (approximately 10–26%) was observed in the media of osteoblasts cultures in the presence of Apatos and Gen-Os, resulting in similar OC levels (Figure 7b–e). Vitamin

D3-treated osteoblasts cultured in the presence of hydroxyapatite had enhanced OC levels compared to controls and non-treated cells (hydroxyapatite only). Finally, exposing cells to D3+K1, D3+MK-4 or D3+MK-7 resulted in higher OC levels compared to controls and non-treated cells (hydroxyapatite only). The greatest enhancements were observed with vitamin D3 (55–120%) and MK-7 (40–120%).

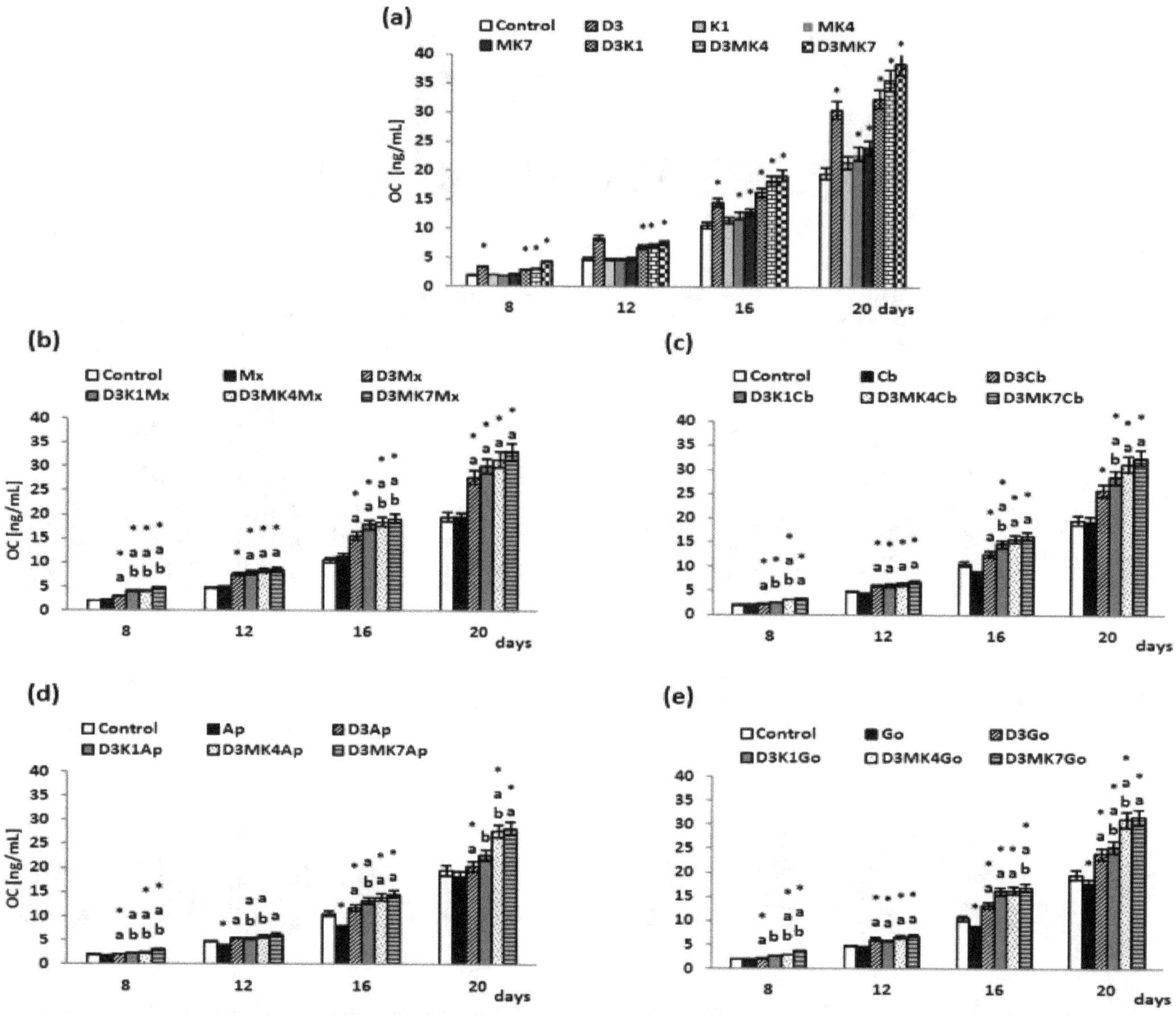

Figure 7. OC levels in osteoblasts after incubation with hydroxyapatites and treated with vitamins D3, K1, MK4, and MK7 after 8, 12, 16, and 20 days. The results are expressed as the ng/mL of medium and are shown as the mean ± SD ($n = 5$). The values for the control cells and the treated cells were significantly different according to unpaired Student's t-test. * Statistically significant differences versus control, $p < 0.05$; (**a**) statistically significant differences versus group hydroxyapatites (Mx (Maxgraft), Cb (Cerabone), Ap (Apatos), Go (Gen-Os), respectively for graphs (**b**–**e**)), $p < 0.05$. (**a**) statistically significant differences versus group vitamin D_3 with hydroxyapatites (Mx (Maxgraft), Cb (Cerabone), Ap (Apatos), Go (Gen-Os), respectively for graphs (**b**–**e**)), $p < 0.05$.

4. Discussion

Hydroxyapatite-based biomaterials are often implanted to support bone regeneration or are used as a scaffold for culturing osteogenic cells, notably osteoblasts, which are essential for bone regeneration producing protein and mineral components of bone [50]. During the bone regeneration process, osteoblasts are recruited to the site of the bone defect, where they proliferate, differentiate, and contribute to extracellular matrix mineralization [51]. Biomaterials can cause inflammation and prolong wound healing as a result of increased ROS generation [52]. Vitamin D3, as well as K1, MK-4, and MK-7, possess osteoinductive, anti-inflammatory, and antioxidant properties, and may be able to prevent oxidative stress to improve bone regeneration [18,33,53].

Redox homeostasis results from the balance between ROS generation and the activity of cellular antioxidants [54]. Under physiological conditions, osteoblasts maintain redox homeostasis; however, interaction between osteoblasts and biomaterials can disturb this balance leading to oxidative stress. Results of this study showing that osteoblasts cultured in the presence of xenogeneic and allogeneic biomaterials exhibit increased ROS generation and reduced antioxidant capacity. Increased ROS production has been shown to directly affect the biological activity of osteoblasts via inhibition of Runx2 phosphorylation, the process responsible for transcription of type I collagen, osteopontin, osteoprotegin, sialoprotein, OC, and ALP [55].

Results of this study indicate that the presence of xenogeneic and allogeneic biomaterials can lead to antioxidant-related disorders in osteoblasts, including those associated with the GSH-dependent system. In addition to performing regulatory functions that ensure redox homeostasis, GSH tripeptide is also a cofactor of glutathione peroxidase, the main enzyme involved in breaking down peroxides including lipid peroxides. Thus, GSH is responsible for protecting cell membrane phospholipids and as a consequence, its deficiency can affect the function of biological membranes [56]. Lower GSH levels prevent glutathione peroxidase from performing its function; therefore, allogeneic biomaterials, in particular Apatos and Gen-Os, which lower GSH levels, can affect the activity of glutathione peroxidase, as observed in these studies. As a consequence, the GSH-dependent antioxidant system cannot properly perform its protective role and increased peroxidation of unsaturated fatty acids results in the formation of reactive aldehydes, such as 4-HNE, as confirmed by our study. Biomaterials based on bioactive glass have also been shown to alter the redox homeostasis in osteoblasts' cells, manifested as increased 4-HNE production [57]. It should be mentioned that 4-HNE belongs to the α,β-unsaturated aldehydes, which include carbonyl groups and unsaturated bonds, thus, rendering them electrophilic and highly biologically reactive compounds mostly targeting nucleophilic compounds, such as GSH or proteins, leading to a reduction in biological activity including reduced GSH-Px phosphorylation [58,59]. This is also important because lipid peroxidation products decrease the expression of genes encoding for bone morphogenetic proteins (BMPs), including BMP2, BMP7, and BMP4, and therefore inhibit osteoblast maturation [55,60]. On the other hand, 4-HNE is known to act as growth regulating factor, acting in the concentration dependent and cell-type-specific manner influencing proliferation, differentiation, and apoptosis [61,62]. Mesenchymal cells, notably osteoblast-like cells, are especially sensitive to the growth regulating effects of 4-HNE, which interfere with complex bioactivities of numerous other growth-regulating factors (cytokines, hormones, etc.) and might eventually be involved in pathophysiology of altered bone growth, as in case of hypertrophic callus formation or otosclerosis [63,64]. Since 4-HNE was found to be produced by osteoblast-like cells, the growth of which was enhanced in vitro by bioactive glass, we may assume that production of 4-HNE by the osteoblasts observed in current study (also in the control cell cultures) supports previously described growth regulating effects of 4-HNE for the bone cells, while hydroxyapatite biomaterials resemble effects of bioactive glass [57,65]. However, to maintain functional growth of the bone cells it is necessary to enhance not only their proliferation but also differentiation, while oxidative stress and in particular 4-HNE were already proposed to be key mediators that might define the overall bioactivities of the implantable bioactive materials that require maintenance of the redox balance and the presence of supportive factors, like cytokines and vitamins [66,67].

Thus, the use of vitamins D3 and K to prevent redox imbalance could be beneficial for the bone growth and regeneration. Vitamin D3 did not have any effect on ROS level in this study, whereas ROS levels were significantly decreased with the addition of vitamin K to culture media, particularly the MK-7 homologue. The mechanism of vitamin K action may be associated with the inhibition of 12-LOX (lipoxygenase 12) activity [33], which can decrease ROS generation. Moreover, inhibiting ROS production may help maintain the metabolic activity of osteoblasts, but at the same time, could increase osteoclast activity, resulting in increased bone resorption [60]. This could reduce bone mineral density and the mechanical strength of newly formed and existing bone [68]. Therefore, vitamin K treatment can have associated consequences including oxidation of reduced glutathione to glutathione disulphide and reduced GSH-Px activity [69]. This is critical since GSH-Px activity protects membrane phospholipids and prevents the formation of 4-HNE. At the cellular level, oxidative damage disturbs cellular metabolism, thereby inhibiting osteoblast proliferation. This could be related to the oxidative damage to fibronectin, a glycoprotein that regulates certain osteoblast activities including adhesion, proliferation, migration, and cell morphology [70].

Here, we report decreased GSH levels in osteoblast cultures grown in the presence of allogeneic and xenogenic bone derivatives and treated with vitamin D3 and/or K. The decrease in GSH levels was particularly evident in cultures treated with the MK-7 homologue in combination with D3, which could be related to the vitamin D3-mediated increase in cysteine glutamate ligase (GCLC) activity which catalysis GSH biosynthesis [71]. Owing to the antioxidant properties of K vitamins, they can also affect GSH concentration, as confirmed by previous studies on nerve cells [34]. Higher GSH levels are accompanied by elevated GSH-Px activity, which enhances antioxidant activity in osteoblasts. Cells with enhanced antioxidant activity are less sensitive to oxidative damage and cell membrane damage.

Our results suggest that vitamin D3 and vitamin K protect osteoblasts from oxidative stress and have beneficial effects on proliferation, differentiation, and mineralization of osteoblasts cultured both with and without hydroxyapatite-based biomaterials. This study suggests that vitamin K enhances the proliferative activity of osteoblasts (measured by DNA concentration). The MK-7 homologue is the most effective, either alone or combined with vitamin D3. Enhanced proliferation of MK-7-treated osteoblasts has been confirmed by other authors [72,73]. However, in previous studies, osteoblasts were not cultured in the presence of hydroxyapatite-based biomaterials. Osteoblast DNA levels have been associated with OC expression, osteoprotegerin, NFκB ligand (RANKL), and RANK ligand promoter genes facilitated by vitamin K [74,75]. Furthermore, the vitamin K family participates in sphingolipids synthesis, in addition to their structural role, sphingolipids take part in proliferation, differentiation, and intercellular recognition activities of neurons [76]. Vitamin D3 cannot affect proliferation on its own, but does support osteoblast differentiation, as demonstrated by increased expression of typical markers such as runt-related transcription factor 2 (RUNX2), type I collagen, alkaline phosphatase, and OC [77–79]. Our results show that unlike vitamin D3, the K vitamins do not influence osteoblast differentiation. However, the combination of vitamin K and D3 can significantly enhance osteoblast differentiation with increasing efficiency: K1 < MK-4 < MK-7. This can clearly be observed when osteoblasts are cultured in the presence of xenogeneic biomaterials (Maxgraft) and was previously observed by others [79]. Results of this study confirm previous reports of the participation of vitamin D3 and K vitamins in the bone regeneration process [75]. Human bone marrow stromal cells (hBMSC) and 2T3 osteoblast lines treated with vitamin D3 showed a concentration-dependent increase in ALP activity [80], whereas vitamin K did not affect ALP activity [42].

Osteocalcin is responsible for depositing calcium in bone tissue [81] and both vitamin D3 and vitamin K participate in the OC biosynthesis and activation processes [82]. In our study, vitamin D3 was shown to significantly affect OC levels in osteoblasts cultured without hydroxyapatite. A significant increase in OC synthesis was observed on day 12, indicating the start of the mineralization process. Other studies have confirmed the influence of vitamin D3 on OC production by osteoblasts [80]. However, we have shown that the presence of biomaterials, in particular, allogeneic bone-derived biomaterials (Apatos and Gen-Os), significantly decreases OC levels in the medium of osteoblast cultures. Furthermore, hydroxyapatite has been shown to bind 60–90% of OC [83]. Therefore, observed differences in the effects of different biomaterials on OC levels may be the result of varying levels of absorption related to the porosity and composition of the material. The lowest OC levels observed for osteoblasts cultured in the presence of Gen-Os is most likely owing to the presence of collagen in the biomaterial structure. Collagen content favours the absorption of OC since OC binds to the surface of collagen and further mediates the binding and accumulation of hydroxyapatite crystals on collagen fibers [40].

The K vitamin family supports biosynthesis of OC by participating in post-translational carboxylation of Gla residues of OC propeptides [26]. In contrast, carboxylated OC binds calcium to bone hydroxyapatite, thus supporting bone metabolism [81]. The results of our study indicate the effects of K vitamins are less significant than those of vitamin D3; however, the combination results in synergistic effects, even in the presence of allogeneic biomaterials (Apatos and Gen-Os). This could be because post-translational modification of OC depends on both vitamin D3 and vitamin K [37]. Vitamin D3 activates γ-glutamylcarboxylase for which vitamin K is a cofactor [26,84]. Moreover, vitamin K intensifies the synthesis and accumulation of carboxylated OC [82], thereby supporting hydroxyapatite deposition on collagen fibers. This process is regulated by the carboxylated form of OC [85]. Since vitamins D3 and K have a high affinity for hydroxyapatite, they are absorbed onto its surface and remain available to cells for a long time, despite the relatively short half-life of both vitamins [79]. Our results show intensified biosynthesis of OC between day 16 and 20 of the culture. Clinical studies have also confirmed that vitamin K (MK-4) and vitamin D3 treatments increase bone mass [86].

5. Conclusions

This study suggests that proliferation and differentiation of human osteoblasts in vitro was strongly influenced by redox balance, which might be altered in the presence of hydroxyapatite-based biomaterials. While physiological ROS production-induced lipid peroxidation generating 4-HNE might be important for the redox homeostasis of the growing osteoblasts, vitamins D3 and/or K could prevent their excessive production through oxidative stress of the osteoblasts induced by the biomaterials, thereby preventing their adverse effects on cellular proliferation and osteoblast differentiation. The most desirable effects on the maintenance of redox homeostasis were in our in vitro study observed for vitamins D3 and K2, especially the homologue MK-7.

It is suggested that proliferation and differentiation of human osteoblasts in vitro is strongly influenced by redox balance, which is altered in the presence of hydroxyapatite-based biomaterials. ROS production and thus induced lipid peroxidation with 4-HNE generation might be important for the redox homeostasis of the growing osteoblasts, whereas vitamins D3 and/or K could prevent oxidative stress of the osteoblasts induced by the biomaterials, thereby preventing their adverse effects on proliferation and osteoblasts differentiation. The most desirable effects were in our in vitro study observed for vitamins D3 and K2, especially the MK-7 homologue.

Author Contributions: Conceptualization, E.S. and G.G.; methodology, E.A.; validation, M.M.; formal analysis, M.M.; investigation, E.A.; resources, G.T.; data curation, E.A.; writing—original draft preparation, E.A. and G.T.; writing—review and editing, N.Z.; supervision, E.S.; project administration, E.S.; funding acquisition, G.G.

References

1. Ono, H.; Sase, T.; Tanaka, Y.; Takasuna, H. Histological assessment of porous custom-made hydroxyapatite implants 6 months and 2.5 years after cranioplasty. *Surg. Neurol. Int.* **2017**, *8*, 8. [CrossRef]
2. Yamada, M.; Ueno, T.; Minamikawa, H.; Sato, N.; Iwasa, F.; Hori, N.; Ogawa, T. N-Acetyl cysteine alleviates cytotoxicity of bone substitute. *J. Dent. Res.* **2010**, *89*, 411–416. [CrossRef] [PubMed]
3. Anusavice, K.J. *Phillip's Science of Dental Materials*, 11th ed.; Elsevier: Saunders, FL, USA, 2003; pp. 170–190.
4. Anusavice, K.J.; Shen, C.; Rawls, H.R. *Phillip's Science of Dental Materials*, 12th ed.; Elsevier: Saunders, FL, USA, 2013; pp. 170–190.

5. Yamada, M.; Kojima, N.; Att, W.; Minamikawa, H.; Sakurai, K.; Ogawa, T. Improvement in the osteoblastic cellular response to a commercial collagen membrane and demineralized freeze-dried bone by an amino acid derivative: An in vitro study. *Clin. Oral Implants Res.* **2011**, *22*, 165–172. [CrossRef] [PubMed]
6. Mouthuy, P.A.; Snelling, S.J.B.; Dakin, S.G.; Milković, L.; Gašparović, A.Č.; Carr, A.J.; Žarković, N. Biocompatibility of implantable materials: An oxidative stress viewpoint. *Biomaterials* **2016**, *109*, 55–68. [CrossRef]
7. Lenz, R.; Mittelmeier, W.; Hansmann, D.; Brem, R.; Diehl, P.; Fritsche, Y.; Bader, R. Response of human osteoblasts exposed to wear particles generated at the interface of total hip stems and bone cement. *J. Biomed. Mater. Res. A* **2009**, *89*, 370–378. [CrossRef]
8. Oryan, A.; Alidadi, S.; Moshiri, A.; Maffulli, N. Bone regenerative medicine: classic options, novel strategies, and future directions. *J. Orthop. Surg. Res.* **2014**, *9*, 18. [CrossRef]
9. Fujioka-Kobayashi, M.; Schaller, B.; Zhang, Y.; Pippenger, B.E.; Miron, R.J. In vitro evaluation of an injectable biphasic calcium phosphate (BCP) carrier system combined with recombinant human bone morphogenetic protein (rhBMP)-9. *Biomed. Mater. Eng.* **2017**, *28*, 293–304. [CrossRef] [PubMed]
10. Hoffmann, T.; Al-Machot, E.; Meyle, J.; Jervøe-Storm, P.M.; Jepsen, S. Three-year results following regenerative periodontal surgery of advanced intrabony defects with enamel matrix derivative alone or combined with a synthetic bone graft. *Clin. Oral Investig.* **2016**, *20*, 357–364. [CrossRef] [PubMed]
11. Van Ballegooijen, A.J.; Pilz, S.; Tomaschitz, A.; Grübler, M.R.; Verheyen, N. The Synergistic Interplay between Vitamins D and K for Bone and Cardiovascular Health: A Narrative Review. *Int. J. Endocrinol.* **2017**, *2017*, 7454376. [CrossRef]
12. Norman, P.E.; Powell, J.T. Vitamin D and Cardiovascular Disease. *Circ. Res.* **2014**, *114*, 379–393. [CrossRef]
13. Neve, A.; Corrado, A.; Cantatore, F.P. Osteocalcin: Skeletal and extra-skeletal effects. *J. Cell. Physiol.* **2013**, *228*, 1149–1153. [CrossRef]
14. Nakai, K.; Fujii, H.; Kono, K.; Goto; Kitazawa, R.; Kitazawa, S.; Hirata, M.; Shinohara, M.; Fukagawa, M.; Nishi, S. Vitamin D activates the Nrf2-Keap1 antioxidant pathway and amelioratesn ephropathy in diabetic rats. *Am. J. Hypertens.* **2014**, *27*, 586–595. [CrossRef] [PubMed]
15. Salum, E.; Kals, J.; Kampus, P.; Salum, T.; Zilmer, K.; Aunapu, M.; Arend, A.; Eha, J.; Zilmer, M. Vitamin D reduces deposition of advanced glycation end-products in the aortic wall and systemic oxidative stress in diabetic rats. *Diabetes Res. Clin. Pract.* **2013**, *100*, 243–249. [CrossRef]
16. Sardar, S.; Chakraborty, A.; Chatterjee, M. Comparative effectiveness of vitamin D3 and dietary vitamin E on peroxidation of lipids and enzymes of the hepatic antioxidant system in Sprague-Dawley rats. *Int. J. Vitam. Nutr. Res.* **1996**, *66*, 39–45. [PubMed]
17. George, N.; Kumar, T.P.; Antony, S.; Jayanarayanan, S.; Paulose, C.S. Effect of vitamin D3 in reducing metabolic and oxidative stress in the liver of streptozotocin-induced diabetic rats. *Br. J. Nutr.* **2012**, *2012*, 1–9. [CrossRef]
18. Hamden, K.; Carreau, S.; Jamoussi, K.; Miladi, S.; Lajmi, S.; Aloulou, D.; Ayadi, F.; Elfeki, A. Alpha,25 dihydroxyvitamin D3: therapeutic and preventive effects against oxidative stress, hepatic, pancreatic and renal injury in alloxan-induced diabetes in rats. *J. Nutr. Sci. Vitaminol.* **2009**, *55*, 215–222. [CrossRef]
19. Zhang, H.Q.; Teng, J.H.; Li, Y.; Li, X.X.; He, Y.H.; He, X.; Sun, C.H. Vitamin D status and its association with adiposity and oxidative stress in schoolchildren. *Nutrition* **2014**, *30*, 1040–1044. [CrossRef]
20. Codoner-Franch, P.; Tavarez-Alonso, S.; Simo-Jorda, R.; Laporta-Martin, P.; Carratala- Calvo, A.; Alonso-Iglesias, E. Vitamin D status is linked to biomarkers of oxidative stress, inflammation, and endothelial activation in obese children. *J. Pediatr.* **2012**, *161*, 848–854. [CrossRef]
21. Haas, M.J.; Jafri, M.; Wehmeier, K.R.; Onstead-Haas, L.M.; Mooradian, A.D. Inhibition of endoplasmic reticulum stress and oxidative stress by vitamin D in endothelial cells. *Free Radic. Biol. Med.* **2016**, *99*, 1–10. [CrossRef]
22. Uberti, D.; Lattuada, D.; Morsanuto, V.; Nava, U.; Bolis, G.; Vacca, G.; Squarzanti, D.F.; Cisari, C.; Molinari, C. Vitamin D protects human endothelial cells from oxidative stress through the autophagic and survival pathways. *J. Clin. Endocrinol. Metab.* **2014**, *99*, 1367–1374. [CrossRef] [PubMed]
23. Borges, M.C.; Martini, L.A.; Rogero, M.M. Current perspectives on vitamin D, immune system, and chronic diseases. *Nutrition* **2011**, *27*, 399–404. [CrossRef]
24. Hamidi, M.; Gajic-Veljanoski, O.; Cheung, A. Vitamin K and bone health. *J. Clin. Densitom.* **2013**, *16*, 409–413. [CrossRef]

25. Lanham-New, S.A. Importance of calcium, vitamin D and vitamin K for osteoporosis prevention and treatment. *Proc. Nutr. Soc.* **2008**, *67*, 163–176. [CrossRef]
26. Shiraki, M.; Tsugawa, N.; Okano, T. Recent advances in vitamin K-dependent Gla-containing proteins and vitamin K nutrition. *Osteoporos. Sarcopenia* **2015**, *1*, 22–38. [CrossRef]
27. Iwamoto, J.; Sato, Y.; Takeda, T.; Matsumoto, H. High-dose vitamin K supplementation reduces fracture incidence in postmenopausal women: A review of the literature. *Nutr. Res.* **2009**, *29*, 221–2288. [CrossRef]
28. Urayama, S.; Kawakami, A.; Nakashima, T.; Tsuboi, M.; Yamasaki, S.; Hida, A.; Ichinose, Y.; Nakamura, H.; Ejima, E.; Aoyagi, T.; et al. Effect of vitamin K2 on osteoblast apoptosis: Vitamin K2 inhibits apoptotic cell death of human osteoblasts induced by Fas, proteasome inhibitor, etoposide, and staurosporine. *J. Lab. Clin. Med.* **2000**, *136*, 181–193. [CrossRef]
29. Ichigawa, T.; Horie-Inoue, K.; Ikeda, K.; Blumberg, B.; Inoue, S. Vitamin K2 induces phosphorylation of protein kinase A and expression of novel target genes in osteoblastic cells. *J. Mol. Endocrinol.* **2007**, *39*, 239–247. [CrossRef]
30. Villa, J.K.D.; Diaz, M.A.N.; Pizziolo, V.R.; Martino, H.S.D. Effect of vitamin K in bone metabolism and vascular calcification: A review of mechanisms of action and evidences. *Crit. Rev. Food Sci. Nutr.* **2017**, *57*, 3959–3970. [CrossRef]
31. Westhofen, P.; Watzka, M.; Marinova, M.; Hass, M.; Kirfel, G.; Müller, J.; Bevans, C.G.; Müller, C.R.; Oldenburg, J. Human vitamin K 2,3-epoxide reductase complex subunit 1-like 1 (VKORC1L1) mediates vitamin K-dependent intracellular antioxidant function. *J. Biol. Chem.* **2011**, *286*, 15085–15094. [CrossRef]
32. Mukai, K.; Itoh, S.; Morimoto, H. Stopped-flow kinetic study of vitamin E regeneration reaction with biological hydroquinones (reduced forms of ubiquinone, vitamin K, and tocopherolquinone) in solution. *J. Biol. Chem.* **1992**, *267*, 22277–22281.
33. Li, J.; Wang, H.; Rosenberg, P.A. Vitamin K prevents oxidative cell death by inhibiting activation of 12-lipoxygenase in developing oligodendrocytes. *J. Neurosci. Res.* **2009**, *87*, 1997–2005. [CrossRef]
34. Li, J.; Lin, J.C.; Wang, H.; Peterson, J.W.; Furie, B.C.; Furie, B.; Booth, S.L.; Volpe, J.J.; Rosenberg, P.A. Novel role of vitamin K in preventing oxidative injury to developing oligodendrocytes and neurons. *J. Neurosci.* **2003**, *23*, 5816–5826. [CrossRef]
35. Ohsaki, Y.; Shirakawa, H.; Hiwatashi, K.; Furukawa, Y.; Mizutani, T.; Komai, M. Vitamin K suppresses lipopolysaccharide-induced inflammation in the rat. *Biosci. Biotechnol. Biochem.* **2006**, *70*, 926–932. [CrossRef]
36. Shea, M.K.; Cushman, M.; Booth, S.L.; Burke, G.L.; Chen, H.; Kritchevsky, S.B. Associations between vitamin K status and haemostatic and inflammatory biomarkers in community-dwelling adults. The Multi-Ethnic Study of Atherosclerosis. *Thromb. Haemost.* **2014**, *112*, 438–444.
37. Poon, C.C.; Li, R.W.; Seto, S.W.; Kong, S.K.; Ho, H.P.; Hoi, M.P.; Lee, S.M.; Ngai, S.M.; Chan, S.W.; Leung, G.P.; et al. In vitro vitamin K(2) and 1α,25-dihydroxyvitamin D(3) combination enhances osteoblasts anabolism of diabetic mice. *Eur. J. Pharmacol.* **2015**, *767*, 30–40. [CrossRef]
38. Miyake, N.; Hoshi, K.; Sano, Y.; Kikuchi, K.; Tadano, K.; Koshihara, Y. 1,25- Dihydroxyvitamin D3 promotes vitamin K2 metabolism in human osteoblasts. *Osteoporos. Int.* **2001**, *12*, 680–687. [CrossRef]
39. Przekora, A.; Ginalska, G. Enhanced differentiation of osteoblastic cells on novel chitosan/β-1, 3-glucan/bioceramic scaffolds for bone tissue regeneration. *Biomed. Mater.* **2015**, *10*, 015009. [CrossRef]
40. Neve, A.; Corrado, A. Osteoblast physiology in normal and pathological conditions. *Cell Tissue Res.* **2011**, *343*, 289–302. [CrossRef]
41. Atkins, G.J.; Welldon, K.J.; Wijenayaka, A.R.; Bonewald, L.F.; Findlay, D.M. Vitamin K promotes mineralization, osteoblast-to-osteocyte transition, and an anticatabolic phenotype by gamma-carboxylation-dependent and -independent mechanisms. *Am. J. Physiol. Cell. Physiol.* **2009**, *297*, C1358–C1367. [CrossRef]
42. Matsumoto, T.; Igarachi, C.; Taksushi, Y.; Harada, S.; Kikuchi, T.; Yamato, H.; Ogata, E. Stimulation by 1,25 dihydroxyvitamin D3 of in vitro mineralization induced by osteoblast-like MCT3-E1 cells. *Bone* **1991**, *12*, 27–32. [CrossRef]
43. Ducy, P.; Zhang, R.; Geoffroy, V.; Ridall, A.L.; Karsenty, G. OSF2/Cbfal: a transcriptional activator of osteoblast differentiation. *Cell* **1997**, *89*, 742–754. [CrossRef]
44. Koshihara, Y.; Hoshi, K. Vitamin K2 enhances osteocalcin accumulation in the extracellular matrix of human osteoblasts in vitro. *J. Bone Miner. Res.* **1997**, *12*, 431–438. [CrossRef]

45. Kuzkaya, N.; Weissmann, N.; Harrison, D.G.; Dikalov, S. Interactions of peroxynitrite, tetrahydrobiopterin, ascorbic acid, and thiols: implications for uncoupling endothelial nitricoxide synthase. *J. Biol. Chem.* **2003**, *278*, 22546–22554. [CrossRef]
46. Maeso, N.; Garcia-Martinez, D.; Ruperez, F.J.; Cifuentes, A.; Barbas, C. Capillary electrophoresis of glutathione to monitor oxidative stress and response to antioxidant treatments in an animal model. *J. Chromatogr. B* **2005**, *822*, 61–69. [CrossRef]
47. Paglia, D.E.; Valentine, W.N. Studies on the quantitative and qualitative characterization of erythrocyte glutathione peroxidase. *J. Lab. Clin. Med.* **1967**, *70*, 158–169.
48. Luo, X.P.; Yazdanpanah, M.; Bhooi, N.; Lehotay, D.C. Determination of aldehydes and other lipid peroxidation products in biological samples by gas chromatography-mass spectrometry. *Anal. Biochem.* **1995**, *228*, 294–298. [CrossRef]
49. Gęgotek, A.; Rybałtowska-Kawałko, P.; Skrzydlewska, E. Rutin as a mediator of lipid metabolism and cellular signaling pathways interactions in fibroblasts altered by UVA and UVB radiation. *Oxid. Med. Cell. Longev.* **2017**, *2017*, 4721352. [CrossRef]
50. Schirmer, K.; Ganassin, R.C.; Brubacher, J.L.; Bols, N.C. A DNA fluorometric assay for measuring fish cell proliferation in microplates with different well sizes. *J. Tissue Cult. Methods* **1994**, *16*, 133–142. [CrossRef]
51. Sabokbar, A.; Millett, P.J.; Myer, B.; Rushton, H. A rapid, quantitative assay for measuring alkaline phosphatase activity in osteoblastic cells in vitro. *Bone Miner.* **1994**, *27*, 57–67. [CrossRef]
52. Zimmermann, G.; Moghaddam, A. Allograft bone matrix versus synthetic bone graft substitutes. *Injury* **2011**, *2*, S16–S21. [CrossRef]
53. Rutkovskiy, A.; Stensløkken, K.O.; Vaage, I.J. Osteoblast Differentiation at a Glance. *Med. Sci. Monit. Basic Res.* **2016**, *22*, 95–106. [CrossRef] [PubMed]
54. Morais, J.M.; Papadimitrakopoulos, F.; Burgess, D.J. Biomaterials/tissue interactions: possible solutions to overcome foreign body response. *AAPS J.* **2010**, *12*, 188–196. [CrossRef]
55. Gigante, A.; Torcianti, M.; Boldrini, E.; Manzotti, S.; Falcone, G.; Greco, F.; Mattioli-Belmonte, M. Vitamin K and D association stimulates in vitro osteoblast differentiation of fracture site derived human mesenchymal stem cells. *J. Biol. Regul. Homeost. Agents* **2008**, *22*, 35–44.
56. Tan, D.Q.; Suda, T. Reactive Oxygen Species and Mitochondrial Homeostasis as Regulators of Stem Cell Fate and Functio. *Antioxid. Redox Signal.* **2018**, *29*, 149–168. [CrossRef]
57. Jung, W.W. Protective effect of apigenin against oxidative stress-induced damage in osteoblastic cells. *Int. J. Mol. Med.* **2014**, *33*, 1327–1334. [CrossRef]
58. Morris, G.; Anderson, G.; Dean, O.; Berk, M.; Galecki, P.; Martin-Subero, M.; Maes, M. The glutathione system: A new drug target in neuroimmune disorders. *Mol. Neurobiol.* **2014**, *50*, 1059–84. [CrossRef]
59. Mrakovcic, L.; Wildburger, R.; Jaganjac, M.; Cindric, M.; Cipak, A.; Borovic-Sunjic, S.; Waeg, G.; Milankovic, A.M.; Zarkovic, N. Lipid peroxidation product 4-hydroxynonenal as factor of oxidative homeostasis supporting bone regeneration with bioactive glasses. *Acta Biochim. Pol.* **2010**, *57*, 173–8. [CrossRef]
60. Castro, J.P.; Jung, T.; Grune, T.; Siems, W. 4-Hydroxynonenal (HNE) modified proteins in metabolic diseases. *Free Radic. Biol. Med.* **2017**, *111*, 309–315. [CrossRef]
61. uczaj, W.; Gęgotek, A.; Skrzydlewska, E. Antioxidants and HNE in redox homeostasis. *Free Radic. Biol. Med.* **2017**, *111*, 87–101. [CrossRef] [PubMed]
62. Koerdt, S.; Siebers, J.; Bloch, W.; Ristow, O.; Kuebler, A.C.; Reuther, T. Role of oxidative and nitrosative stress in autogenous bone grafts to the mandible using guided bone regeneration and a deproteinized bovine bone material. *J. Craniomaxillofac. Surg.* **2014**, *42*, 560–567. [CrossRef] [PubMed]
63. Borović Šunjić, S.; Čipak, A.; Rabuzin, F.; Wildburger, R.; Žarković, N. The Influence of 4-Hydroxy-2-nonenal on Proliferation, Differentiation and Apoptosis of Human Osteosarcoma Cells. *Biofactors* **2005**, *24*, 141–148. [CrossRef]
64. Borović, S.; Čipak, A.; Meinitzer, A.; Kejla, Z.; Perovic, D.; Waeg, G.; Žarković, N. Differential effect of 4-hydroxynonenal on normal and malignant mesenchymal cells. *Redox Rep.* **2007**, *207*, 50–54. [CrossRef]

65. Rudić, M.; Milković, L.; Žarković, K.; Borović-Šunjić, S.; Sterkers, O.; Ferrary, E.; Bozorg Grayeli, A.; Žarković, N. The effects of angiotenzin II and oxidative stress mediator 4-hydroxynonenal on the human osteoblast-like cell growth: Possible relevance for otosclerosis. *Free Radic. Biol. Med.* **2013**, *57*, 22–28. [CrossRef]
66. Milkovic, L.; Cipak Gasparovic, A.; Zarkovic, N. Overview on major lipid peroxidation bioactive factor 4-hydroxynonenal as pluripotent growth regulating factor. *Free Radic. Res.* **2015**, *49*, 850–860. [CrossRef]
67. Milkovic, L.; Hoppe, A.; Detsch, R.; Boccaccini, A.R.; Zarkovic, N. Effects of Cu-doped 45S5 bioactive glass on the lipid peroxidation-associated growth of human osteoblast-like cells in vitro. *J. Biomed. Mater. Res. Part A* **2014**, *102*, 3556–3561. [CrossRef]
68. Egea, J.; Fabregat, I.; Frapart, Y.M.; Ghezzi, P.; Görlach, A.; Kietzmann, T.; Kubaichuk, K.; Knaus, U.G.; Lopez, M.G.; Olaso-Gonzalez, G.; et al. European Contribution to the study of ROS: A Summary of the Findings and Prospects for the Future from the COST Action BM1203 (EU-ROS). *Redox Biol.* **2017**, *13*, 94–162. [CrossRef]
69. Almeida, M.; O'Brien, C.A. Basic biology of skeletal aging: Role of stress response pathways. *J. Gerontol. A Biol. Sci. Med. Sci.* **2013**, *68*, 1197–1208. [CrossRef]
70. Ke, C.Y.; Yang, F.L.; Wu, W.T.; Chung, C.H.; Lee, R.P.; Yang, W.T.; Subeq, Y.M.; Liao, K.W. Vitamin D3 reduces tissue damage and oxidative stress caused by exhaustive exercise. *Int. J. Med. Sci.* **2016**, *13*, 147–153. [CrossRef]
71. Abdollahi, M.; Larijani, B.; Rahimi, R.; Salari, S. Role of oxidative stress in osteoporosis. *Therapy* **2005**, *2*, 787–796. [CrossRef]
72. Jain, S.K.; Micinski, D.; Huning, L.; Kahlon, G.; Bass, P.F.; Levine, S.N. Vitamin D and L-cysteine levels correlate positively with GSH and negatively with insulin resistance levels in the blood of type 2 diabetic patients. *Eur. J. Clin. Nutr.* **2014**, *68*, 1148–1153. [CrossRef]
73. Lancaster, C.E.; Harrison, R.E. Effects of Vitamin D, K1, and K2 Supplementation on Bone Formation by Osteoblasts In Vitro: A Meta-analysis. *J. Biom. Biostat.* **2017**, *8*, 365. [CrossRef]
74. Zhu, M.; Ma, J.; Lu, S.; Zhu, Y.; Cui, Y.; Tan, H.; Wu, J.; Yongqing, X. Vitamin K2 analog menaquinone-7 shows osteoblastic bone formation activity in vitro. *Biomed. Res.* **2017**, *28*, 1364–1369.
75. Katsuyama, H.; Otsuki, T.; Tomit, M.; Fukunaga, M.; Fukunaga, T.; Suzuki, N.; Saijoh, K.; Fushimi, S.; Sunami, S. Menaquinone-7 regulates the expressions of osteocalcin, OPG, RANKL and RANK in osteoblastic MC3T3E1 cells. *Int. J. Mol. Med.* **2005**, *15*, 231–236. [CrossRef]
76. Yamaguchi, M. Role of nutritional factor menaquinone-7 in bone homeostasis and osteoporosis prevention. *Integr. Mol. Med.* **2014**, *1*, 1–6. [CrossRef]
77. Ferland, G. Vitamin K and the nervous system: an overview of its actions. *Adv. Nutr.* **2012**, *3*, 204–212. [CrossRef]
78. Posa, F.; Di Benedetto, A.; Colaianni, G.; Cavalcanti-Adam, E.A.; Brunetti, G.; Porro, C.; Trotta, T.; Grano, M.; Mori, G. Vitamin D Effects on Osteoblastic Differentiation of Mesenchymal Stem Cells from Dental Tissues. *Stem Cells Int.* **2016**, *2016*, 9150819. [CrossRef]
79. Van Driel, M.; Pols, H.A.; van Leeuwen, J.P. Osteoblast differentiation and control by vitamin D and vitamin D metabolites. *Curr. Pharm. Des.* **2004**, *10*, 2535–2555. [CrossRef]
80. Ozeki, K.; Aoki, H.; Fukui, Y. The effect of adsorbed vitamin D and K to hydroxyapatite on ALP activity of MC3T3-E1 cell. *J. Mater. Sci. Mater. Med.* **2008**, *19*, 1753–1757. [CrossRef]
81. Zarei, A.; Hulley, P.A.; Sabokbar, A.; Javaid, M.K.; Morovat, A. 25-Hydroxy- and 1α,25-dihydroxycholecalciferol have greater potencies than 25-Hydroxy- and 1α,25-dihydroxyergocalciferol in modulating cultured human and mouse osteoblast activities. *PLoS ONE* **2016**, *11*, e0165462. [CrossRef]
82. Buranasinsup, S.; Bunyaratavej, N. The Intriguing Correlation between Undercarboxylated Osteocalcin and Vitamin, D. *J. Med. Assoc. Thai* **2015**, *98*, 16–20.
83. Gigante, A.; Brugè, F.; Cecconi, S.; Manzotti, S.; Littarru, G.P.; Tiano, L. Vitamin MK-7 enhances vitamin D3-induced osteogenesis in hMSCs: Modulation of key effectors in mineralization and vascularization. *J. Tissue Eng. Regen. Med.* **2015**, *9*, 691–701. [CrossRef]
84. Chen, L.; Jacquet, R.; Lowder, E.; Landis, W.J. Refinement of collagen-mineral interaction: A possible role for osteocalcin in apatite crystal nucleation, growth and development. *Bone* **2015**, *71*, 7–16. [CrossRef]

11

Nutritional Stress in Head and Neck Cancer Originating Cell Lines: The Sensitivity of the NRF2-NQO1 Axis

Lidija Milković [1,†], Marko Tomljanović [2,†], Ana Čipak Gašparović [1], Renata Novak Kujundžić [2], Dina Šimunić [2], Paško Konjevoda [2], Anamarija Mojzeš [2], Nikola Đaković [3,4], Neven Žarković [1] and Koraljka Gall Trošelj [2,*]

1 Laboratory for Oxidative Stress, Division of Molecular Medicine, Ruđer Bošković Institute, 10000 Zagreb, Croatia
2 Laboratory for Epigenomics, Division of Molecular Medicine, Ruđer Bošković Institute, 10000 Zagreb, Croatia
3 University Hospital Centre Sisters of Charity, Institute for Clinical Medical Research and Education, 10000 Zagreb, Croatia
4 Department of Clinical Oncology, School of Medicine, University of Zagreb, 10000 Zagreb, Croatia
* Correspondence: Koraljka.Gall.Troselj@irb.hr
† These authors contributed equally to this work.

Abstract: Nutritional stress disturbs the cellular redox-status, which is characterized by the increased generation of reactive oxygen species (ROS). The NRF2-NQO1 axis represents a protective mechanism against ROS. Its strength is cell type-specific. FaDu, Cal 27 and Detroit 562 cells differ with respect to basal NQO1 activity. These cells were grown for 48 hours in nutritional conditions (NC): (a) Low glucose–NC2, (b) no glucose, no glutamine–NC3, (c) no glucose with glutamine–NC4. After determining the viability, proliferation and ROS generation, NC2 and NC3 were chosen for further exploration. These conditions were also applied to IMR-90 fibroblasts. The transcripts/transcript variants of *NRF2* and *NQO1* were quantified and transcript variants were characterized. The proteins (NRF2, NQO1 and TP53) were analyzed by a western blot in both cellular fractions. Under NC2, the NRF2-NQO1 axis did not appear activated in the cancer cell lines. Under NC3, the NRF2-NQO1axis appeared slightly activated in Detroit 562. There are opposite trends with respect to TP53 nuclear signal when comparing Cal 27 and Detroit 562 to FaDu, under NC2 and NC3. The strong activation of the NRF2-NQO1 axis in IMR-90 resulted in an increased expression of catalytically deficient NQO1, due to NQO1*2/*2 polymorphism (rs1800566). The presented results call for a comprehensive exploration of the stress response in complex biological systems.

Keywords: glucose deprivation; glutamine deprivation; viability; proliferation; ROS; NRF2-NQO1 axis; IMR-90; NQO1 transcript variants; rs1800566; TP53 mutation

---

1. Introduction

To date, the roles of glucose and glutamine in the biology of transformed cells both in vitro and in vivo, have been evaluated in various cellular systems, most often as separate entities. It is well-established that cancer cells need glucose as a source of carbon. They also need glutamine. Not only as an alternative substrate for the Krebs cycle and ATP production, but also as a source of carbon and nitrogen, glutamine is needed for various biosynthetic reactions and glutathione production to support antioxidant defense [1].Under normoglycemic conditions, the generation of the fundamental metabolite nicotinamide adenine dinucleotide phosphate (NADPH) is secured via the glucose catabolism pathway—the pentose phosphate pathway (PPP). Glucose starvation results in decreased ATP production and could induce oxidative stress by downregulating NADPH production by PPP. Under these conditions, metabolic reprogramming and redox regulation are closely related to the activation of 5′ AMP-activated protein kinase (AMPK) pathway. It is a protective mechanism aimed at prolonging cell survival by preventing excessive NADPH consumption in fatty acid synthesis and increasing NADPH generation in the process of fatty acid oxidation [2]. When the protective capacity of the AMPK pathway is exceeded, the regeneration rate of glutathione decreases, and there is an increase in ROS, an indicator of the disbalanced cellular redox status. It is well-known that ROS are involved in the nutrient deprivation-induced Warburg effect [3]. Glutamine has also been shown as a source of NADPH. In pancreatic cancer cells with a K-RAS-regulated metabolic pathway, the glutamine-derived malate can be converted to pyruvate by malic enzyme. This reaction is associated with NADPH generation [4]. It has also been shown recently that a lack of glutamine may promote rapid and transient activation of AMPK [5].

During glucose deprivation, activated AMPK phosphorylates the wild type (WT) TP53 at serine 15, leading to G1/S cell cycle arrest and cellular senescence [6]. WT TP53 has an important role in cellular metabolism. It inhibits the monomeric form of the enzyme glucose-6-phosphate dehydrogenase (G6PD), which is present in the cytoplasm. This event results in one more instance of NADPH depletion. The effect seems to be characteristic of WT TP53, but not its mutant forms. It was proposed as a main function of cytoplasmic WT TP53 in resting cells [7].

When deprived of oxygen and glucose, the cells activate the AMPK by NAD(P)H:quinone oxidoreductase 1 (NQO1) [8]. This enzyme was purified and characterized for the first time in 1988 [9]. It was originally considered only as a flavin adenine dinucleotide (FAD)—dependent, two-electron reductase. There are numerous proofs of its effectiveness, associated with reducing quinones to hydroquinones through a two-electron transfer. The catalytically active form of the enzyme is a homodimeric protein. It has two identical active sites located at the interface between monomers and with one FAD bound per monomer. Each of these two sites is shared by both reduced pyridine nucleotide cofactors, NADH and NADPH [10]. The model of the NQO1 mode of action (ping-pong-bi-bi kinetic mechanism), proposed in 1974 [11], is still considered valid. The catalytic cycle is initiated by the binding of reduced pyridine nucleotide in the active site, followed by a hydride transfer to FAD. It leads to a conformational change expelling the oxidized pyridine nucleotide, nicotinamide adenine dinucleotide (NAD+), and creating an environment for quinone binding. The generation of NAD+ makes a strong, functional, yet indirect link between NQO1 and two very important cellular enzymes relevant for metabolism and metabolic reprogramming in cancer. These are NAD+ dependent sirtuin 1 (SIRT1) and PARP-1, a major NAD+-consuming enzyme [12].

The enzymatic activity of NQO1 can be detected in the cytosol and in the nucleus [13]. It has an important role in eliminating free radicals [14] which increase during nutritional stress. According to the most recent data, NQO1 is a central unit of the redox-dependent switch. It depends on NQO1 conformational change, in which NADH has strong protective role against tryptic digestion and loss of the C-terminal NQO1 domain. To a lesser extent, a protective role was also obtained with NADPH [15].

Altered pyridine nucleotide ratios could induce a switch in protein conformation. This results in binding of NQO1 to a different set of proteins and RNA under oxidative conditions [16]. Thus, NQO1 action influences the activity of other proteins indirectly, through generating NAD+ (SIRT1, PARP-1) [17] and through direct binding (hypoxia-inducible factor, alpha subunit, HIF1-α, TP53) [18,19].

NQO1 stabilizes both wild-type (WT) [18] and mutant-types (MT) TP53 protein [20] by protecting them from the ubiquitin-independent 20S proteasomal degradation. This stabilizing effect is most prominent under oxidative stress. However, the presence of the single nucleotide variation (SNV) rs1800566 that occurs in NQO1 exon 6, strongly decreases the enzymatic activity of NQO1 and abolishes TP53 stability mediated by NQO1 [21,22]. This polymorphism, also known as NQO1*2 (heterozygote)/NQO1*2/*2 (homozygote), was shown to be an important factor in a poor clinical response to quinone (mitomycin C, β-lapachone)-based chemotherapy. This is due to a lack of drug bioactivation [23,24].

TP53, which is traditionally considered a tumor suppressor, is currently an emerging research topic relating to nutritional stress [25]. Its connection to NQO1 may be a critical factor for cellular adaptive stress response, especially during nutrient deprivation. The most recent data have shown that the withdrawal of glutamine activates TP53 [26]. In a glutamine deprived cell, TP53 binds to the promoter of the solute-like carrier family 7, member 3 (SLC7A3). It promotes cancer cell adaptation to glutamine deprivation by upregulating SLC7A3to increase arginine uptake [27].

The state of oxidative stress is of utmost importance for activating *NQO1* transcription, which is mediated by NFE2L2 (Nuclear Factor, Erythroid 2 Like 2: NRF2). When there is an excess of ROS, NRF2 dissociates from its cytoplasmic partner Kelch-Like ECH-Associated Protein 1 (KEAP-1). It enters the nucleus and binds to the *cis*-acting elements in an array of NRF2 target genes called antioxidant response elements (AREs) [28]. These are present in the *NQO1* promoter [29]. Consequentially, this event leads to an increased transcriptional activity of the *NQO1* gene. This phenomenon has been shown in various models as a part of a strong antioxidative cellular response.

One very interesting molecular-genetic aspect of *NQO1* mRNA is associated with the deposit of four NQO1 transcript variants (TVs) in the GeneBank. The gene itself contains six exons (Figure 1). All of them are part of the longest transcript (TV1; NM_000903.3, N = 2521 nt). Another three transcripts are characterized, as follows: TV2: NM_001025433.2; exon 5 excluded (N = 2419 nt); TV3: NM_001025434.2, exon 4 excluded (N = 2407 nt); TV4: NM_001286137.2, exons 4 and 5 excluded (N = 2305 nt).

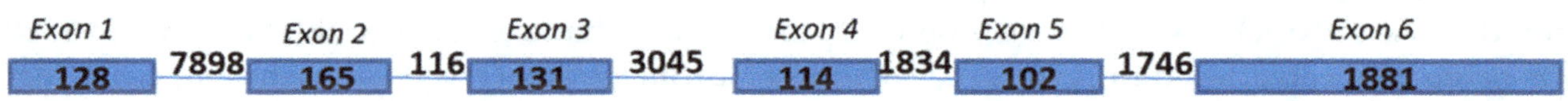

Figure 1. Structure of the *NQO1* gene, NG_011504.2.

In 1995, Gasdaska et al. described the *NQO1* transcript lacking exon 4 (TV3) in cancer cell lines SW 480 and HT-29. The existence of the corresponding protein was not confirmed [30]. Seven years later, it was proposed that the polymorphism present at the end of exon 4, rs1131341 (Arg137Trp, also known as NQO1*3*), has a strong influence on NQO1 splicing. As a consequence, the ratio TV1/TV3 (shown by end-point PCR to be around 9.0 in NQO1*1/*1, NQO1*2/*1, NQO1*2/2*), significantly decreases (TV1/TV3 = 2) [31]. The ratio of TVs may vary depending on stressful conditions [30]. This was shown only once, in the mononuclear cells obtained from patients before and at various times following treatment with mitomycin [32]. According to SwissProt, there is only one experimentally verified NQO1 protein variant which is coded by NQO1 TV1. It consists of 274 amino acids (30.868 kDa).

As recently shown in a yeast model, introns negatively regulate growth in a rich medium. They are clearly required for maintaining cellular viability during the deprivation of nutrients (dextrose and phosphates) [33]. In 2007, Pleiss et al. showed that two different stress-inducers (ethanol exposure and amino acids deprivation) induce unique splicing profiles. This suggests that in yeast at least two independent pathways connect the spliceosome with the cellular environment [34].

Alternative splicing was shown to take place during nutrient depletion in an organoid model system derived from murine intestinal epithelial cells. This included exon skipping events and events involving full intron retention (IR-S; intron retention simple) and complex intron retention (IR-C; intron retention complex) [35].

Based on these facts, we wanted to explore selected cellular parameters (cellular viability and proliferation rate, ROS generation) and molecular events included in the axis NRF2-NQO1/TP53, under two different forms of nutritional stress. The transcripts (quantitatively - NRF2, NQO1 and qualitatively - NQO1 splice variants) and proteins (NRF2, NQO1, TP53) in cytoplasmic and nuclear cellular fractions were validated. Three cell lines originating from the head and neck squamous cell carcinomas (HNSSC) were used: FaDu; Cal 27; and Detroit 562. These cells significantly differ with respect to basal NQO1 activity (FaDu > Detroit 562 > Cal 27) [36]. IMR-90 fibroblasts, which are considered as NQO1 non-expressing cell lines [37], were used as representative of an untransformed cell line.

2. Materials and Methods

2.1. Cell Lines and Cell Culture Conditions

The cells originating from metastatic pharyngeal cancer (pleural effusion-Detroit 562) and human fetal lung fibroblasts (IMR-90) were purchased from Sigma-Aldrich (St. Louis, MO, USA). The human tongue squamous carcinoma cells (Cal 27) and human hypopharyngeal squamous carcinoma cells (FaDu) were purchased from the American Type Culture Collection (ATCC, LGC Standards GmbH, Wesel, Germany). The cells were cultured in T75 cell culture flasks (Sarstedt AG&Co.KG, Nümbrecht, Germany), in Dulbecco's Modified Eagle's Medium (DMEM, D5796; Sigma-Aldrich, St. Louis, MO, USA), supplemented with a 10% fetal bovine serum (FBS, Sigma-Aldrich, St. Louis, MO, USA), without antibiotics, at 37 °C in a humidified atmosphere and in the presence of 5% CO_2. Prior to the experiments, the cells were harvested with 0.25% (*w/v*) Trypsin-0.53 mM EDTA (Ethylenediaminetetraacetic acid) solution and counted with the trypan blue exclusion assay in Bürker-Türk hemocytometer (Brand, Wertheim, Germany). For the experiments performed, the cells were cultured in DMEM, under the following four nutritional conditions (NCs) with respect to glucose and glutamine: NC1-high glucose (4.5 g/L) with L-glutamine (0.584 g/L) (D5796; Sigma-Aldrich, St. Louis, MO, USA); NC2-low glucose (1 g/L) with L-glutamine (0.584 g/L) (D6046, Sigma-Aldrich, St. Louis, MO, USA); NC3-no glucose and no glutamine (A14430, Gibco, Life Technologies Corporation, Grand Island, NY, USA); NC4-no glucose (A14430), but with 0.584 g/L of L-glutamine (Sigma Aldrich, St. Louis, MO, USA).

2.2. Cell Viability Assay

Cellular viability was measured using EZ4U assay (Biomedica, Vienna, Austria), which assesses cellular viability through reducing tetrazolium salts to colored formazan derivatives in the mitochondria of living cells. The cells were seeded in 96-well plates (TPP, Trasadingen, Switzerland) at a density of 1×10^4 cells per well and cultivated in 200 μL of previously described media formulations (NC1-NC4), supplemented with 10% FBS. After a cultivation period of 48 hours, 20 μL of the dye substrate (tetrazolium salts) was added to each well. After a 2 h incubation, formazan derivatives were quantified by measuring the absorbance using the microplate reader Multiskan EX (Thermo Electron Corporation, Shanghai, China) at 450 nm, with 620 nm as a reference wavelength. Cellular viability under tested conditions was expressed as a percentage of the viability of the control cells (cells grown in a high glucose + glutamine, NC1, medium).

2.3. Cell Proliferation Assay

The rate of cellular proliferation was estimated through incorporating pyrimidine analogue BrdU (5-bromo-2′-deoxyuridine), in place of thymidine, into the DNA of proliferating cells, using the Cell Proliferation ELISA, BrdU (colorimetric) Kit (Roche Applied Science, Mannheim, Germany). The antibody conjugated anti-BrdU-peroxidase binds incorporated BrdU. The complex BrdU/anti-BrdU-peroxidase was detected by the reaction between peroxidase conjugated to the BrdU antibody and the substrate (3,3',5,5'-tetramethylbenzidine). After reaching a satisfactory color intensity (after incubating between 5 and 30 min), the reaction was stopped with 1 M H_2SO_4 solution.

The cells were seeded in 96-well plates (TPP, Trasadingen, Switzerland) at a density of 1×10^4 cells per well and were maintained in 200 μL of previously described media formulations (NC1-NC4), supplemented by 10% FBS. After 48 h of incubation, the assay was performed according to the manufacturer's instructions. The reaction product (3,3',5,5'-tetramethyl-benzidine diimine) was quantified by measuring absorbance using a microplate reader Multiskan EX (Thermo Electron Corporation, Shanghai, China) set at 450 nm (reference wavelength: 620 nm). Cell proliferation was expressed as a percentage of the cells grown under condition NC1 (high glucose + L-glutamine medium).

2.4. ROS Measurement

The intracellular levels of reactive oxygen species (ROS) were detected by DCFH-DA (2′,7′-Dichlorofluorescin Diacetate; Sigma-Aldrich, St. Louis, MO, USA). The cells were seeded in white 96-well plates (Thermo Fisher Scientific, Nunc A/S, Roskilde, Denmark) at a density of 1×10^4 cells per well and maintained in 200 μL of the previously described media formulations (NC1-NC4), supplemented by 10% FBS. After growing for 48 hours, the cells were incubated with 20 μL of 100 μM DCFH-DA, which was added to the culture media. After 45 min of incubation, the medium containing DCFH-DA was replaced with 200 μL of fresh medium. The fluorescence intensity was measured immediately (zero point) and after one hour, on a plate reader Infinite 200 PRO (Tecan Group Ltd., Männedorf, Switzerland). The excitation/emission wavelengths for DCFH-DA were set at 500/529 nm. The values of the emitted fluorescence were expressed as arbitrary units, which represent the difference between the two points of measurement (one hour and zero point). Additionally, the values were corrected with respect to the cell numbers, which varied in relation to the treatment applied.

2.5. Nucleic Acids Extraction

The cells were cultured for 48 h at a density of 1×10^6 in T25 flasks (Sarstedt AG&Co.KG, Nümbrecht, Germany), in 5 mL of the previously described media formulations (NC1-NC3), supplemented by 10% FBS.

The total RNA was extracted from the cells cultivated and treated in 25 cm^2 flasks (Sarstedt AG&Co.KG, Nümbrecht, Germany). The medium was removed and extraction was performed by TRIzol (Invitrogen, Carlsbad, CA, USA), according to the manufacturer's instructions. The integrity of isolated RNA was determined by electrophoresis, on 1% agarose gel stained with ethidium bromide (EtdBr) (Sigma-Aldrich, St. Louis, MO, USA). As there were no issues relating to the integrity of the extracted RNA, all samples were further purified with gDNA Removal Kit (Jena Bioscience, Jena, Germany), according to the manufacturer's instructions. The concentration and purity of extracted RNA was determined spectrophotometrically (BioSpec-nano, Shimadzu Biotech, Japan) by measuring the absorbance at the following wavelengths: 230, 260 and 280 nm. The samples were stored at −80 °C.

The genomic DNA was extracted by phenol-chloroform extraction, after an overnight incubation with Proteinase K (QIagen, Holden, Germany), as previously described [38]. After successful precipitation, the samples of extracted DNA were re-suspended in TE buffer (10 mM Tris, 1 mM EDTA, pH 7.4). The concentration and quality of the extracted DNA was determined spectrophotometrically and electrophoretically, in 1% gel agarose stained with EtdBr. The samples were stored at +4 °C.

2.6. Construction of Primers

For all primers used in this research, with the exception of GAPDH1/GAPDH2 which are commonly used, the modeling through combining the programs Primer-BLAST and Primer3Plus were performed. Table 1 shows the primer sequences, their exact position on the RefSeq and the expected amplicon sizes.

The composition of the nucleotides of the primers used allowed the authors to perform the polymerase chain reaction under almost identical conditions. The primers for gDNA were selected to anneal to the template at 58 °C, while the primers for cDNA annealed to the template at 59 °C.

The three primers for GAPDH were constructed in a way which allowed combining the primer GAPDH2 with primers GAPDH1—for determining cDNA quality and GAPDH3—for a rigorous check of the potential gDNA contamination. These strict precautionary measures were undertaken because the TaqMan probe used for quantifying NRF2 may bind to the gDNA, at least according a statement provided.

The selection of primers which would allow for the amplification of all four NQO1 TVs in one reaction was based on the primary structure of the NQO1 TV1 mRNA (Figure 1). The primers were complementary to the stretch of nucleotides positioned in the 3′ region of the exon 2 (NQO1F) and 5′ region of the exon 6 (NQO1R), respectively. All primers which allowed for examining the exon/intron boundaries sequences were used only on genomic DNA.

Table 1. The positions and composition of primers used for analyses of *NQO1*.

Primers	Primer Sequences	Ref. Seq.	Primer Position	Amplicon Size
	GAPDH primers			
GAPDH1	5′AACGGATTTGGTCGTATTGGGC3′	NM_002046.7	101–122	600 bps
GAPDH2	5′AGGGATGATGTTCTGGAGAGCC3′		679–700	
GAPDH2	5′AGGGATGATGTTCTGGAGAGCC3′	NG_007073.2	3145–3166	644 bps
GAPDH3	5′AAGCTGACTCAGCCCGCAAAGG3′		2523–2545	
	NQO1 primers			
	cDNA			
NQO1 F	5′GTCGGACCTCTATGCCATGA3′	NM_000903.3	238–257	TV1—685 bps TV2—583 bps TV3—571 bps
NQO1 R	5′GTCAGTTGGGATGGACTTGC3′		905–922	TV4—469 bps
NQO1389460 F	5′CAGCTCACCGAGAGCCTAGT3′		3756	
NQO1389460 R	5′CATGGCATAGAGGTCCGACT3′		237–257	221 bps
	Genomic DNA			
NQO1 g1F	5′CACACACACCCCTACAATCCCC3′	NG_011504.2	(−246)–(−225)	509 bps
NQO1 g1R	5′CCAGGTCCCTAATCTCTTCCC3′		243263	
NQO1 g2F	5′ACATTTCTGGCTACAGGAGATGGA3′		78827905	704 bps
NQO1 g3	5′GTCAGTTGGGATGGACTTGC3′		8573–8594	
NQO1 g4F	5′CAGCTCACCGAGAGCCTAGT3′		11299–11318	361 bps
NQO1 g4R	5′GAAATCCATGTAATACTGCACCT3′		11641–11659	
NQO1 g5F	5′AGTTGGCTGACCAAGGACAA3′		13285–13304	591 bps
NQO1 g5R	5′CCCTGCATCAGGACAGACC3′		13855–13875	
NQO1 g6F	5′TAGCTCAGGGGAGCCAAAGT3′		15104–15124	693 bps
NQO1 g6R	5′TGAATTCCCCTGAAGGTTCGT3′		1577715796	
NQO1 g1F	5′TGGTAACGGCTAGGTAGAGGG3′		(−246)–(−225)	509 bps
NQO1 g1R	5′AGCCCAGTCGGATTTTGGTT3′		243–263	

2.7. Reverse Transcription, RT – PCR, and PCR

The reverse transcription was performed with a High-Capacity cDNA Reverse Transcription Kit (Thermo Fisher Scientific, Waltham, MA, USA), with anchored Oligo(dT)$_{23}$ primers (Sigma-Aldrich, St. Louis, MO, USA) and 1 µg of total RNA in a 20 µL volume, according to the manufacturer's instructions. The reaction conditions were: 25 °C/10 min; 37 °C/120 min; 85 °C/5 min; 4 °C/indefinite. After finalization of the reverse transcription, 80 µL of sterile, deionized water was added to the tubes to achieve a total volume of 100 µL of cDNA, which was used for subsequent reactions.

The efficacy of reverse transcription was assessed with the end-point polymerase chain reaction (PCR) using the primer pair GAPDH 1/GAPDH 2 and 1 µL of diluted cDNA. This template volume was used as a standard in all end-point PCR reactions. For discovering the potentially present traces of contaminating DNA, the primer pair GAPDH2/GAPDH3 was used, as the sequence of the GAPDH3 primer is complementary to the nucleotides in intron 5. The polymerase chain reaction was carried out in GeneAmp PCR System 2400 (Applied Biosystems, Foster City, CA, USA). The reaction mixture (12.5 µL) contained AmpliTaq 360 Gold Master Mix and GC Enhancer (Thermo Fisher Scientific, Waltham, MA, USA), home-made nuclease free-water and primers (final concentration: 400 nM).

The genomic DNA was amplified with the same sets of chemicals and in the same volume, with 200 ng of gDNA. The reaction conditions were: Predenaturation 95 °C/5 min, followed by 35 cycles: 9 °C/30 s; 58 °C and 59 °C for gDNA and cDNA, respectively/30 s, 72 °C/30 s. The final elongation was at 72 °C, for 7 min.

2.8. Densitometry, Purification of PCR Products from Agarose Gel and Sequencing

The PCR products were separated by electrophoresis in 2% agarose gel and photographed for future densitometric analysis, which was performed with ImageJ [39]. Bands of interest were cut out of the gel and purified using a GenElute Gel Extraction Kit (Sigma-Aldrich, St. Louis, MO, USA), according to the manufacturer's instructions. After being eluted from the column, 10 μL of eluate was loaded in 2% agarose gel in order to determine the purity and amount of eluate that was going to be used in the sequencing reaction. Only the amplicon corresponding to TV4 needed to be purified, re-amplified one more time with NQO1F/NQO1R and purified again. A sufficient amount of the template was obtained for the sequencing reaction only after this additional procedure.

The sequencing reaction contained 12 μL mixture of the purified PCR product (approximately 50 ng per 100 bps), nuclease-free water and 1 μL of the forward and reverse primer used in PCR, respectively (primer concentration 3.2 pmol/μL), for a total volume of 13 μL. The samples were sequenced at the DNA Sequencing Core Facility of the Rudjer Boskovic Institute.

2.9. Real-Time RT-PCR and Rationale for Specific TaqMan's Probe Selection

For quantification of NQO1 and NRF2, TaqMan chemistry was used. The reactions were performed in a 7300 Real-Time PCR System (Applied Biosystems, Foster City, CA, USA). We used always 1.5 μL of cDNA template, 10 μL of TaqMan Fast Advanced Master Mix (Thermo Fisher Scientific, Waltham, MA, USA), 1 μL of the probe and 7.5 μL of sterile, deionized water. The reactions were performed in triplicate for each template and for each probe and in at least three biological replicates, which we tried to associate with three consecutive passages. Microamp 96-well rxn plates (Thermo Fisher Scientific, Waltham, MA, USA) were used, and amplification was performed under the following conditions: Incubation 50 °C/2 min + 95 °C/10 min, followed by 40 cycles; 95 °C/15 s, 60 °C/60 s. The following probes were used for house-keepers: GAPDH—Hs99999905_m1 (as GAPDH was used in the end-point PCR reactions); TBP—Hs00920495_m1 (as the TBP was used in the WBs); HPRT1—Hs02800695_m1 (as it was shown to be relatively stable in some published studies) [40].

For relative quantification, delta-delta Ct 2–ΔΔCt developed by Applied Biosystems was used, which presumes identical amplification efficiencies of the target and reference gene and the Pfaffl method [41], which served as a correctional factor in subsequent calculations.

2.10. Protein Extraction and Western Blot Analyses

The cells were cultured for 48 hours at a density of 1×10^6 in T25 flasks (Sarstedt AG&Co.KG, Nümbrecht, Germany), in 5 mL of the previously described media formulations (NC1-NC3), supplemented by 10% FBS. The proteins were extracted with NE-PER nuclear and cytoplasmic extraction reagents (Thermo Scientific -Pierce Biotechnology, Rockford, IL, USA), supplemented by the protease inhibitor (Complete Mini Protease Inhibitor Cocktail Tablets; Roche Applied Science, Mannheim, Germany). The protein amount was estimated using the Bradford method [42]. The absorbance was measured at 595 nm using the microplate reader Multiskan EX (Thermo Electron Corporation, Shanghai, China). The protein samples were mixed with Laemmli buffer and boiled for 5 min at 95 °C. The equal amounts of protein (10 μg) were loaded on the gel (9% resolving and 5% stacking), separated electrophoretically and transferred to nitrocellulose membranes (Roti®-NC, Carl Roth, Karlsruhe, Germany). The transfer efficacy was evaluated by staining the membranes with Ponceau S solution (Sigma Aldrich, St. Louis, MO, USA). The stained membranes were scanned. After incubating the membranes with 5% nonfat milk (Cell Signaling Technology, Danvers, MA, USA) in Tris-buffered saline (TBS; 50 mM Tris-Cl, 150 mM NaCl, pH 7.6) containing 0.1% Tween-20 for 1 h, the membranes were probed overnight with the following primary antibodies: (all rabbit monoclonal, Cell Signaling Technology, Danvers, MA, USA): anti-NQO1 (1:1000; CST: #62262); anti-NRF2 (1:1000; CST:#12721); anti-TP53 (1:1000; CST:#2527); anti-TBP (1:1000; CST:#44059); anti-β-actin (1:1000, CST:#8457). The last two antibodies were used as the loading controls for nuclear and cytoplasmic fractions, respectively.

The expected molecular weights of the detected proteins were: NQO1—29 kDa; NRF2—97–100 kDa; p53—53 kDa; TBP 35–45 kDa; β-actin 45 kDa. After three washings of the membranes with TBST (0.1% Tween 20 in 1× TBS), the immunoreactive bands were detected with an HRP- linked anti-rabbit IgG secondary antibody (1:2000; CST: #7074). The immunological complexes were visualized using SuperSignal^{TM} West Pico PLUS Chemiluminescent Substrate (Thermo Scientific, Rockford, IL, USA) and Alliance 4.7 (UVITEC, Cambridge, UK). The protein expression levels were quantified using ImageJ and/or Image Studio Lite (LI-COR, Lincoln, NE, USA) analysis software. The relative change of signals obtained was calculated after normalization according to the loading controls and Ponceau S signals.

2.11. Statistical Analyses

Each experiment related to cellular biology (viability, proliferation, ROS generation) was performed in technical triplicates or quadriplicates and repeated three times (as specified in the Figure legend). The data obtained was analyzed with 1-way ANOVA and Tukey post-hoc test, as indicated in the figure legends. The same principle was applied for producing and analyzing the data obtained with molecular biology methods in biological triplicates. For both analyses and visualization, GraphPad 6.0 was used. The statistical significance of the differences obtained for all data analyzed was considered significant at $p < 0.05$.

3. Results

Nutritional conditions were first profiled using the cancer cell lines originating from HNSCC: FaDu; Cal 27; Detroit 562. In this study, highly proliferating cells were used for exploringbasic cellular parameters before including a very slowly proliferating cell line—IMR-90 fibroblasts.

3.1. Viability, Proliferation and Generation of ROS

The choice of experimental conditions needed for nutritional stress induction, in relation to the concentration of glucose and glutamine in the medium, was combined with measuring cellular viability (Figure 2A), cellular proliferation (Figure 2B), and the amount of ROS generated at hour 48 (Figure 2C). Initially, the four nutritional conditions (NC1-NC4) were established, as described in the Material and Methods section, to which only the cancer cells were exposed. Then, based on the data obtained, this research was extended to IMR-90, through applying nutritional conditions NC2 and NC3. NC1 should be considered the control condition.

As presented in Figure 2A, the viability of the cancer cell lines, regardless of their genetic background, was similar under the given conditions. Predictably, the most intensive decrease in cellular viability (up to 70%) was recorded for all three cancer cell lines in the medium without glucose and glutamine (NC3), when compared to both NC1 and NC2 ($p < 0.0001$).

The presence of glutamine in a medium without glucose (NC4) was beneficial for the viability of all cancer cell lines (Figure 2A). It was also beneficial for the cellular proliferative capacity (Figure 2B) of Detroit 562 ($p = 0.0007$) and Cal 27 ($p < 0.0001$), but not FaDu. The strongest ROS generation was associated with condition NC3. The presence of glutamine in a medium without glucose (NC4) led to a decreased ROS generation in all three cancer cell lines (Figure 2C). Under NC2, the generation of ROS in Cal 27 and FaDu was stronger ($p = 0.0055$ and $p < 0.0001$, respectively) than in Detroit 562 (Figure 2C).

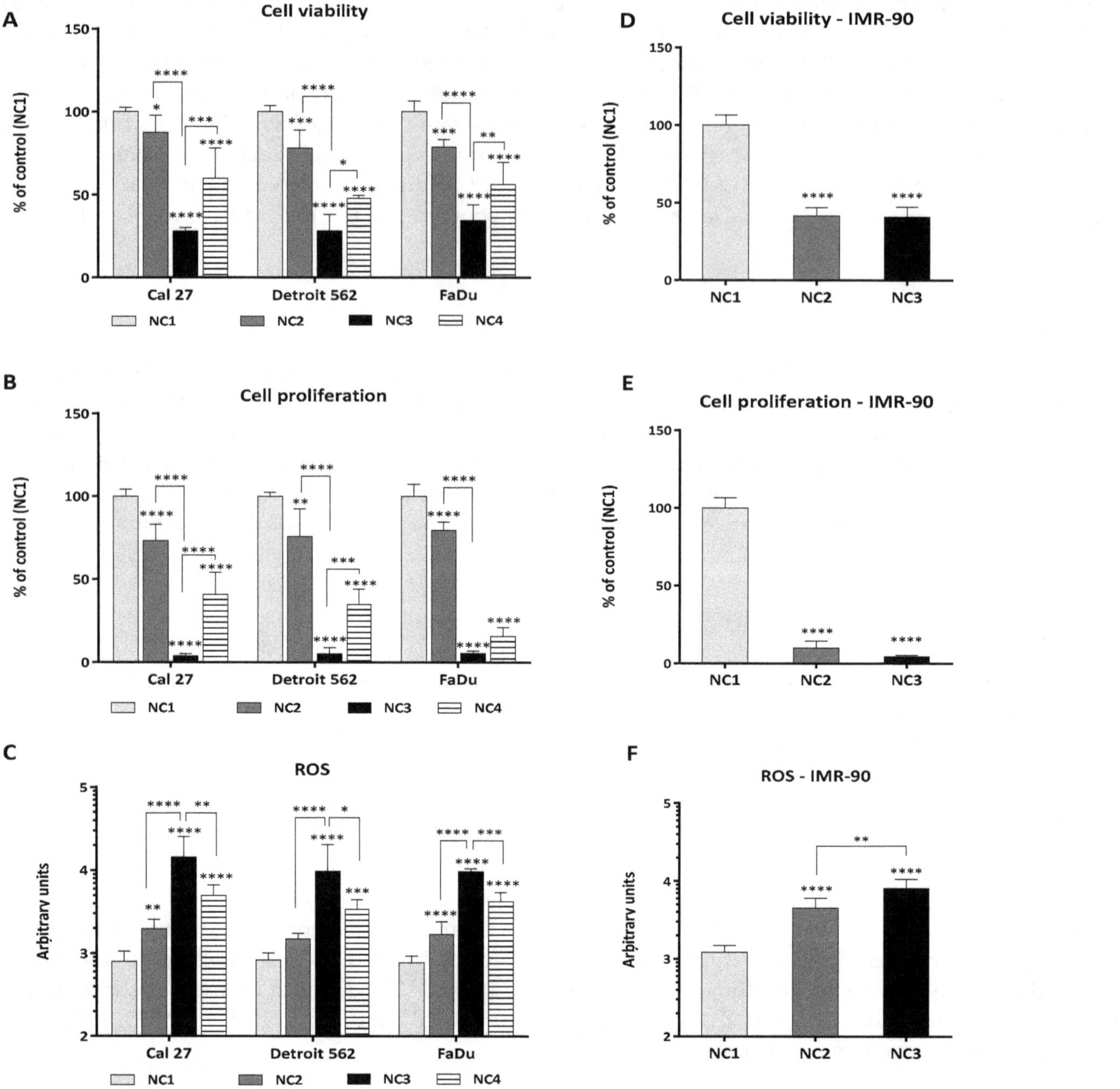

Figure 2. The viability, proliferation and generation of ROS in cancer cell lines (**A** **C**) and IMR-90 (**D** **F**) after exposure to NC1-NC4 and NC1-NC3, respectively, for 48 hours. One-way ANOVA with Tukey post-hoc test was used to test the differences with regard to nutrient conditions. The values are shown as the mean ± 95% CI. N = 3. * $p < 0.05$; ** $p < 0.01$; *** $p < 0.001$; **** $p < 0.0001$.

When reviewed, these data indicated that the cancer cell lines showed some interesting and unique features. Under NC2, the generation of ROS was not significantly increased only in Detroit 562. Under NC4, when there is a lack of glucose, FaDu was far less sensitive to the rescuing effect of glutamine on proliferative capacity.

These data are also very indicative regarding the degree of cellular sensitivity to glutamine deprivation, showing that Cal 27 and Detroit 562 were more dependent on glutamine than was FaDu.

Knowing that non-transformed cells are highly dependent on glucose, and relying on the data obtained with the cancer cell lines (Figure 2A–C), we chose to continue the experiments using the mildest (NC2) and the most robust condition (NC3), now including the IMR-90 fibroblasts.

They were considered a good control system, to compare with the cancer cell lines. The Figure 2D–F represent IMR-90 response to NC2 and NC3. The viability of IMR-90 was unique in the extreme sensitivity of this cell line to the mild glucose deprivation, during 48 hours (NC2; $p < 0.0001$) (Figure 2D). The viability and proliferative capacity after 48 hours in NC2 (Figure 2D,E) seems to be a maximal effect of a critical nutrient deprivation (glucose) because 48 h of cultivation in the medium without glucose and glutamine (NC3) did not influence these cellular parameters further. However, the generation of ROS did differ between NC2 and NC3 ($p = 0.0019$) (Figure 2F), although not as strong as in the cancer cell lines ($p < 0.0001$) (Figure 2C). The lack of a significant change of IMR-90 viability after 48 hours of exposure to mild and extreme starvation, is in clear contrast with the cancer cell lines and needs to be further explored.

3.2. *Quantification of NQO1 and NRF2 in a Real-Time*

According to the majority of literature data, the NRF2-NQO1 axis should be highly active under conditions which induce the generation of ROS. As a first step forward, the transcriptional activation of *NRF2* and *NQO1* in real-time was explored. Three different house-keeping genes were used, as we were aware that under the experimental conditions we chose to explore, we might not be able to make an accurate quantification of our targets. Under the given conditions, none of the three housekeepers (*GAPDH*, *TBP*, *HPRT1*) used in this research were universally stable in all four cell lines. The changes in their transcription rates were clearly both condition specific and cell-type specific, as already presented [40]. Finally, under the given conditions, a quantification was made by combining these three housekeepers in a particular cell line.

The general picture shows that in all cell lines, the transcriptional activity of *NRF2* increases under NC2 and NC3 (Figure 3A,C,E,G). The quantification for FaDu, under NC3, was estimated based on the Ct values for NRF2 and NQO1 under NC1 and NC2, in respect to NC3 (Figure 3E,F). However, under NC2, the increasing trends of *NQO1* were not present in Cal 27 and FaDu. The statistically significant differences in transcription rates were reached only in Cal 27, for both genes: *NRF2*: NC1 versus NC2; $p = 0.0008$; NC2 versus NC3; $p = 0.0102$: *NQO1*: NC1 versus NC3; $p = 0.0153$; NC2 vs. NC3; $p = 0.0072$ and IMR-90, but for *NQO1* only: NC1 versus NC3; $p = 0.0276$. Notwithstanding the mathematical calculation, it is visible that only in Cal 27 and already at NC2 does the Fold Change of *NRF2* reach a value above 2 (Figure 3A). Thus, it is quite obvious that the transcriptional level of *NRF2*, with the exception of Cal 27, does not exhibit as strong changes as expected, especially not with respect to the increased generation of ROS (all cancer cell lines: NC2 versus NC3 $p < 0.0001$; Figure 2C). An increase of *NQO1* mRNA was recorded under NC3 in Cal 27, Detroit 562 and IMR-90. Due to significant variations in the transcriptional level of house-keepers in FaDu under NC3, the target transcripts were not accurately quantified, under that condition and only in that cell line.

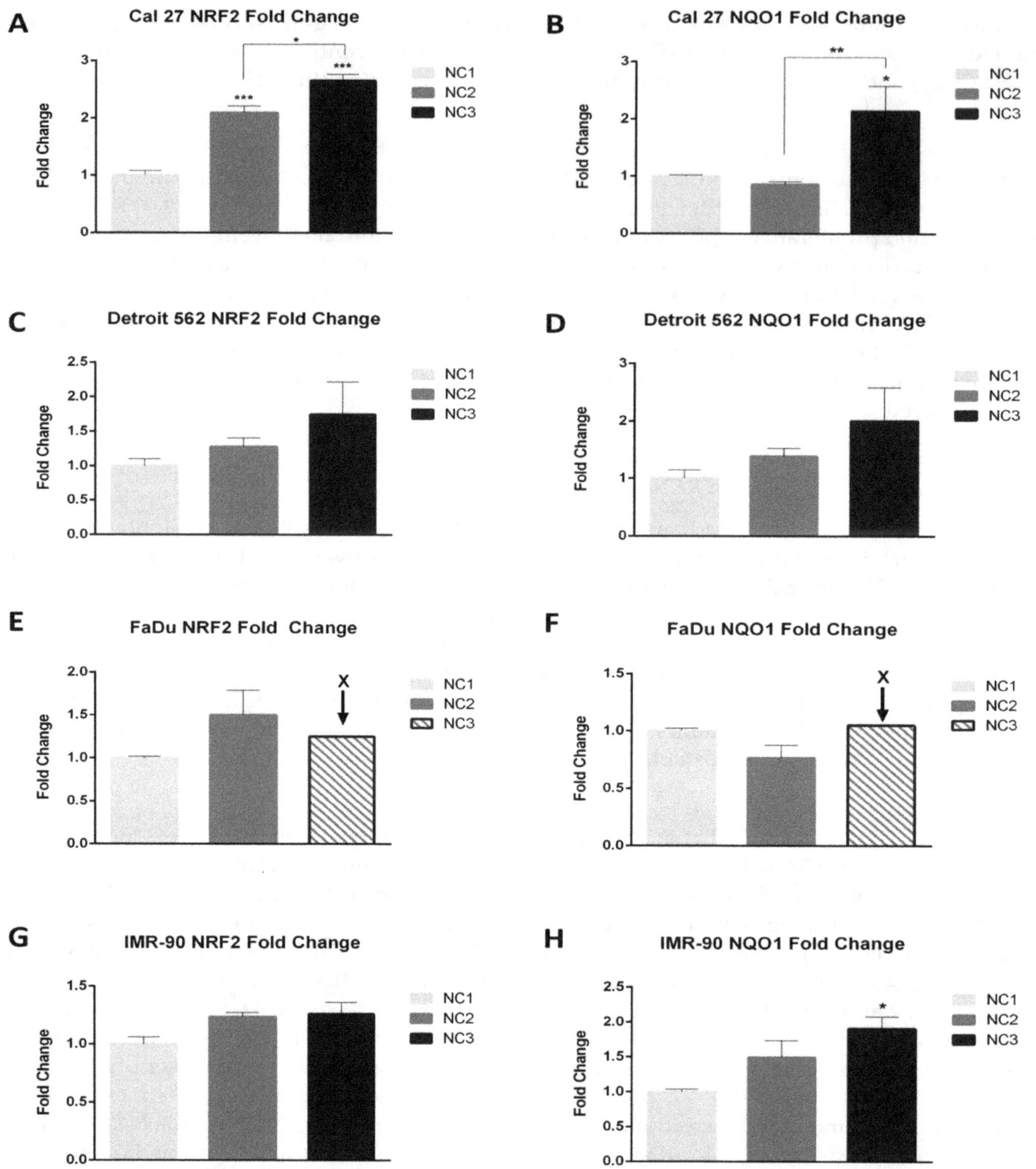

Figure 3. The quantification of target transcripts, *NRF*2 and *NQO*1, in real time (RT-qPCR). One-way ANOVA with Tukey post hoc test was used to test the differences with respect to the quantity of *NRF*2 and *NQO*1 mRNA, under different nutritional conditions. The values are shown as the mean ± SD. N = 3. * $p < 0.05$; ** $p < 0.01$; *** $p < 0.001$; **** $p < 0.0001$. * The Fold Change for NRF2 and NQO1 in FaDu under NC3 was estimated according to the Ct values.

3.3. Three Splice Variants of NQO1 are Present in All Four Cell Lines, Notwithstanding the Experimental Conditions Applied

In all four cell lines, three reproducible bands occurred in the end-point PCR (35 cycles), regardless of the cellular background and/or the type of nutritional stress (Figure 4A–C). Based on their size, we were certain of the existence of TV1 and TV4. The nature of the amplicon which was in the middle position (as the length of TV2 and TV3 differ for only 13 nucleotides, Figure 1) was revealed only after analyzing the sequencing data (Figure 5). Under all experimental conditions and in all cell lines, TV1 gave the most prominent signal and TV3 was visible. However, in Detroit 562, TV4 varied and was almost undetectable. There was an additional 400 bps long amplicon present in all cell lines under all conditions, which was not characterized further. After a densitometric analysis of the signals after 28 cycles (when the signal was less saturated) using the ImageJ, the changes in the ratio TV1/TV3 were recorded (Figure 6).

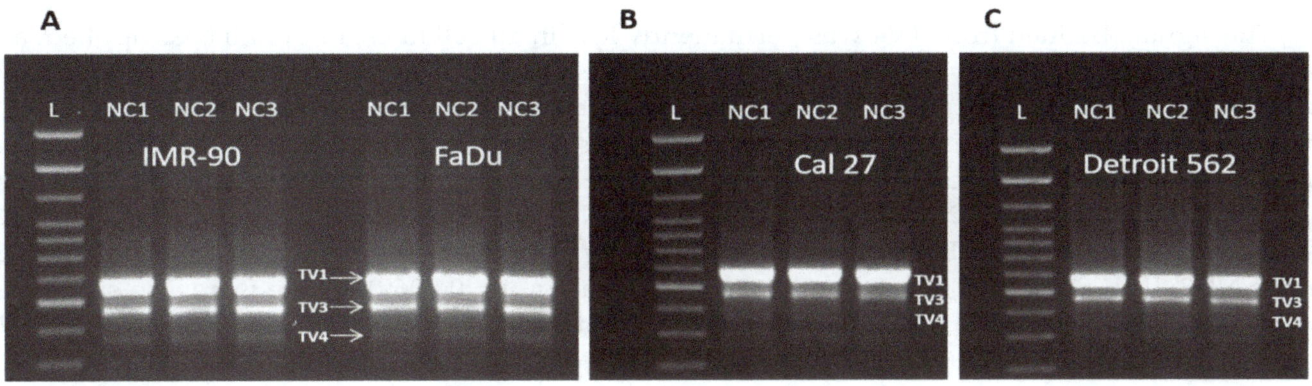

Figure 4. The presence of three NQO1 splice variants in all four cell lines, under three nutritional conditions (35 PCR cycles). L: 100 bp DNA ladder (Invitrogen). **A**. IMR-90 and FaDu; **B**. Cal 27; **C**. Detroit 562. TV1: 685 bps, TV3: 571 bps and TV4: 469 bps.

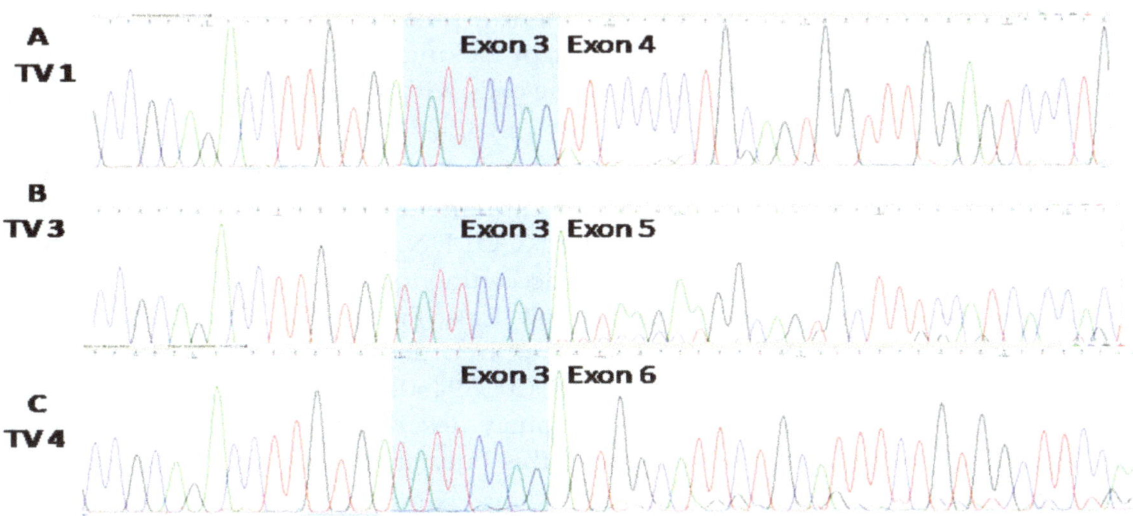

Figure 5. The sequence analyses of amplicons obtained from the Cal 27 NQO1-cDNA amplified with the primer pair NQO1F/NQO1R (Table 1), presented in Figure 4, confirm the presence of transcript variants TV1 (**A**), TV3 (**B**), and TV4 (**C**). Eight nucleotides in the terminal part of the 3′ exon 3 are common to all three amplicons and are shown in blue. The sequences of TVs were identical in all cell lines.

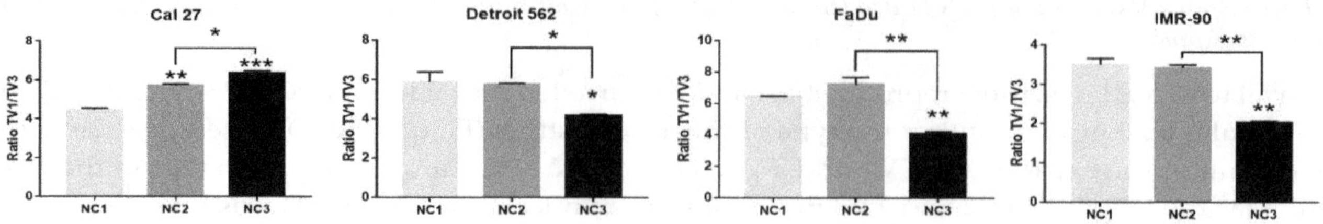

Figure 6. The dependence of the TV1/TV3 ratio on NCs applied. One-way ANOVA with Tukey post-hoc test. The ascending trends were recorded in Cal 27 as contrasted with the other three cell lines. The values are shown as the mean ± SD. N = 3. * $p < 0.05$; ** $p < 0.01$; *** $p < 0.001$; **** $p < 0.0001$.

The signal obtained from TV4 was permanently low in all cell lines, under all NCs applied, and was excluded from the analyses. The intensity of the densitometric signal originating from TV1 and TV3 was expressed as their ratio (TV1/TV3) obtained after 28 PCR cycles (Supplementary Figure S1). The opposite trends in Cal 27 versus three other cell lines were observed. In Cal 27, the TV1/TV3 ratio increased from 4.47 (NC1) to 5.7 (NC2) and 6.36 (NC3) (Figure 6). The differences were statistically significant: NC1 versus NC2: $p = 0.0020$; NC1 versus NC3: $p = 0.0006$; NC2 versus NC3: $p = 0.0123$. In IMR-90, Detroit 562 and FaDu, the TV1/TV3 ratio was significantly decreased under NC3 (NC1 versus NC3; IMR-90: $p = 0.0014$; Detroit 562: 0.0213; FaDu: $p = 0.0069$). There were also significant differences between the ratio of TVs which was observed between NC2 and NC3 (IMR-90: $p = 0.0017$; Detroit 562: $p = 0.0261$; FaDu: 0.0031).

3.4. Detection of SNVs in Intron/Exon Boundaries

3.4.1. rs 689460, G+C, Is Present in IMR-90 and Does Not Influence the NQO1 Splicing

Based on the one paper showing that the polymorphism present at the very end of *NQO1* exon 4 (rs1131341) favors occurrence of the transcript lacking exon 4 [31], and based on the cDNA end-point PCR and sequencing data, the status of the cells with respect to rs1131341 was determined.

All exon/intron boundaries with primers shown in Table 1 were analyzed, using the gDNA as a template. All primers, except primer NQO1 g6R (complementary to nucleotides in the 5′ part of the exon 6), were complementary to intronic/non-coding NQO1 DNA sequence. With these sets of analyses, we discovered a polymorphism in intron 1 (nt #248, according to NG_011504.2), present only in IMR-90 genomic DNA (Figure 7A), but not in the cancer cell lines (Figure 7B). To date, there is no data on the potential influence of this SNV on *NQO1* splicing. In order to explore the potential influence of this polymorphism on splicing, novel primers, NQO1389460F and NQO1389460R, (Table 1) were constructed and analyses were performed as previously described. The existence of any alternative new splice variants which could be associated with this polymorphism, under the applied conditions could not be confirmed (Figure 7C).

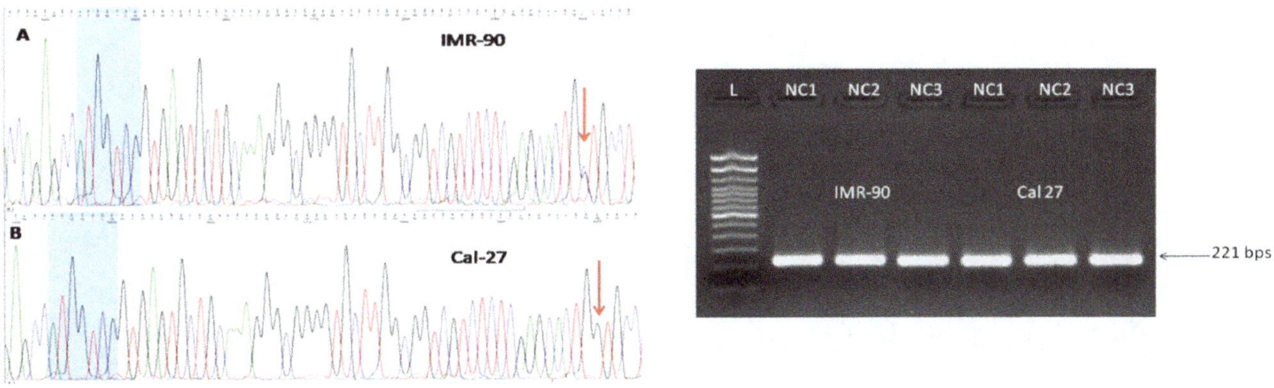

(C)

Figure 7. A. Single nucleotide variation C+G wass present in IMR-90, in NQO1 intron 1. The terminal part of the 3' exon 1 is shown in blue. The distance between the polymorphic locus, revealed to be rs 689,460 according to NCBI, and 3' of exon 1, is only 50 nts. **B**. All three cancer cell lines were homozygos, rs 689,460 G+G, as shown here, for Cal 27. **C**. Single nucleotide variant C+G in IMR-90 did not influence the NQO1 splicing. The amplicons obtained in IMR-90 under given conditions did not differ from those obtained in homozygos (G+G) cancer cell lines (only Cal 27 is shown). L: 100 bp DNA ladder. Lines 1–3 and lines 4–6: IMR-90 and Cal 27 under NC1, NC2 and NC3, respectively.

3.4.2. rs 689452, C+G, Is Present in Detroit 562 and Cal 27

The presence of one more SNV in intron 1 (nt# 8070, according to NG_011504.2) was further discovered, in Detroit 562 and Cal 27 (Figure 8). The sequence variant (C+G) corresponds to rs 689452. In IMR-90 and FaDu, this position was homozygous, C+C, as shown on Figure 8. Based on the splice variants analyses, rs 689,452 does not influence the splicing of *NQO*1.

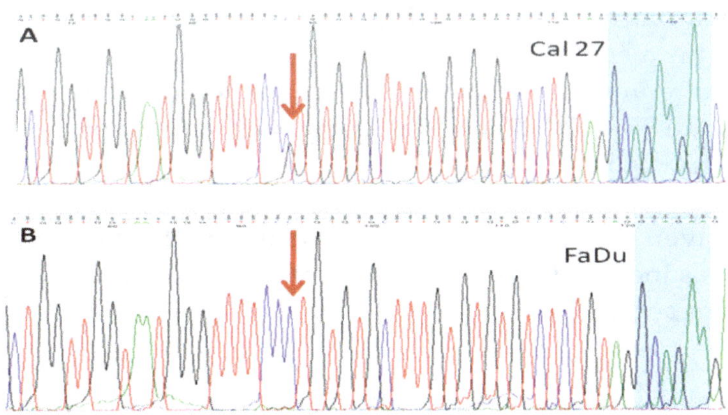

Figure 8. A. Single nucleotide variant, rs 689452, C+G in Cal 27 (also present in Detroit 562) is in intron 1, separated from the 5' part of exon 2 (labeled blue) by only 27 nts. **B**. In FaDu (shown by an arrow) and IMR-90 (not shown), the sequence was homozygous, C + C.

3.4.3. Presence of rs1800566 in IMR-90 but Not in Cancer Cell Lines

The rs1800566 was detected only in IMR-90. This well-known polymorphism was present as a homozygous SNV–NQO1*2/*2 (nt #15389, according to NG_011504.2), leading to a change of the triplet CCT into TCT (Figure 9A). This is highly consequential, because this SNV missense variant, as described earlier, leads to replacing proline with serine at position 187. This polymorphism was shown as the one which influenced the structure of the NQO1 protein, making it highly unstable and catalytically compromised [43]. While it is known that IMR-90 expresses only traces of NQO1 protein, the data related to IMR-90 genotype, with respect to rs1800566, could not be found.

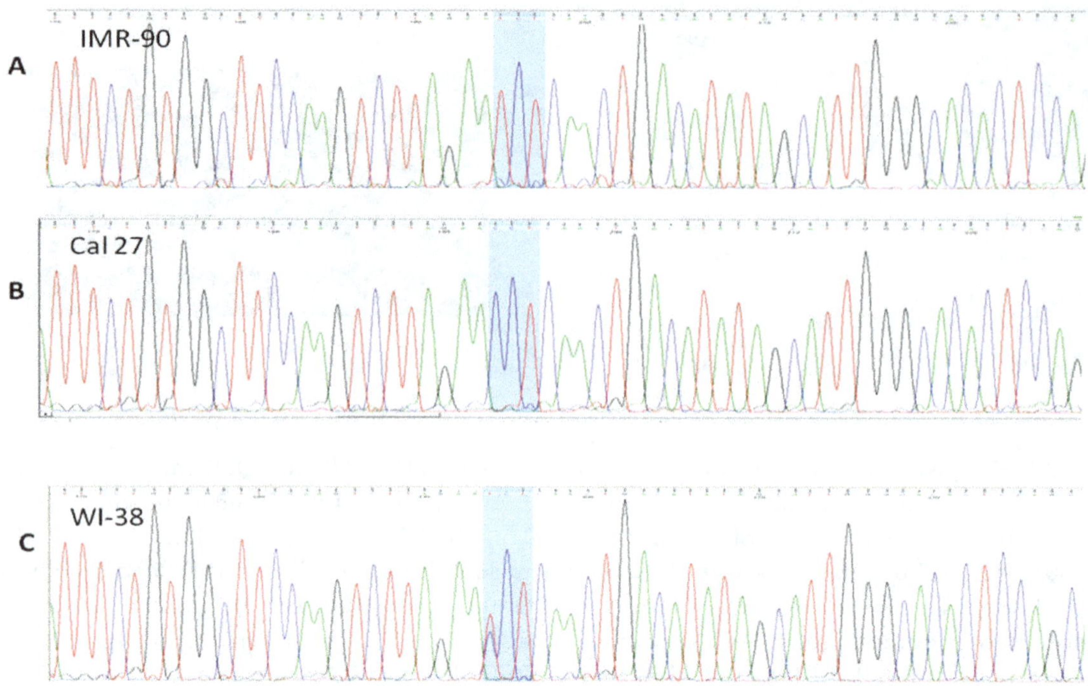

Figure 9. The presence of SNV rs1800566, in IMR-90 (NQO1*2/*2; (**A**)) and WI-38 (NQO1*1/*2; (**C**)). The homozygous triplet TCT, coding for Serine, replaces the CCT triplet coding for Proline, which was present in all cancer cell lines, as shown for Cal 27 (**B**).

As human fibroblasts WI-38 express NQO1 but are not able to develop the NQO1-mediated redox cycle that could lead to HSP90 (Heat Shock Protein) inhibition upon β-Lapachone induction [44], it was presumed that these cells may have a genotype NQO1*1/*2. When we worked with WI-38 [45], the DNA was preserved from an early passage and kept frozen at −20 °C. Indeed, WI-38 are heterozygotes, NQO1*1/*2 (Figure 9C). This means that both commonly used cell lines are, regarding NQO1 activity, severely compromised and represent two totally different biological systems. All cancer cell lines were homozygous, P187P (Figure 9B – Cal 27) and the enzymatic activity of their NQO1 should be intact, at least with respect to rs1800566.

Finally, under the given conditions, we explored the NRF2/NQO1 axis on the protein level. In these analyses, TP53 was included. For the reasons explained earlier, western blots on nuclear and cytoplasmic fractions were performed.

3.5. Western Blot Analyses

3.5.1. Analyses of NQO1

Under NC1, the NQO1 signal was least visible in IMR-90. Although all three cancer cell lines had a considerably high level of NQO1 in both cellular fractions, a lower basal amount of NQO1 in Cal 27 nuclear fraction, as compared with FaDu and Detroit 562, was obvious on all our blots (Figure 10A,B). The level of NQO1 in the cellular fractions did not significantly change under the NC2 condition. However, some trends can be observed. There was a slight decrease of NQO1 in both cellular fractions of Cal 27 and Detroit 562 and an increase in both cellular IMR-90 fractions. The signal in FaDu was slightly increased in the cytoplasm, and, at the same time, there was a mild decrease in the nucleus (Figure 10C,D). Although under NC2 densitometry registers an obvious increase of NQO1 in both IMR-90 fractions, the intensity of the NQO1 signal remained very weak and far below the signal obtained in all cancer cell lines and under all tested conditions.

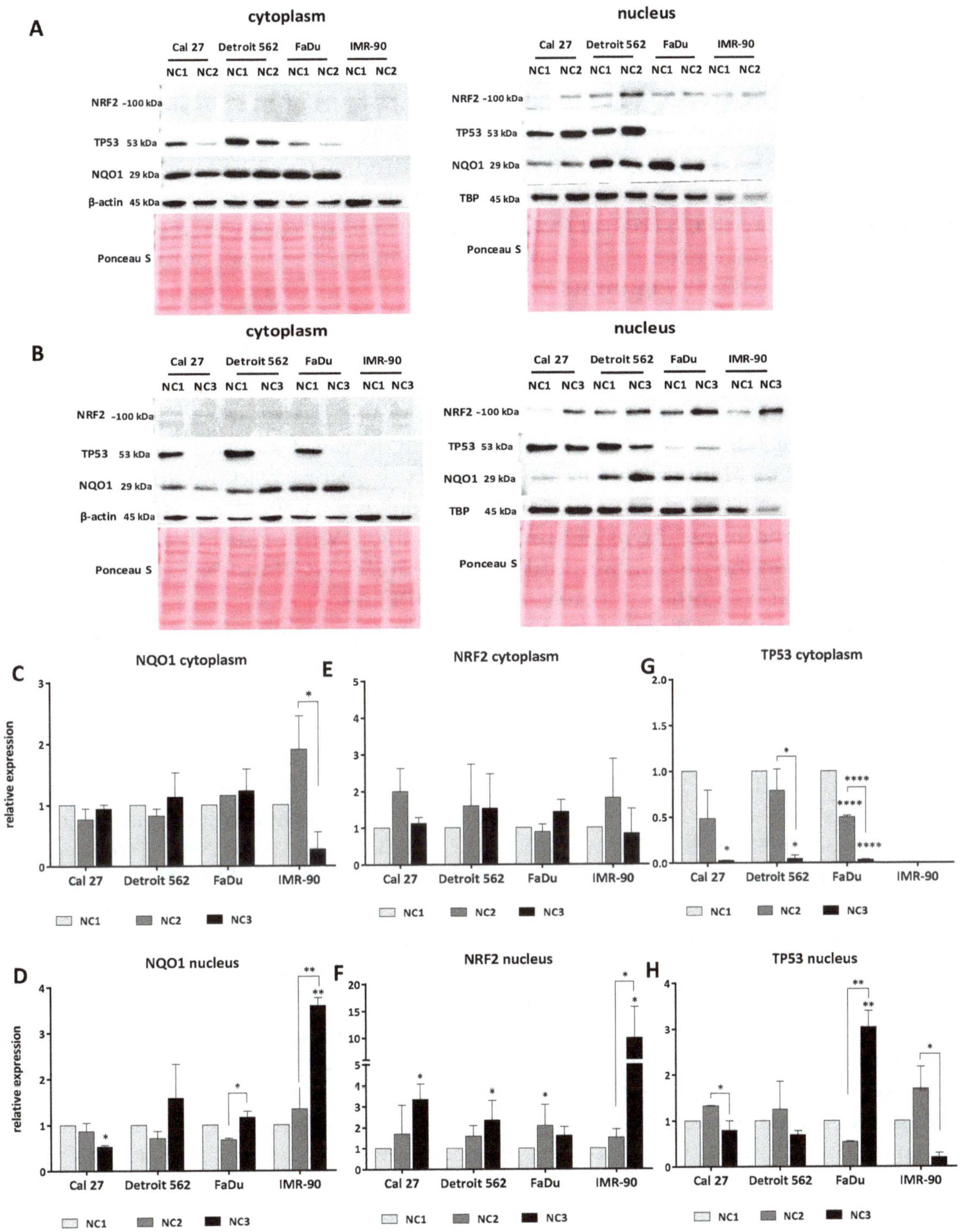

Figure 10. The expression of NQO1, NRF2 and TP53 in cancer cell lines and IMR-90. The representative western blots of cytoplasmic and nuclear NQO1, NRF2 and TP53 content under different nutritional conditions: (**A**) NC2 (low glucose + L-glutamine); (**B**) NC3 (no glucose and no L-glutamine). The relative expression is calculated as compared to the control condition (NC1 – high glucose + L-glutamine) for: **C**-cytoplasmic and **D**-nuclear NQO1; **E**-cytoplasmic and **F**-nuclear NRF2; **G**-cytoplasmic and **H**-nuclear TP53. One-way ANOVA with Tukey post hoc test was used to test the differences in relative expression of selected proteins under different nutrient conditions. The values are shown as the mean ± SD. n = 3. * $p < 0.05$; ** $p < 0.01$; *** $p < 0.001$; **** $p < 0.0001$.

In all cancer cell lines, the more harsh condition of NC3 did not significantly affect the NQO1 cytoplasmic levels which remained close to the NQO1 control levels (NC1). In IMR-90, cytoplasmic level of NQO1 decreased (Figure 10C). Although not statistically significant as compared with the control (NC1), this decrease was quite intense when compared with cytoplasmic NQO1 expressed with respect to NC2 ($p = 0.0107$). IMR-90 responded with increased nuclear accumulation of NQO1 when exposed to NC3 ($p = 0.0056$). Under the same conditions, Cal 27 showed a nuclear decrease of NQO1 ($p = 0.0457$). Although the NC3 condition did not significantly affect the nuclear level of NQO1 in FaDu when compared with the control (NC1), a significant increase was recorded when compared to NC2 ($p = 0.0178$). An increased nuclear NQO1 level in NC3, when compared to NC2, was also observed in IMR-90 ($p = 0.0085$) (Figure 10D).

3.5.2. Analyses of NRF2

The cytoplasmic expression of NRF2 in all four cell lines was weak and unaffected by nutritional conditions (Figure 10E). The nuclear NRF2 expression showed some differences. Although a nuclear increase of NRF2 in all cell lines under NC2 was recorded, it was statistically significant only in FaDu ($p = 0.0414$). Under NC3, a significant increase of NRF2 was recorded in the nucleus of the remaining three cell lines (NC1 vs. NC3: Cal 27 $p = 0.039$, Detroit 562 $p = 0.0306$, IMR-90 $p = 0.0103$). However, under these extreme conditions, it was not present in FaDu. With regard to all other cell lines, there was a difference between the nuclear amount of NRF2 under NC2 and NC3 in IMR-90 ($p = 0.0142$), which was manifested as a significant increase (Figure 10F).

It was proposed that NQO1 stabilizes both the wild-type and mutant-type TP53. There was a mixed situation—wild-type NQO1 and mutant TP53 in the cancer cell lines and mutant-type NQO1 and wt TP53 in IMR-90.

3.5.3. Analyses of TP53

As expected, the wild type TP53 was not detected in the IMR-90 cytoplasm (Figure 10A,B). There was a significantly high level of the expressed mutant types of TP53 in the cytoplasm of the cancer cells in NC1. Cytoplasmic TP53 decreased in all cell lines under both NC2 and NC3. However, under NC2, the decrease reached a statistically significant level only in FaDu ($p < 0.0001$) (Figure 10G). The nuclear level of TP53 in Cal 27 and Detroit 562 exposed to NC2 for 48 h, was arround or slightly above the control (NC1) values (Figure 10H). Under NC3, the strong TP53 signal in the cytoplasm of the cancer cell lines literally disappeared (Figure 10B,G). It was present in the nucleus and showed very interesting trends in relation to the conditions applied. In Cal 27, IMR-90 and Detroit 562, the NC3 related TP53 signal was lower than in NC2. The decrease was significant for Cal 27 and IMR-90; NC2 versus NC3; $p = 0.0411$ and $p = 0.0258$. In FaDu, the signal of nuclear TP53 strongly increased and reached statistical significance (NC2 versus NC3; $p = 0.0025$) (Figure 10H).

The images on Figure 10 were obtained with 10 micrograms of proteins, under the same conditions. For improving the visibility of the bands corresponding to TP53 and NQO1 in IMR-90, a separate blot was made with the same amount of protein loaded (10 μg). For detection of the signals, software-controlled prolonged time exposure was used (Supplementary Figure S2).

3.5.4. The General Picture of the NRF2-NQO1 Axis and TP53 at the Protein Level, in the Experimental Model

Under NC2, the NRF2-NQO1 axis did not seem to be activated in the cancer cell lines. This was because: a) Only FaDu exhibited a significant increase of NRF2 in the nucleus, which was joined with decreased NQO1 and TP53 signals in the nucleus; and b) Cal 27 and Detroit 562 had no significant increase of NRF2 in their nuclear fraction, nor did they exhibit a significant change in NQO1 expression. However, they had a slight increase of TP53 in their nuclei.

Under NC3, the NRF2-NQO1 axis seemed to have moderate activity in Detroit 562. This was because: (a) There was a significant increase of nuclear NRF2, associated with a slight increase of

NQO1 in both cellular fractions; (b) Cal 27 had the strongest increase of nuclear NRF2 among all the cancer cell lines, which was not associated with the corresponding increase of NQO1; and (c) FaDu seems to be unresponsive.

There are opposite trends with respect to the TP53 nuclear signal when comparing Cal 27 and Detroit 562 to FaDu, under NC2 and NC3. Under NC2, the TP53 nuclear signal increased in Cal 27 and Detroit 562 and decreased in FaDu. Under NC3, the TP53 nuclear signal increased in FaDu and decreased in Cal 27 and Detroit 562.

The axis NRF2-NQO1 was active in IMR-90, the cell line with a wt TP53. This was because: (a) a slight and strong accumulation of NRF2 was observed in the nucleus under NC2 and NC3, respectively (Figure 10A,B,F); and (b) and was joined by a recordable accumulation of the NQO1 (S187S) protein, however unstable it was shown to be [43].

The simplified graphic presentation of phenomena observed at the protein level is presented in Figure 11.

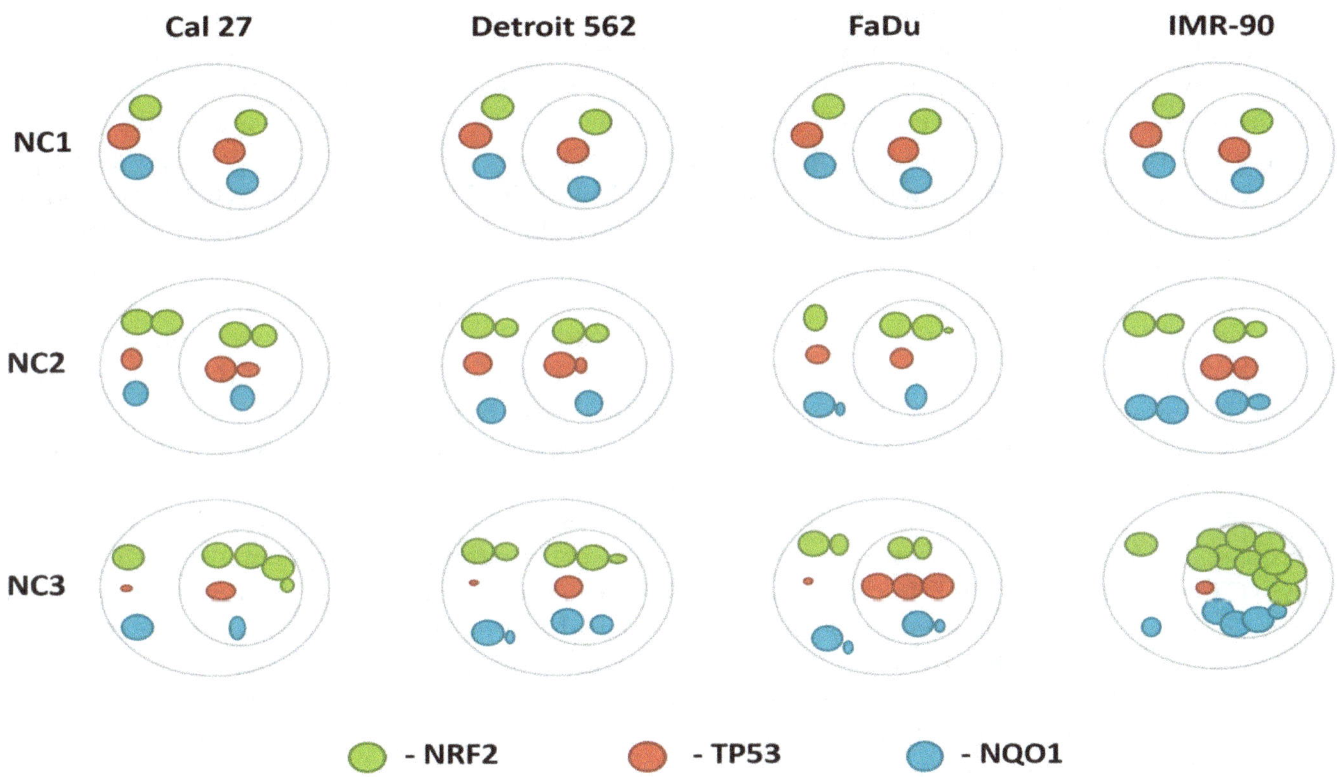

Figure 11. The cellular distribution and the rate of change of the NRF2, NQO1 and TP53 proteins in all four cell lines, under NC1-NC3.

4. Discussion

The lack of nutrients leads to cellular oxidative responses. We wanted to explore the NRF2-NQO1 axis by inducing stress through incubating cancer cell lines in the following: A medium with a decreased concentration of glucose (1 g/L) (NC2); a medium with a complete lack of glucose (NC4) but supplemented with glutamine; and, in order to induce the strongest level of ROS, by creating an artificial situation, using a medium deprived of glutamine and glucose (NC3). However, due to trace amounts of glucose and glutamine in FBS, these cells were still exposed to a minimal concentration of glucose (~0.125 g/L $\approx$ 0.69 mM) [46] and glutamine (0.05 mM) [47].

The cellular viability, ROS generation and rate of cellular proliferation, were measured and compared under these four conditions, in three cancer cell lines, originating from HNSCC: FaDu, Detroit 562 and Cal 27. According to COSMIC and published papers, these three cell lines can be classified according to their TP53 mutation status [48]: 1. FaDu—heterozygous mutation leading to substitution of arginine with leucine at aa248 (R→L); 2. Cal 27—mutation (unknown zigosity status) leading to substitution of histidine with leucine at aa193 (H→L); 3. Detroit 562 mutation (unknown zigosity status) leading to the substitution of arginine with histidine at aa 175 (R→H). All these mutations are in the DNA-binding domain of the TP53 protein. Based on the cellular effects observed in the cancer cells, molecular-genetic experiments were chosen to be performed under conditions NC2 and NC3, to which fetal lung fibroblasts IMR-90 were also exposed. This cell line was used to show how MYC, after being transduced in IMR-90, makes IMR-90 addicted to glutamine [49]. Terashima et al. were able to stimulate NRF2 entering in the nucleus of these cells under the low glucose condition [50]. Thus, it was assumed that IMR-90 is likely to be sensitive to NC2 [49]. Many research groups were not able to show the expressed NQO1 protein in these cells in their native state [36]. However, there are also data showing the expressed, mature NQO1 in IMR-90 [51].

4.1. Cancer Cells and Fibroblasts IMR-90 Differ in Their Sensitivity to Nutrient Deprivation

We were aware that the chosen conditions were likely to reduce cellular viability and increase the level of ROS, but we could not predict to what extent that would happen. What was visible immediately was that the viability of the cancer cells under NC2 decreased significantly, but far less than under NC3. With respect to NC2 and the cancer cell lines, our results were in the strongest agreement with results published by Terashima et al., who presented a similar decrease of HepG2 viability cultured in a medium with low glucose (1 g/L) [50] In IMR-90 only, the NC2 condition had extremely strong effects with respect to the examined cellular parameters. In IMR-90 only, the effect of the NC2 condition was much stronger than in the cancer cell lines, confirming that IMR-90 are primarily dependent on glucose. The result obtained on IMR-90 is in agreement with data presented by Yunova et al., who also convincingly showed the sensitivity of IMR-90 to glucose deprivation [49]. With respect to glutamine, van den Heuvel et al. showed the insensitivity to glutamine depletion of normal human lung fibroblasts (NHLF)) cells originating from lung fibroblasts [52].

As shown on Figure 2A,D, among all four tested cell lines exposed to NC2, the viability of fibroblasts decreased below 50% (app. 80% in cancer cell lines). It did not decrease further when there was a lack of both glucose and glutamine (NC3). One possible explanation of this effect may be the high expression of glutamate/cystine antiporter solute carrier family 7 member 11 (SLC7A11, also called xCT) in IMR-90 cells [53]. In cancer cells, SLC7A11 mediates the efflux of intracellular glutamate, thereby rendering them metabolically less adaptable and more reliant on glucose for survival [54]. The drop in cellular viability of all cancer cell lines, contrary to IMR-90, was less pronounced under NC2 than under NC3. The proliferation rate of FaDu under NC4 was far below the proliferation rate in Cal 27 and Detroit 562. This indicates that there is a differential sensitivity to glutamine between FaDu and the other two cancer cell lines, when there is a lack of glucose.

Both FaDu and Cal 27 cells have been reported to express a high level of SLC7A11 [55]. The relative unresponsiveness of FaDu observed under NC4 may be explained, at least in part, by the observed high increase of nuclear NRF2 ($p = 0.0414$) (Figure 10F), joined with increased *NRF*2 transcription and the significantly increased generation of ROS ($p < 0.0001$) (Figure 2C), in response to NC2. Considering that NRF2 has been reported to induce *SLC7A11* expression in response to glucose starvation [51], it is conceivable that FaDu cells, in which exposure to NC2 upregulated NRF2, would under NC4 (complete lack of glucose, albeit supplemented with glutamine) have a more pronounced expression of *SLC7A*11 and efflux of glutamate that would prevent the recovery of cellular proliferation. Cal 27 cells, which were also reported to express a high level of *SLC7A*11 [55] but did not respond to NC2 by statistically significant upregulation of nuclear NRF2, significantly recovered cellular proliferation under NC4NC4), as contrasted with FaDu cells.

Under NC3, FaDu was the only cancer cell line which had a significant increase of NQO1 in the nucleus (NC2 vs. NC3 $p = 0.0178$). However, at the transcriptional level, significant changes relating to

*NQO*1 transcriptional activity in FaDu were not detected, neither under NC2 (accurately measured), nor under NC3 (approximated).

Cal 27 was the cell line with the lowest basal level of NRF2 (Figure 10A,B, NC1—nuclear fraction). Probst et al. were able to show that the cell lines with high basal NRF2 activity exhibited little or no increase in *NQO1* mRNA levels following NRF2 activation with the compound RTA 405 [56]. We think this fact can explain part of our results obtained on NQO1 transcripts.

4.2. Importance of Defining the SNVs, in a Given Experimental Model

When *NQO*1 transcripts were quantified, a significant increase of the *NQO*1 transcript was shown only in IMR-90 and Cal 27 but not in FaDu (under NC2) and Detroit 562. Unfortunately, however high was the expression in IMR-90, there was a minimal amount of NQO1 protein. The level of the *NQO*1 transcript in IMR-90 was strong and it did not differ from the mRNA NQO1 signal in the cancer cell lines. A majority of research groups had not detected NQO1 protein in IMR-90 (at least not under restful conditions). However, it seems that nobody reported the data on the IMR-90 *NQO*1 genotype. After performing the DNA sequencing, it was concluded that this cell line is a NQO1*2/*2 homozygote (rs180566). Some inconclusive data obtained on another human fibroblast cell line, WI-38, encouraged us to analyze it in the same way. It was shown that WI-38 also bears rs180566 but is a heterozygote (NQO1*1/*2) cell line. Thus, at least with respect to NQO1, IMR-90 and WI-38 are not normal fibroblasts. Accordingly, any conclusion obtained on one cell line cannot be automatically translated to another (or any other) cell line. Regarding rs180566, detailed modeling performed by Lienhart et al. [43] did not provide a structure-based explanation for the lower enzymatic activity of NQO1 P187S. A plausible explanation given in an in vivo model by Tsvetkov et al., should be considered in light of discovering that E3 ligase STUB1/CHIP (C terminus of Hsc70-interacting protein) regulates the NQO1 protein level through ubiquitination and degradation [57]. The heterozygote P187S (rs180566) was shown to be a stronger STUB1 interactor with an increased susceptibility to ubiquitination by the E3 ligase STUB. Thus, we concluded that the homozygote, S187S, may have even stronger affinity for STUB1. This finally resulted in an almost undetectable NQO1 in IMR-90, notwithstanding the increase of *NQO*1 transcriptional activity (Figure 3; Figure 4) under NC3 which was associated with a physiological increase of nuclear NRF2 (Figure 10B). An increased sensitivity of IMR-90 to NC2 may be a consequence of lacking a strong mechanism for influencing the NAD(P)+/NAD(P)H redox balance during the stress-related events, due to catalytically insufficient NQO1.

4.3. FaDu Proliferative Potential with Respect to GLS1 and Decreased Sensitivity to Glutamine When Deprived of Glucose

It was very interesting to see that that the effect of glutamine on FaDu, with respect to proliferative potential when deprived of glucose, exerts a less positive effect than with Cal 27 and Detroit 562 (Figure 2B). The enzyme, Glutaminase 1 (GLS1), enables malignant cells to undergo increased glutaminolysis and utilization of glutamine as an alternative nutrient. A recently published study has clearly shown that the expression of *GLS*1 in FaDu is far less prominent than in Detroit 562 cells [58]. Thus, the capacity of FaDu for glutamine utilization seems to be constitutively decreased. Although we did not measure *GLS*1, we hypothesized that the less prominent rescuing effect of glutamine on FaDu than on Detroit 562 and Cal 27, in the absence of glucose, occurs as a result of low basal expression of *GLS*1 In IMR-90, GLS1 is also present at a low level and its pharmacological inhibition does not change the level of intracellular ATP [59]. Sandulache et al. discovered that a majority of HNSCC cancer cell lines show a dependence on glucose and not glutamine [60]. This study also referred to FaDu. However, FaDu, as contrasted with all of the other 14 tested HNSCC (Detroit 562 and Cal 27 were not included in the panel), did not exert a similar rate of sensitivity to non-metabolizable D-glucose analogues. The data was: IC_{50} 10.90 (FaDu) versus IC_{50} 0.79 with UMSCC22B—the most sensitive cell line.

4.4. NQO1 and Its Splice Variants—TaqMan Probes Validation

Regardless of the conditions applied, three *NQO1* transcript variants were detected. Based on WB analyses, only the longest one, TV1, as compared with the other two, TV3 and TV4, seemed to translate in the mature protein. The change of the ratio TV1/TV3 was measured. We have shown that only in Cal 27, under NC3, did this ratio increase. It decreased in the other three cell lines. It is unfortunate that the *NQO1* transcript fold change under NC3 in FaDu could not be accurately measured (Figure 3F), where there was a significant decrease of the TV1/TV3 ratio (Figure 6). Based on the Ct values, we estimated that the *NQO1* transcription, under NC3, does not significantly vary from the control condition (NC1) (Figure 3). The change of TV1/TV3 ratio was not influenced by rs1131341 [31], as all cell lines had the same genotype with respect to this SNV. Thus, we concluded that the change TV1/TV3 depends on both conditions applied and the cell-type specific response, which clearly differentiated Cal 27 from other cell lines [61]. It is possible that, through this mechanism, this cell line compensates NQO1 basal activity, which is significantly lower (90 U) than the NQO1 activity in Detroit 562 (790 U) and FaDu (1400 U) [36]. It also raises a question with respect to compensatory mechanisms associated not only with the level of transcription, but also with splicing, when there is a high ROS generation. Cal 27, the cell line with the lowest NQO1 activity under basal conditions, had the strongest increase of NQO1 transcript and adjusted the TV1/TV3 ratio in favor of TV1. FaDu, the cell line with the highest NQO1 activity under basal conditions, had no increase of NQO1 transcript and reduced the TV1/TV3 ratio.

4.5. Activation of NRF2 and Sensitivity to Glutamine

In the Results section, we presented how some cellular parameters are indicative of a different sensitivity to glutamine, which is higher in Cal 27 and Detroit 562, and lower in FaDu. How is that to be explained, in addition to the previously given explanation relating to the cellular transporters? First, it should be noted that the intensity of the nuclear and cytoplasmic NQO1 signal was far less prominent in Cal 27 under NC1, than in the other two cancer cell lines (Figure 10A,B). The same phenomenon was shown in Li's paper [36]. The reason for less NQO1 protein is not the presence of rs18066, as Cal 27 are NQO1*1/*1. One possibility may be that these cells have a genuinely less active NRF2-KEAP1 pathway. Indeed, Romero et al. have suggested NQO1 as a suitable biomarker for NRF2 activation, when researching a human KRAS-mutant lung adenocarcinoma (LUAD) [62]. In our experimental model, this was not shown. However, one can speculate that the NRF2-KEAP1 axis, weaker in Cal 27 than in FaDu and Detroit 562, can be strongly activated. Based on the results presented, Cal 27 cells indeed have a stronger potential for activating the NRF2-NQO1 axis than the other two cancer cell lines. The activation of the NRF2-KEAP1 axis, which was shown to be, according to the majority of the parameters measured, highly dependent on glutamine [63], needs to be explored further in dynamic and not end-point experiments.

4.6. TP53 and Its Potential to Influence Phenomena Observed

Regarding the influence of MT TP53 on NRF2, there are many data, but there are no conclusions. Lisek et al. showed that mutant TP53 increases NRF2 localization to the nucleus of cancer cells, where it redirects NRF2 to ARE elements of specific genes to activate their transcription. Conversely, it sequesters NRF2 from other targets, leading to their downregulation [64]. Kalo et al. have shown that induction of stress in HCT116 cells bearing TP53 mutant R273H results in NRF2 nuclear accumulation. However, the transcription of target genes was induced to a much lesser extent than in HCT116 without TP53 activity (TP53−/−). They also showed that the down-regulation of endogenous mutant TP53, results in increased mRNA levels of NQO1 and Hem Oxygenase (HMOX-1). Thus, they proposed that MT TP53 promotes the survival of cells with high level ROS [65]. Under NC3, a decrease of nuclear TP53 in Cal 27 is related to an increased transcription rate of *NQO1*. At the same time, the strong increase of TP53 in the nuclei of FaDu under NC3 may be associated with its silencing effect on *NQO1* transcription (as estimated). This would be in accord with Kalo's results, at least at the level of transcriptional activity of only *NQO*1. When we measured the transcriptional activity of *HMOX*-1, a tremendous transcriptional activity of *HMOX*-1 in IMR-90 and FaDu under NC3 was recorded (data not shown). This does not necessarily mean that this increased transcription relies on NRF2, as we need to perform chromatin immunoprecipitations and selective silencing in order to understand the molecular mechanisms involved.

One important aspect of the events which seems connected to accumulating phenomena related to nutritional stress and TP53, may be on an entirely different level. The ubiquitin ligase Mdm-2, which mediates TP53 degradation in the proteasome [66], is a transcriptional target of TP53 [67]. Qie et al. [68] reported the upregulated transcription of MDM2 in Hep3B cells cultured in a glutamine free medium, despite the homozygous deletion of TP53, pointing to the existence of alternative regulatory mechanisms. Considering the role of MDM2 in TP53 degradation, if the reported increase in MDM2 transcripts translates to an increased protein level, one would expect a decrease in the TP53 upon glutamine deprivation in TP53 expressing cells. Although it is seemingly surprising, a redistribution of TP53 among cellular compartments (cytoplasm and nucleus) was observed, rather than its decrease in the cells deprived of glutamine (NC3; Figure 10B). A possible explanation for this effect may be the nuclear retention of TP53 due to its poly(ADP-ribosyl)ation that prevents TP53 interaction with the nuclear export receptor CRM1 [69]. Chiodi et al. recently reported that glucose and/or glutamine deprivation causes very rapid PARP-1 activation and protein poly(ADP-ribosyl)ation [70]. This is consistent with the intracellular distribution of TP53 that was observed in all the tested cells grown in a low glucose medium (NC2) or glucose- and glutamine-free medium (NC3) (Figure 10C). Namely, in all cell lines, TP53 was less abundant in the cytoplasm of the cells grown under NC2 and almost completely absent in the cytoplasm of the cells simultaneously deprived of glucose and glutamine (NC3). Contrary to cytoplasm, a significant amount of TP53 was present in the nuclear fraction. The retention of TP53 in the nuclei of all cells exposed to NC2 and NC3, regardless of their TP53 mutational status, is consistent with the report that three amino acids (E258, D259 and E271) are the targets of poly(ADP-ribosyl)ation, and that TP53 failed to get poly(ADP-ribosyl)ated only when all three of them were replaced with alanine [69]. Therefore, poly(ADP-ribosyl)ation can retain both mutant and WT TP53 in the nucleus.

The importance of NQO1 activity for TP53 accumulation in the cell strongly argues in favor of the involvement of poly(ADP-ribosy)lation in TP53 stabilization. This is because it was shown that the inhibition of NQO1 activity by dicumarol induces proteasomal degradation of WT and MT p53 [18]. The enzymatic activity of NQO1 is needed for generating NAD^+, which is a co-substrate for PARP-1 that transfers ADP-ribose moieties from NAD^+ to proteins including TP53 [71]. Experimentally induced PARP-1 or NAD^+ deficiency has been reported to result in a significantly reduced level and activity of TP53 [72]. Therefore, the accumulation of TP53 in the nuclei of the cells exposed to glucose and glucose/glutamine deprivation (NC2 and NC3), may be partly mediated by NQO1 (and other oxidoreductases like WOX1 [73]. Asher et al. [18] suggested that, considering the presence of several putative TP53-binding elements in *NQO*1 promoter, *NQO*1 may belong to TP53-inducible genes involved in a positive autoregulatory loop that regulates the level of TP53. Therefore, the mutational status of TP53 may have a profound influence on NQO1 expression in cells that are exposed to nutritional stress.

5. Conclusions

After modulating FaDu, Cal 27 and Detroit 562 for the vital cellular parameters of viability, proliferation and generation of ROS while cultivating them under four different nutritional conditions (NC1-NC4), some general conclusions can be drawn: (a) In relation to all three parameters analyzed, these cell lines showed sensitivity to glucose deprivation; (b) when having available minimal amounts of glucose and glutamine (NC2) FaDu, Cal 27, Detroit 562 responded strongly with respect to all three parameters; (c) only FaDu cells showed an increased need for glucose, not glutamine (NC3 versus NC4), for sustaining replication activity. A strong increase of ROS influences the NRF2-NQO1 axis in these cells in a fashion which is apparently cell-type specific. When considering the activation of the axis through NRF2 nuclear accumulation, the strongest response under a milder condition (NC2) was recorded in FaDu, associated with a decrease of the nuclear TP53 signal. Under harsh conditions (NC3), Cal 27 and Detroit 562 responded with NRF2 nuclear accumulation, associated again, with a decrease of the nuclear TP53 signal. Obtaining the same phenomena under different conditions (Cal 27 and Detroit 562 versus FaDu) pointed out the differences in response to identical stress, which correlates with the fact that only FaDu cells did not recover their replicative potential when deprived of glucose, in the presence of glutamine.

When considering the activation of the axis through an increase in *NQO1* transcription, only Cal 27 responded adequately, through an increase in the *NQO1* transcription rate and a modulation of alternative splicing, in favor of TV1. FaDu responded in an entirely different fashion, with a decrease in *NQO1* transcript and a modulation of alternative splicing, in favor of TV3. These responses may be consequential with respect to NQO1 enzymatic activity in Cal 27 and FaDu, which was previously shown to be 15 times higher in FaDu. Thus, the whole response of the NRF2-NQO1 axis to stress should be considered in the broader context of the cellular background.

Detroit 562 is the cell line which moderately activated its NRF2-NQO1 axis, on both the transcriptional and protein level. It is the only cancer cell line which had no significant increase of ROS, under NC2.

Fibroblasts IMR-90 were entirely dependent on glucose. These cells exhibited a physiological cellular response relating to the activation of NRF2-NQO1 axis during nutritional stress, which resulted with hardly detectable NQO1 signals when compared to the cancer cell lines. IMR-90 are homozygous – NQO1*2/*2, with respect to rs1800566.

Without making the genotyping in respect to rs1800566, we would still be confident that IMR-90 indeed are *NQO1* non-expressing cells. According to all available data, their rs1800566 genotype directs an extremely high rate of the NQO1 protein degradation, although the NRF2-NQO1 axis in these cells activates during nutritional stress.

Thus, when making important conclusions on the strength of the NRF2-NQO1 axis through NQO1 protein level/enzymatic activity, the status of rs1800566, as well as specifics of the cellular background, are always relevant and must be considered.

Author Contributions: Conceptualization, K.G.T.; L.M.; methodology, L.M., M.T., K.G.T., D.Š., A.Č.G., A.M., R.N.K.; software, L.M., M.T., P.K., K.G.T.; validation, K.G.T., L.M., M.T.; formal analysis, K.G.T., L.M., M.T.; R.N.K.; investigation, K.G.T., L.M., M.T.; resources, K.G.T.; data curation, K.G.T.; writing—original draft preparation, K.G.T.; writing—K.G.T., R.N.K., L.M., P.K., N.Đ., N.Ž.; visualization, L.M., M.T., K.G.T.; supervision, K.G.T.; project administration, K.G.T.; funding acquisition, K.G.T.

Acknowledgments: The authors are grateful to Aaron Etra for his work on English editing.

Abbreviations

AMPK	5′AMP-Activated Protein Kinase
ARE	Antioxidant Response Element
ATCC	American Type Culture Collection
ATP	Adenosine Triphosphate
BrdU	5-Bromo-2′-Deoxyuridine
COSMIC	Catalogue of Somatic Mutations In Cancer
DCFH-DA	2′,7′-Dichlorofluorescin Diacetate
DMEM	Dulbecco's Modified Eagle's Medium

EDTA	Ethylenediaminetetraacetic Acid
ER	Endoplasmic Reticulum
EtdBr	Ethidium Bromide
FAD	Flavin Adenine Dinucleotide
FBS	Fetal Bovine Serum
G6PD	Glucose-6-Phosphate Dehydrogenase
GAPDH	Glyceraldehyde-3-Phosphate Dehydrogenase
gDNA	Genomic DNA
GLS1	Glutaminase 1
HCT116	Human Colorectal Carcinoma
HIF-1α	Hypoxia-Inducible Factor, Alpha Subunit
HMOX-1	Heme Oxygenase 1
HNSCC	Head and Neck Squamous Cell Carcinoma
HPRT1	Hypoxanthine Phosphoribosyltransferase 1
HSP90	Heat Shock Protein 90
IR-C	Intron Retention Complex
IR-S	Intron Retention Simple
KEAP-1	Kelch-Like ECH-Associated Protein 1
LUAD	Lung Adenocarcinoma
Mdm-2	Mouse Double Minute 2 Homolog
MT	Mutated
NADH/NAD+	Nicotinamide Adenine Dinucleotide
NADPH/NADP+	Nicotinamide Adenine Dinucleotide Phosphate
NC	Nutritional Condition
NC1	Nutritional Condition 1
NC2	Nutritional Condition 2
NC3	Nutritional Condition 3
NC4	Nutritional Condition 4
NFE2L2 (NRF2)	Nuclear Factor (Erythroid-Derived 2)-Like 2
NHLF	Normal Human Lung Fibroblasts
NQO1	NAD(P)H:Quinone Oxidoreductase 1
PARP-1	Poly (ADP-Ribose) Polymerase 1
PCR	Polymerase Chain Reaction
PPP	Pentose Phosphate Pathway
ROS	Reactive Oxygene Species
RT-qPCR	Reverse Transcription-QuantitativePolymerase Chain Reaction
SIRT1	Sirtuin 1
SLC7A11	Solute-Like Carrier Family 7, Member 11
SLC7A3	Solute-Like Carrier Family 7, Member 3
SNP	Single-Nucleotide Polymorphism
SNV	Single Nucleotide Variant
TBP	TATA-Box Binding Protein
TBS	Tris-Buffered Saline
TBST	Tris Buffered Saline with Tween-20
TE buffer	Tris-EDTA Buffer
TP53	Tumor Protein P53
TV	Transcript Variant
WOX1	WUSCHEL Related Homeobox 1
WT	Wild-Type

References

1. Bott, A.J.; Maimouni, S.; Zong, W.-X. The pleiotropic effects of glutamine metabolism in cancer. *Cancers* **2019**, *11*, 770. [CrossRef] [PubMed]

2. Jeon, S.-M.; Chandel, N.S.; Hay, N. AMPK regulates NADPH homeostasis to promote tumour cell survival during energy stress. *Nature* **2012**, *485*, 661–665. [CrossRef]
3. Wu, C.-A.; Chao, Y.; Shiah, S.-G.; Lin, W.-W. Nutrient deprivation induces the Warburg effect through ROS/AMPK-dependent activation of pyruvate dehydrogenase kinase. *Biochim. Biophys. Acta* **2013**, *1833*, 1147–1156. [CrossRef] [PubMed]
4. Son, J.; Lyssiotis, C.A.; Ying, H.; Wang, X.; Hua, S.; Ligorio, M.; Perera, R.M.; Ferrone, C.R.; Mullarky, E.; Shyh-Chang, N.; et al. Glutamine supports pancreatic cancer growth through a KRAS-regulated metabolic pathway. *Nature* **2013**, *496*, 101–105. [CrossRef] [PubMed]
5. Lie, S.; Wang, T.; Forbes, B.; Proud, C.G.; Petersen, J. The ability to utilise ammonia as nitrogen source is cell type specific and intricately linked to GDH, AMPK and mTORC1. *Sci. Rep.* **2019**, *9*, 1461. [CrossRef] [PubMed]
6. Jones, R.G.; Plas, D.R.; Kubek, S.; Buzzai, M.; Mu, J.; Xu, Y.; Birnbaum, M.J.; Thompson, C.B. AMP-activated protein kinase induces a p53-dependent metabolic checkpoint. *Mol. Cell* **2005**, *18*, 283–293. [CrossRef] [PubMed]
7. Jiang, P.; Du, W.; Wang, X.; Mancuso, A.; Gao, X.; Wu, M.; Yang, X. p53 regulates biosynthesis through direct inactivation of glucose-6-phosphate dehydrogenase. *Nat. Cell Biol.* **2011**, *13*, 310–316. [CrossRef] [PubMed]
8. Lee, H.; Oh, E.-T.; Choi, B.-H.; Park, M.-T.; Lee, J.-K.; Lee, J.-S.; Park, H.J. NQO1-induced activation of AMPK contributes to cancer cell death by oxygen-glucose deprivation. *Sci. Rep.* **2015**, *5*, 7769. [CrossRef]
9. Jaiswal, A.K.; McBride, O.W.; Adesnik, M.; Nebert, D.W. Human dioxin-inducible cytosolic NAD(P)H:menadione oxidoreductase. cDNA sequence and localization of gene to chromosome 16. *J. Biol. Chem.* **1988**, *263*, 13572–13578. [PubMed]
10. Lind, C.; Cadenas, E.; Hochstein, P.; Ernster, L. DT-diaphorase: Purification, properties, and function. *Methods Enzymol.* **1990**, *186*, 287–301. [CrossRef] [PubMed]
11. Hosoda, S.; Nakamura, W.; Hayashi, K. Properties and reaction mechanism of DT diaphorase from rat liver. *J. Biol. Chem.* **1974**, *249*, 6416–6423. [PubMed]
12. Bai, P.; Cantó, C.; Oudart, H.; Brunyánszki, A.; Cen, Y.; Thomas, C.; Yamamoto, H.; Huber, A.; Kiss, B.; Houtkooper, R.H.; et al. PARP-1 inhibition increases mitochondrial metabolism through SIRT1 activation. *Cell Metab.* **2011**, *13*, 461–468. [CrossRef] [PubMed]
13. Winski, S.L.; Koutalos, Y.; Bentley, D.L.; Ross, D. Subcellular localization of NAD(P)H:quinone oxidoreductase 1 in human cancer cells. *Cancer Res.* **2002**, *62*, 1420–1424. [PubMed]
14. Criddle, D.N.; Gillies, S.; Baumgartner-Wilson, H.K.; Jaffar, M.; Chinje, E.C.; Passmore, S.; Chvanov, M.; Barrow, S.; Gerasimenko, O.V.; Tepikin, A.V.; et al. Menadione-induced reactive oxygen species generation via redox cycling promotes apoptosis of murine pancreatic acinar cells. *J. Biol. Chem.* **2006**, *281*, 40485–40492. [CrossRef] [PubMed]
15. Siegel, D.; Dehn, D.D.; Bokatzian, S.S.; Quinn, K.; Backos, D.S.; Di Francesco, A.; Bernier, M.; Reisdorph, N.; de Cabo, R.; Ross, D. Redox modulation of NQO1. *PLoS ONE* **2018**, *13*, e0190717. [CrossRef] [PubMed]
16. Ross, D.; Siegel, D. NQO1 in protection against oxidative stress. *Curr. Opin. Toxicol.* **2018**, *7*, 67–72. [CrossRef]
17. Ross, D.; Siegel, D. Functions of NQO1 in cellular protection and CoQ10 metabolism and its potential role as a redox sensitive molecular switch. *Front. Physiol.* **2017**, *8*, 595. [CrossRef]
18. Asher, G.; Lotem, J.; Cohen, B.; Sachs, L.; Shaul, Y. Regulation of p53 stability and p53-dependent apoptosis by NADH quinone oxidoreductase 1. *Proc. Natl. Acad. Sci. USA* **2001**, *98*, 1188–1193. [CrossRef]
19. Oh, E.-T.; Kim, J.-W.; Kim, J.M.; Kim, S.J.; Lee, J.-S.; Hong, S.-S.; Goodwin, J.; Ruthenborg, R.J.; Jung, M.G.; Lee, H.-J.; et al. NQO1 inhibits proteasome-mediated degradation of HIF-1α. *Nat. Commun.* **2016**, *7*, 13593. [CrossRef]
20. Asher, G.; Lotem, J.; Tsvetkov, P.; Reiss, V.; Sachs, L.; Shaul, Y. P53 hot-spot mutants are resistant to ubiquitin-independent degradation by increased binding to NAD(P)H:quinone oxidoreductase 1. *Proc. Natl. Acad. Sci. USA* **2003**, *100*, 15065–15070. [CrossRef]
21. Han, Y.; Shen, H.; Carr, B.I.; Wipf, P.; Lazo, J.S.; Pan, S. NAD(P)H:quinone oxidoreductase-1-dependent and -independent cytotoxicity of potent quinone Cdc25 phosphatase inhibitors. *J. Pharmacol. Exp. Ther.* **2004**, *309*, 64–70. [CrossRef] [PubMed]
22. Asher, G.; Lotem, J.; Kama, R.; Sachs, L.; Shaul, Y. NQO1 stabilizes p53 through a distinct pathway. *Proc. Natl. Acad. Sci. USA* **2002**, *99*, 3099–3104. [CrossRef] [PubMed]

23. Fleming, R.A.; Drees, J.; Loggie, B.W.; Russell, G.B.; Geisinger, K.R.; Morris, R.T.; Sachs, D.; McQuellon, R.P. Clinical significance of a NAD(P)H: Quinone oxidoreductase 1 polymorphism in patients with disseminated peritoneal cancer receiving intraperitoneal hyperthermic chemotherapy with mitomycin C. *Pharmacogenetics* **2002**, *12*, 31–37. [CrossRef] [PubMed]
24. Oh, E.-T.; Park, H.J. Implications of NQO1 in cancer therapy. *BMB Rep.* **2015**, *48*, 609–617. [CrossRef] [PubMed]
25. Itahana, Y.; Itahana, K. Emerging roles of p53 family members in glucose metabolism. *Int. J. Mol. Sci.* **2018**, *19*, 776. [CrossRef] [PubMed]
26. Reid, M.A.; Wang, W.-I.; Rosales, K.R.; Welliver, M.X.; Pan, M.; Kong, M. The B55α subunit of PP2A drives a p53-dependent metabolic adaptation to glutamine deprivation. *Mol. Cell* **2013**, *50*, 200–211. [CrossRef] [PubMed]
27. Lowman, X.H.; Hanse, E.A.; Yang, Y.; Ishak Gabra, M.B.; Tran, T.Q.; Li, H.; Kong, M. p53 promotes cancer cell adaptation to glutamine deprivation by upregulating Slc7a3 to increase arginine uptake. *Cell Rep.* **2019**, *26*, 3051–3060. [CrossRef] [PubMed]
28. Krajka-Kuźniak, V.; Paluszczak, J.; Baer-Dubowska, W. The Nrf2-ARE signaling pathway: An update on its regulation and possible role in cancer prevention and treatment. *Pharmacol. Rep.* **2017**, *69*, 393–402. [CrossRef] [PubMed]
29. Dhakshinamoorthy, S.; Porter, A.G. Nitric oxide-induced transcriptional up-regulation of protective genes by Nrf2 via the antioxidant response element counteracts apoptosis of neuroblastoma cells. *J. Biol. Chem.* **2004**, *279*, 20096–20107. [CrossRef]
30. Gasdaska, P.Y.; Fisher, H.; Powis, G. An alternatively spliced form of NQO1 (DT-diaphorase) messenger RNA lacking the putative quinone substrate binding site is present in human normal and tumor tissues. *Cancer Res.* **1995**, *55*, 2542–2547.
31. Pan, S.-S.; Han, Y.; Farabaugh, P.; Xia, H. Implication of alternative splicing for expression of a variant NAD(P)H:quinone oxidoreductase-1 with a single nucleotide polymorphism at 465C>T. *Pharmacogenetics* **2002**, *12*, 479–488. [CrossRef] [PubMed]
32. Yao, K.S.; Godwin, A.K.; Johnson, C.; O'Dwyer, P.J. Alternative splicing and differential expression of DT-diaphorase transcripts in human colon tumors and in peripheral mononuclear cells in response to mitomycin C treatment. *Cancer Res.* **1996**, *56*, 1731–1736. [PubMed]
33. Parenteau, J.; Maignon, L.; Berthoumieux, M.; Catala, M.; Gagnon, V.; Abou Elela, S. Introns are mediators of cell response to starvation. *Nature* **2019**, *565*, 612–617. [CrossRef] [PubMed]
34. Pleiss, J.A.; Whitworth, G.B.; Bergkessel, M.; Guthrie, C. Rapid, transcript-specific changes in splicing in response to environmental stress. *Mol. Cell* **2007**, *27*, 928–937. [CrossRef] [PubMed]
35. Tsalikis, J.; Pan, Q.; Tattoli, I.; Maisonneuve, C.; Blencowe, B.J.; Philpott, D.J.; Girardin, S.E. The transcriptional and splicing landscape of intestinal organoids undergoing nutrient starvation or endoplasmic reticulum stress. *BMC Genom.* **2016**, *17*, 680. [CrossRef] [PubMed]
36. Li, L.-S.; Reddy, S.; Lin, Z.-H.; Liu, S.; Park, H.; Chun, S.G.; Bornmann, W.G.; Thibodeaux, J.; Yan, J.; Chakrabarti, G.; et al. NQO1-mediated tumor-selective lethality and radiosensitization for head and neck cancer. *Mol. Cancer Ther.* **2016**, *15*, 1757–1767. [CrossRef] [PubMed]
37. Punganuru, S.R.; Madala, H.R.; Arutla, V.; Zhang, R.; Srivenugopal, K.S. Characterization of a highly specific NQO1-activated near-infrared fluorescent probe and its application for in vivo tumor imaging. *Sci. Rep.* **2019**, *9*, 8577. [CrossRef] [PubMed]
38. Green, M.R.; Sambrook, J. Isolation of high-molecular-weight DNA using organic solvents. *Cold Spring Harb. Protoc.* **2017**, *2017*, pdb.prot093450. [CrossRef] [PubMed]
39. Rasband, W.S. ImageJ. 1997–2018; U. S. National Institutes of Health: Bethesda, MD, USA. Available online: https://imagej.nih.gov/ij/ (accessed on 11 July 2019).
40. Krzystek-Korpacka, M.; Hotowy, K.; Czapinska, E.; Podkowik, M.; Bania, J.; Gamian, A.; Bednarz-Misa, I. Serum availability affects expression of common house-keeping genes in colon adenocarcinoma cell lines: Implications for quantitative real-time PCR studies. *Cytotechnology* **2016**, *68*, 2503–2517. [CrossRef] [PubMed]
41. Pfaffl, M.W. A new mathematical model for relative quantification in real-time RT-PCR. *Nucleic Acids Res.* **2001**, *29*, e45. [CrossRef] [PubMed]
42. Bradford, M.M. A rapid and sensitive method for the quantitation of microgram quantities of protein utilizing the principle of protein-dye binding. *Anal. Biochem.* **1976**, *72*, 248–254. [CrossRef]

43. Lienhart, W.-D.; Gudipati, V.; Uhl, M.K.; Binter, A.; Pulido, S.A.; Saf, R.; Zangger, K.; Gruber, K.; Macheroux, P. Collapse of the native structure caused by a single amino acid exchange in human NAD(P)H:quinone oxidoreductase(1.). *FEBS J.* **2014**, *281*, 4691–4704. [CrossRef] [PubMed]
44. Wu, Y.; Wang, X.; Chang, S.; Lu, W.; Liu, M.; Pang, X. β-Lapachone induces NAD(P)H:quinone oxidoreductase-1- and oxidative stress-dependent heat shock protein 90 cleavage and inhibits tumor growth and angiogenesis. *J. Pharmacol. Exp. Ther.* **2016**, *357*, 466–475. [CrossRef] [PubMed]
45. Novak Kujundzić, R.; Grbesa, I.; Ivkić, M.; Katdare, M.; Gall-Troselj, K. Curcumin downregulates H19 gene transcription in tumor cells. *J. Cell. Biochem.* **2008**, *104*, 1781–1792. [CrossRef] [PubMed]
46. Kanska, J.; Aspuria, P.-J.P.; Taylor-Harding, B.; Spurka, L.; Funari, V.; Orsulic, S.; Karlan, B.Y.; Wiedemeyer, W.R. Glucose deprivation elicits phenotypic plasticity via ZEB1-mediated expression of NNMT. *Oncotarget* **2017**, *8*, 26200–26220. [CrossRef]
47. Bannai, S.; Kitamura, E. Adaptive enhancement of cystine and glutamate uptake in human diploid fibroblasts in culture. *Biochim. Biophys. Acta* **1982**, *721*, 1–10. [CrossRef]
48. Kim, M.P.; Lozano, G. Mutant p53 partners in crime. *Cell Death Differ.* **2018**, *25*, 161–168. [CrossRef]
49. Yuneva, M.; Zamboni, N.; Oefner, P.; Sachidanandam, R.; Lazebnik, Y. Deficiency in glutamine but not glucose induces MYC-dependent apoptosis in human cells. *J. Cell Biol.* **2007**, *178*, 93–105. [CrossRef]
50. Terashima, J.; Habano, W.; Gamou, T.; Ozawa, S. Induction of CYP1 family members under low-glucose conditions requires AhR expression and occurs through the nuclear translocation of AhR. *Drug Metab. Pharmacokinet.* **2011**, *26*, 577–583. [CrossRef]
51. Liu, K.; Jin, B.; Wu, C.; Yang, J.; Zhan, X.; Wang, L.; Shen, X.; Chen, J.; Chen, H.; Mao, Z. NQO1 stabilizes p53 in response to oncogene-induced senescence. *Int. J. Biol. Sci.* **2015**, *11*, 762–771. [CrossRef]
52. van den Heuvel, A.P.J.; Jing, J.; Wooster, R.F.; Bachman, K.E. Analysis of glutamine dependency in non-small cell lung cancer: GLS1 splice variant GAC is essential for cancer cell growth. *Cancer Biol. Ther.* **2012**, *13*, 1185–1194. [CrossRef] [PubMed]
53. Okuno, S.; Sato, H.; Kuriyama-Matsumura, K.; Tamba, M.; Wang, H.; Sohda, S.; Hamada, H.; Yoshikawa, H.; Kondo, T.; Bannai, S. Role of cystine transport in intracellular glutathione level and cisplatin resistance in human ovarian cancer cell lines. *Br. J. Cancer* **2003**, *88*, 951–956. [CrossRef] [PubMed]
54. Koppula, P.; Zhang, Y.; Shi, J.; Li, W.; Gan, B. The glutamate/cystine antiporter SLC7A11/xCT enhances cancer cell dependency on glucose by exporting glutamate. *J. Biol. Chem.* **2017**, *292*, 14240–14249. [CrossRef] [PubMed]
55. Wu, Y.; Sun, X.; Song, B.; Qiu, X.; Zhao, J. MiR-375/SLC7A11 axis regulates oral squamous cell carcinoma proliferation and invasion. *Cancer Med.* **2017**, *6*, 1686–1697. [CrossRef] [PubMed]
56. Probst, B.L.; McCauley, L.; Trevino, I.; Wigley, W.C.; Ferguson, D.A. Cancer cell growth is differentially affected by constitutive activation of NRF2 by KEAP1 deletion and pharmacological activation of NRF2 by the synthetic triterpenoid, RTA 405. *PLoS ONE* **2015**, *10*, e0135257. [CrossRef] [PubMed]
57. Tsvetkov, P.; Adamovich, Y.; Elliott, E.; Shaul, Y. E3 ligase STUB1/CHIP regulates NAD(P)H:quinone oxidoreductase 1 (NQO1) accumulation in aged brain, a process impaired in certain Alzheimer disease patients. *J. Biol. Chem.* **2011**, *286*, 8839–8845. [CrossRef] [PubMed]
58. Yang, J.; Guo, Y.; Seo, W.; Zhang, R.; Lu, C.; Wang, Y.; Luo, L.; Paul, B.; Yan, W.; Saxena, D.; et al. Targeting cellular metabolism to reduce head and neck cancer growth. *Sci. Rep.* **2019**, *9*, 4995. [CrossRef]
59. Lee, J.-S.; Kang, J.H.; Lee, S.-H.; Hong, D.; Son, J.; Hong, K.M.; Song, J.; Kim, S.-Y. Dual targeting of glutaminase 1 and thymidylate synthase elicits death synergistically in NSCLC. *Cell Death Dis.* **2016**, *7*, e2511. [CrossRef]
60. Sandulache, V.C.; Ow, T.J.; Pickering, C.R.; Frederick, M.J.; Zhou, G.; Fokt, I.; Davis-Malesevich, M.; Priebe, W.; Myers, J.N. Glucose, not glutamine, is the dominant energy source required for proliferation and survival of head and neck squamous carcinoma cells. *Cancer* **2011**, *117*, 2926–2938. [CrossRef]
61. Gall Trošelj, K.; Novak Kujundzic, R.; Ugarkovic, D. Polycomb repressive complex's evolutionary conserved function: The role of EZH2 status and cellular background. *Clin. Epigenetics* **2016**, *8*, 55. [CrossRef]
62. Romero, R.; Sayin, V.I.; Davidson, S.M.; Bauer, M.R.; Singh, S.X.; LeBoeuf, S.E.; Karakousi, T.R.; Ellis, D.C.; Bhutkar, A.; Sánchez-Rivera, F.J.; et al. Keap1 loss promotes Kras-driven lung cancer and results in dependence on glutaminolysis. *Nat. Med.* **2017**, *23*, 1362–1368. [CrossRef] [PubMed]

63. Sayin, V.I.; LeBoeuf, S.E.; Singh, S.X.; Davidson, S.M.; Biancur, D.; Guzelhan, B.S.; Alvarez, S.W.; Wu, W.L.; Karakousi, T.R.; Zavitsanou, A.M.; et al. Activation of the NRF2 antioxidant program generates an imbalance in central carbon metabolism in cancer. *eLife* **2017**, *6*. [CrossRef] [PubMed]
64. Lisek, K.; Campaner, E.; Ciani, Y.; Walerych, D.; Del Sal, G. Mutant p53 tunes the NRF2-dependent antioxidant response to support survival of cancer cells. *Oncotarget* **2018**, *9*, 20508–20523. [CrossRef] [PubMed]
65. Kalo, E.; Kogan-Sakin, I.; Solomon, H.; Bar-Nathan, E.; Shay, M.; Shetzer, Y.; Dekel, E.; Goldfinger, N.; Buganim, Y.; Stambolsky, P.; et al. Mutant p53R273H attenuates the expression of phase 2 detoxifying enzymes and promotes the survival of cells with high levels of reactive oxygen species. *J. Cell Sci.* **2012**, *125*, 5578–5586. [CrossRef] [PubMed]
66. Haupt, Y.; Maya, R.; Kazaz, A.; Oren, M. Mdm2 promotes the rapid degradation of p53. *Nature* **1997**, *387*, 296–299. [CrossRef] [PubMed]
67. Barak, Y.; Juven, T.; Haffner, R.; Oren, M. Mdm2 expression is induced by wild type p53 activity. *EMBO J.* **1993**, *12*, 461–468. [CrossRef]
68. Qie, S.; Liang, D.; Yin, C.; Gu, W.; Meng, M.; Wang, C.; Sang, N. Glutamine depletion and glucose depletion trigger growth inhibition via distinctive gene expression reprogramming. *Cell Cycle* **2012**, *11*, 3679–3690. [CrossRef]
69. Kanai, M.; Hanashiro, K.; Kim, S.-H.; Hanai, S.; Boulares, A.H.; Miwa, M.; Fukasawa, K. Inhibition of Crm1-p53 interaction and nuclear export of p53 by poly(ADP-ribosyl)ation. *Nat. Cell Biol.* **2007**, *9*, 1175–1183. [CrossRef]
70. Chiodi, I.; Picco, G.; Martino, C.; Mondello, C. Cellular response to glutamine and/or glucose deprivation in in vitro transformed human fibroblasts. *Oncol. Rep.* **2019**, *41*, 3555–3564. [CrossRef]
71. Scovassi, A.I.; Poirier, G.G. Poly(ADP-ribosylation) and apoptosis. *Mol. Cell. Biochem.* **1999**, *199*, 125–137. [CrossRef]
72. Whitacre, C.M.; Hashimoto, H.; Tsai, M.L.; Chatterjee, S.; Berger, S.J.; Berger, N.A. Involvement of NAD-poly(ADP-ribose) metabolism in p53 regulation and its consequences. *Cancer Res.* **1995**, *55*, 3697–3701. [PubMed]
73. Chang, N.S.; Pratt, N.; Heath, J.; Schultz, L.; Sleve, D.; Carey, G.B.; Zevotek, N. Hyaluronidase induction of a WW domain-containing oxidoreductase that enhances tumor necrosis factor cytotoxicity. *J. Biol. Chem.* **2001**, *276*, 3361–3370. [CrossRef] [PubMed]

12

Roles of Toll-Like Receptors in Nitroxidative Stress in Mammals

Yao Li [1,†], Shou-Long Deng [2,†], Zheng-Xing Lian [1,*] and Kun Yu [1,*]

1 Beijing Key Laboratory for Animal Genetic Improvement, National Engineering Laboratory for Animal Breeding, Key Laboratory of Animal Genetics and Breeding of the Ministry of Agriculture, College of Animal Science and Technology, China Agricultural University, Beijing 100193, China; yaoli19881015@126.com

2 CAS Key Laboratory of Genome Sciences and Information, Beijing Institute of Genomics, Chinese Academy of Sciences, Beijing 100101, China; popo84350746@163.com

* Correspondence: lianzhx@cau.edu.cn (Z.-X.L.); young137@163.com (K.Y.)

† These authors contributed equally to this work.

Abstract: Free radicals are important antimicrobial effectors that cause damage to DNA, membrane lipids, and proteins. Professional phagocytes produce reactive oxygen species (ROS) and reactive nitrogen species (RNS) that contribute towards the destruction of pathogens. Toll-like receptors (TLRs) play a fundamental role in the innate immune response and respond to conserved microbial products and endogenous molecules resulting from cellular damage to elicit an effective defense against invading pathogens, tissue injury, or cancer. In recent years, several studies have focused on how the TLR-mediated activation of innate immune cells leads to the production of pro-inflammatory factors upon pathogen invasion. Here, we review recent findings that indicate that TLRs trigger a signaling cascade that induces the production of reactive oxygen and nitrogen species.

Keywords: free radicals; antimicrobial; toll-like receptors

1. Introduction

Reactive oxygen species (ROS) and reactive nitrogen species (RNS) are recognized for their dual role as both deleterious and beneficial species. The overproduction of ROS/RNS results in damage to cell structures, including lipids and membranes, proteins, and DNA, inhibiting their normal function. In contrast, the salutary effects of ROS/RNS occur at low/moderate concentrations in cellular responses, including in defense against infectious agents, in the function of a number of cellular signaling pathways, and the induction of a mitogenic response [1]. The subtle balance between beneficial and harmful effects of ROS/RNS is a very important aspect of living organisms and is achieved by a mechanism called "redox regulation" which maintains cellular "redox homeostasis" using the antioxidants system [2]. ROS and RNS are generated using several different processes including (i) irradiation by UV light, X-rays, and gamma-rays; (ii) as products of metal-catalyzed reactions; (iii) by neutrophils and macrophages during inflammation; and (iv) as products of mitochondria-catalyzed electron transport reactions [3]. An important potential source of oxidizing agents is the phagocytic leukocytes within the body. Neutrophils, monocytes, and macrophages are the most prominent immune cell types that release various pro- and anti-inflammatory mediators for both host defense and inflammatory responses [4]. Oxidation intermediates are also involved in these processes. The cumulative production of ROS/RNS from either endogenous or exogenous sources is termed oxidative stress. Oxidation intermediates are essential activators of oxidative stress. This is because low levels of free radicals, including ROS and RNS, form a stressful oxidative environment that can clear invading pathogens and maintain physiological homeostasis [5]. Although innate immunity is the first line of defense against pathogens, Toll-like receptors (TLRs) also play a crucial role in the early host defense mechanism [6].

In recent years, there have been a significant number of discoveries regarding how the TLR-mediated activation of innate immune cells leads to the production of pro-inflammatory factors upon pathogen invasion. In this review, we discuss recent research findings on TLR activities, with a particular focus on TLR-mediated signaling pathways that are activated during nitroxidative stress.

2. Nitroxidative Stress

2.1. Formation of Nitroxidative Stress

Nitroxidative stress is a cellular condition that reflects a physiological imbalance where excessive reactive nitrogen and reactive oxygen species are present. Under this condition, an excessive production of ROS and RNS occurs, exceeding a level that the body's antioxidant mechanisms can cope with. Examples of ROS species include the superoxide anion ($O_2^{\bullet -}$), hydrogen peroxide (H_2O_2), and hydroxyl radicals (•OH) [7]. RNS include nitric oxide ($NO^{\bullet}$), nitrogen dioxide (NO_2), and the powerful oxidant peroxynitrite ($ONOO^-$) [8]. During microbial infections, excessive $NO^{\bullet}$ produced has varying functions, ranging from anti-microbial and anti-inflammatory host defense and cell protection to proinflammatory and cytotoxic activities [9]. The $NO^{\bullet}$ produced in inflamed tissues during infection or inflammation are affected by the concomitant production of oxygen radicals, particularly $O_2^{\bullet -}$ and H_2O_2. The interaction of $NO^{\bullet}$ with reactive oxygen species causes the formation of several reactive nitrogen oxides, these reactive nitrogen intermediates have a great possibility of causing oxidative and nitrative stress through the oxidation and nitration of biological molecules [10]. A study in mice shows that $O_2^{\bullet -}$ was produced by neutrophils and macrophages in the liver, which is associated with prolonged granulomatous lesions in infected with *Salmonella typhimurium*. The inhibition of $O_2^{\bullet -}$ generation by in vivo superoxide dismutase (SOD) treatment results in a reduction in the area of liver lesions, simultaneously accelerating bacterial growth in the liver. The result suggests that $O_2^{\bullet -}$ may be involved in the host defense mechanism against *S. typhimurium* infections [11]. Granulocyte peroxidases, such as myeloperoxidase, play an important role in oxidative stress [12]. In neutrophils, H_2O_2 produced by $O_2^{\bullet -}$ is metabolized by myeloperoxidase into a strong oxidant, hypochlorous acid (HOCl), in the presence of chloride ions. HOCl plays an important role in host defense and inflammatory tissue injury. In addition, myeloperoxidase generates reactive nitrogen species in vivo only when nitrite and nitrate are available [13]. Overproduced ROS and RNS can easily and rapidly react with intracellular macromolecules, causing oxidative damage to cellular structures that result in progressive physiological dysfunction, which is associated with many pathological conditions. ROS are the major causative factor in steatohepatitis and insulin resistance [14]. Various neurodegenerative diseases, such as Parkinson's disease (PD), Alzheimer's disease (AD), Huntington's disease (HD), and amyotrophic lateral sclerosis (ALS), can be the result of oxidative stress [15].

2.2. Antioxidant Systems

There are two types of antioxidant systems in the body. The first one is formed by antioxidant enzymes, which includes superoxide dismutase (SOD), catalase (CAT), and glutathione peroxidase (GSH-Px). The other is formed by non-enzymatic antioxidants which includes vitamin C, glutathione (GSH), melatonin, and trace elements [16]. Antioxidant substances catalyze the formation of active oxygen intermediates, scavenge free radicals, and terminate peroxidation. ROS is mainly produced by Nicotine adenine dinucleotide phosphate (NADPH) oxidase [17]. When stimulated, NADPH oxidase induces electron transmembrane transport and generates $O_2^{\bullet -}$. As $O_2^{\bullet -}$ is unstable, it is rapidly converted into H_2O_2, which can then be transformed into H_2O and $O_2^{\bullet -}$ or other active oxide derivatives.

RNS family members are derived from $NO^{\bullet}$ through the action of nitric oxide synthase (NOS). NOS is a multidomain enzyme that contains binding sites for the cofactors NADPH, flavin adenine dinucleotide (FAD), flavin mononucleotide (FMN), and (6*R*-)5,6,7,8-tetrahydrobiopterin (BH_4). It catalyzes the production of $NO^{\bullet}$ and L-citrulline from L-arginine. The availability of L-arginine is one of the rate-limiting factors in cellular NO$^\bullet$ synthesis and secretion. There are three members of the NOS family: neuronal NOS (nNOS), endothelial NOS (eNOS), and inducible NOS (iNOS). As nNOS and eNOS are constitutively expressed and not inducible, they are not associated with the inflammatory response [18]. The iNOS isoform is upregulated by inflammatory stimulating factors [19] and can be activated by phagocytes after stimulation with PRR agonists, IFN-γ, and proinflammatory cytokines [20]. In turn, the produced NO$^\bullet$ is able to stimulate the expression of the transcription factor NF-E2-related factor-2 (Nrf2) in macrophages, upregulate ferroportin (FPN), and contribute to nutritional immunity in macrophages [21]. Nrf2 plays a key role in anti-inflammation and oxidation stress and triggers the expression of the downstream target genes catalase (CAT), superoxide dismutase (SOD), glutathione S-transferase alpha 1 (GSTα1), quinone oxidoreductase 1 (NQO1), γ-glutamylcysteine synthetase (γ-GCS), and heme oxygenase-1 (HO-1).

Phosphoinositide 3-kinase (PI3K) and MAPK signaling pathways regulate activation of the Nrf2-antioxidative response element pathway [22]. Oxidative stress activates NF-κB and induces inflammatory responses. Together, Nrf2 and NF-κB cooperatively regulate the oxidative stress response [23]. Oxidation intermediates exert antimicrobial actions against a broad range of pathogens. Indeed, chronic granulomatous disease (CGD) patients that are deficient in oxidation intermediates are susceptible to extracellular bacterial infections [24]. CGD is characterized by inherited defects in the innate immune system resulting from mutations in the genes encoding any of the five components of the NADPH oxidase complex, including gp91-phox, p22-phox, p40-phox, p47-phox, and p67-phox [25]. However, some bacteria have devised strategies to escape killing using oxidation intermediates. *Mycobacterium tuberculosis* binds to macrophage-expressed complement receptor 1 (CR1) and complement receptor 3 (CR3) to overcome reactive oxygen intermediates (ROI) and reactive nitrogen intermediates (RNI) production. This allows for the safe access of pathogens to their intracellular habitat. *Staphylococcus aureus* releases adenosine to overcome oxidative killing in phagocytes and produces the enzymes SOD and CAT to eliminate oxidation intermediates [26]. Mitochondria are major sources of ROS production in cells [27]. Mice with reduced macrophage ROS levels (through the overexpression of CAT in their mitochondria) showed an impaired ability to kill intracellular bacteria following intraperitoneal infection with *Salmonella typhimurium*. This confirms that ROS play a key role in bactericidal activity [28].

3. TLRs and Their Signaling Pathway

3.1. An Introduction on TLRs

Microbial structures and molecules are crucial for their survival and virulence. Pathogen-associated molecular patterns (PAMPs) are highly conserved structural components that are uniquely associated with microorganisms [5]. These include LPS, lipoproteins, carbohydrates, flagellin, and nucleic acids. Upon long-term exposure to pathogens, the innate immune system provides a faithful mechanism to rapidly sense and respond to PAMPs using a number of structurally unrelated host proteins called pattern recognition receptors (PRRs) [6]. PRRs play important intracellular roles to eliminate infection and initiate inflammatory responses [29]. To date, four different classes of PRR families have been identified, including C-type lectin receptors (CLRs), RIG-I-like receptors (RLRs), nucleotide-binding-domain and leucine-rich-repeat-containing receptors (NLRs), and TLRs [30]. PRRs are strategically distributed in the cell to enable the recognition of both extracellular and intracellular pathogens. For example, TLRs and CLRs localize to plasma or endosomal membranes, whereas RLRs and NLRs are cytosolic [31]. TLRs are type I integral membrane glycoproteins that are characterized by an N-terminal extracellular domain with leucine-rich repeats (LRRs) that recognize PAMPs, a transmembrane region, and a cytoplasmic Toll/IL-1R homology (TIR) domain that mediates downstream signaling [30]. TLRs are one of the most-investigated PRR families and are able to detect a wide range of PAMPs, including bacterial lipopolysaccharides (LPS), lipoproteins, flagellin, and nucleic acids. Many structural studies on TLRs, including X-ray crystallography, have been undertaken to investigate the basis of their ligand recognition [5]. In zebrafish (Danio rerio), 19 putative TLR variants, the orthologs of mammalian TLR 2–5 and 7–9, a fish specific receptor type group, and three putative splice variants have been identified [32].

Some studies on oxidative stress in the zebrafish model have been done, including a study evaluating the effects of dietary supplementation with the probiotic *Bacillus amyloliquefaciens* R8, which has a heterologous expression of xylanase from rumen fungi, on zebrafish. The result shows it improved expression levels of oxidative stress-related genes in the fish liver [33]. Zebrafish embryo exposure to Triclocarban (TCC) at environmental concentrations significantly affects the expression of immune-response-related genes following oxidative stress and the release of proinflammatory mediators through the Toll-like receptor signaling pathway [34]. Thirteen TLRs have been discovered in mammals to date [35]. TLR1-9 is present in both humans and mice. Mouse TLR10 is nonfunctional owing to a retroviral insertion. Human TLR11 is a pseudogene, and TLR12 and TLR13 are absent from the human genome. Studies with TLR-knockout mice have demonstrated that each TLR subgroup recognizes distinct PAMPs and initiates a particular immune response (Table 1) [36]. Studies of *Drosophila* TLRs also support the fact that distinct TLRs may differ in their signaling ability. In *Drosophila*, *toll* and *18-wheeler* sharing homologous cytoplasmic domains activate different and nonoverlapping gene expressions when faced with fungi and bacteria. TLRs are expressed in a variety of cells and tissues, including dendritic cells, mononuclear macrophages, and granulocytes. Immune cells are capable of recognizing distinct molecular patterns to elicit a specific response against pathogens or endogenous factors released as a result of cellular damage. TLRs can recognize PAMPs in different cell compartments, including the plasma membrane, endosomes, lysosomes, and endocytic lysosomes. However, the TLRs must be appropriately localized in cells to recognize their ligands. This is also important for self-tolerance and downstream signal transduction. When TLR4 interacts with myeloid differentiation factor 2 (MD2) and CD14, an LPS receptor is then combined with a ligand [37]. This alerts the immune system to the presence of invading organisms, so that an immediate response can be configured to contain the pathogen. Later, during the start of acquired immunity, the recognition event is provided information about the nature of the invading microorganism. From this information, it is determined whether T helper (Th) 1 cells differentiate into Th1 cells that promote cell-mediated immunity or Th2 cells that promote humoral responses [38].

Table 1. TLR Recognition of Microbial Components [36,39,40].

TLR Usage	Expression Patterns in Leucocytes	Cellular Localization	Microbial Component Recognized by the Receptor
TLR1	T-Lymphocytes B-Lymphocytes Natural killer cells PMNs Mononuclear phagocytes Dendritic cells (DCs)	Cell surface	Triacyl lipopeptides
TLR2	T-Lymphocytes PMNs Mononuclear phagocytes DCs	Cell surface	Triacyl lipopeptides Diacyl lipopeptides Lipoteichoic acid Peptidoglycan Porins Lipoarabinomannan Phospholipomannan Glucuronoxylomannan tGPI-mutin Hemagglutinin protein Not determined Zymosan
TLR3	DCs	Endosome	dsRNA
TLR4	PMNs Mononuclear phagocytes DCs T-Lymphocytes	Cell surface	Mannan Glucuronoxylomannan Glycoinositolphospholipids Envelope proteins Heat-shock protein 60,70 Fibrinogen
TLR5	PMNs Mononuclear phagocytes DCs	Cell surface	Flagellin
TLR6	T-Lymphocytes B-Lymphocytes Mononuclear phagocytes DCs	Cell surface	Diacyl lipopeptides lipoteichoic acid Zymosan
TLR7	T-Lymphocytes B-Lymphocytes DCs	Endolysosome	ssRNA Imidazoquinoline
TLR8	T-Lymphocytes Mononuclear phagocytes DCs	Endolysosome	Loxoribine ssRNA Imidazoquinoline
TLR9	T-Lymphocytes B-Lymphocytes Mononuclear phagocytes DCs	Endolysosome	Bropirimin DNA CpG-DNA Hemozoin
TLR10	B-Lymphocytes DCs	Cell surface	not determined
TLR11		Cell surface	Profilin-like molecule not determined
TLR12			not determined
TLR13			not determined

3.2. Signaling Pathway of TLRs

TLRs generally function as homodimers. However, TLR2 exists as a heterodimer with another TLR molecule, for example TLR2/TLR1 and TLR2/TLR6. TLR2 recognizes lipoteichoic acid, lipopeptides, peptidoglycan, and Zymosan [41], whereas TLR4 homodimers specifically recognizes lipopolysaccharides (LPS) from Gram-negative bacteria heat-shock proteins [42] and viral components such as the fusion protein from respiratory syncytial virus (RSV) [24]. TLR5, TLR9, and TLR3 recognize the flagellin component of bacterial flagella, genomic DNA, and the viral replication intermediate dsRNA, respectively [43–46]. All TLRs except TLR3 utilize MyD88 to activate NF-κB and MAPK and induce an overproduction of inflammatory cytokines [47]. This pathway is called the MYD88-dependent pathway. Toll-interleukin1 receptor (TIR) domain containing adaptor protein (TIRAP) assists the recruitment of MyD88 to the surfaces of TLR2 and TLR4. TIRAP has a lipid-binding region that binds to PI (4,5) P in the plasma membrane and PI (3) P on the endosome, allowing the formation of functional signaling complexes at their respective positions [48]. In MYD88-independent pathways, TRIF is recruited to TLR3 and TLR4, which results in the activation of interferon regulatory factor 3 (IRF3), nuclear factor kappa-light-enhancer of activated B cells (NF-κB), and mitogen-activated protein kinase (MAPK) to induce the production of type I interferons and inflammatory cytokines. TRAM is selectively recruited to TLR4 to link TRIF and TLR4. Although TLR3 is able to bind directly to TRIF, this interaction requires the phosphorylation of two tyrosine residues in the cytoplasmic region of TLR3 [49,50].

Two distinct TLR signaling pathways have been identified based on the availability of adaptor molecules. Both pathways activate downstream signaling molecules that lead to the secretion of inflammatory cytokines, type I interferon (IFN), chemokines, and antimicrobial peptides [51]. These instigate neutrophil recruitment, macrophage activation, and the induction of IFN-related genes, which directly kill pathogens. In addition, the activation of the TLR signaling pathway leads to the maturation of dendritic cells (DCs) and contributes to adaptive immunity (Figure 1).

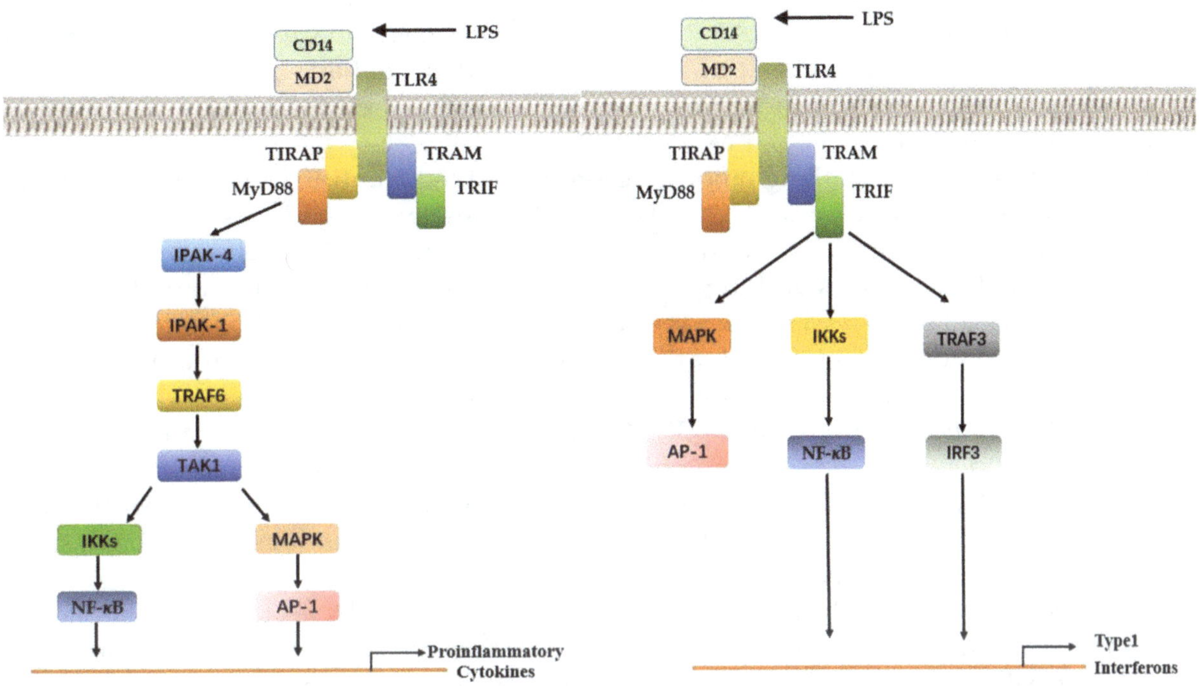

Figure 1. MYD88-dependent pathway (left). MYD88-independent pathways (right) [52].

4. TLRs and Nitroxidative Stress

Under normal conditions, TLRs are maintained at relatively stable levels by intracellular modulation (such as alternative splicing) and degradation by ubiquitination and deubiquitination. Transgenic animals overexpressing TLRs and TLR knockouts have been studied to determine the role of TLRs during bacterial infection in vivo. *S. typhimurium* is a Gram-negative bacterium that replicates in macrophages and has PAMPs that are detectable by at least four TLRs: lipoprotein (TLR2), LPS (TLR4), flagellin (TLR5), and CpG-DNA (TLR9) [53]. In this review, we will focus on data obtained for the TLR2 and TLR4 signaling pathways in mammalian immune cells, which are linked to nitroxidative stress.

Molecules found in nitroxidative stress, such ROS and $NO^{\bullet}$, are found in response to microbial invasion during the neutrophil and macrophage respiratory burst. Both oxidative stress and infective stress can share the same TLR signaling pathways [54]. TLRs interact with adaptor molecules such as MyD88 and TRIF to activate downstream signaling through NF-κB, AP-1, and interferon-regulatory factor-3 (IRF-3). The expression of inflammatory mediators is upregulated; involve notably pro-oxidant enzymes such as NOX and iNOS, producing high levels of ROS [55]. To examine whether TLR signaling could enhance ROS production, a study in RAW 264.7 macrophages shows the use of all kinds of TLRs agonist. The production of ROS was triggered only upon signaling from the cell-surface TLRs (TLR1, TLR2 and TLR4), the same as exposure of cells to rotenone and antimycin A, compounds known to increase mitochondrial $O_2^{\bullet-}$ generation, and increased cellular H_2O_2 [28].

4.1. TLR2 and Nitroxidative Stress

TLR2 recognizes Gram-positive bacteria, including mycobacteria, and viruses and their products. Most Gram-positive bacteria, such as *Bacillus anthracis*, *S. aureus*, and *Clostridium tetani*, cause a wide range of diseases in both immunocompetent and immunocompromised hosts [43]. TLR2 is widely expressed in innate immune cells and epithelial cells and is highly expressed in peripheral blood monocytes. TLR2 was found to induce neutrophil activation and its expression is upregulated when it recognizes its corresponding PAMP. This results in the production of $NO^{\bullet}$ and pro-inflammatory cytokines, including tumor necrosis factor α (TNF-α), interleukin (IL)-1β, IL-6, and chemokines. TNF-α and IL-1β are both able to stimulate the production of monocyte chemotactic protein-1 (MCP-1). MCP-1 expression is also induced by oxidative stress [56]. TLR-2 relies on the MyD88 signaling pathway to activate the NF-κB and MAPK pathways. Phosphorylated MAPK then activates the transcription factor activator protein-1 (AP-1) and the PI3K/protein kinase B (Akt) signaling pathway to induce immune responses. AP-1 is a transcription factor that mediates pro-inflammation through core components including c-Jun and c-Fos. Overexpression of c-Jun induces the production of inflammatory mediators [57].

During pathogen infection, inflammation is accompanied by an anti-inflammation reaction. Native $CD4^{+}$ cells differentiate into Th1 and Th2 cells. The Th1 response produces IL-12 to promote the production of IFN-γ, which functions to inhibit the differentiation of Th2 cells. The Th2 response is characterized by low IFN-γ levels and high IL-4 levels, as well as the production of IL-6 and IL-10. IL-6 produced by Th2 cells acts as a mediator of the acute phase response. It is able to accelerate the infiltration of inflammatory cells [58] but, at later phases, can increase anti-inflammatory cytokine expression to prevent an immune overreaction [59]. IFN-γ levels are low during the Th2 response as it is able to inhibit Th2 cells and the production of IL-10 to decrease the levels of Th1-related cytokines. Mice deficient in TLR2 experienced strong immune reactions and were susceptible to *S. aureus* [60]. Another study revealed that TLR2 signaling contributes to high mortality during polymicrobial intra-abdominal sepsis. This suggests that a novel therapeutic approach to treat severe sepsis would be to specifically target the TLR2 signaling pathway [61].

A previous study reported that mice challenged with Pam3CSK4, a TLR2 agonist, had reduced infiltration of chemokines and inflammatory cells within their tissues [62]. Furthermore, a study by our laboratory revealed that Pam3CSK4 could activate monocytes/macrophages. An overexpression of TLR2 in transgenic goats caused early expression of the pro-inflammatory cytokines' TNF-α and IL-1β, followed by a continuous increase in the expression of the anti-inflammatory factor IL-4. In addition, TLR2 overexpression induced the expression of Th1 type cytokines. A skin inflammation experiment showed that TLR2 overexpression accelerated the inflammatory process in transgenic goats [63].

Free radicals are necessary for the host defense against microbial invasion and inflammatory injury. Following macrophage activation by TLR2, high levels of NOS and ROS are released, in addition to $NO^{\bullet}$ and H_2O_2, to the area of inflammation under oxidative stress. Intriguingly, a study using RAW264.7 cells indicated that NF-κB activation was significantly enhanced upon exposure to NO [64]. Even relatively low levels of $NO^{\bullet}$ are able to trigger downstream pathways and maintain the correct functioning of the defense mechanism. However, NF-κB activation is inhibited by high concentrations of $NO^{\bullet}$ or the overexpression of antioxidants. The inhibition of NF-κB activation has been shown to decrease the inflammatory response and prevent tissue damage. GSH, SOD, and CAT are key anti-oxidative substances in a host organism. Activated NF-κB can induce the secretion of pro-inflammatory factors and also reduce cellular SOD activity. As described earlier, $NO^{\bullet}$ levels are regulated by iNOS. Activated Nrf2 can induce the expression of the proinflammatory gene COX-2 by inhibiting NF-κB activity, which reduces iNOS expression [65]. Furthermore, Nrf2 activation can directly regulate c-Jun signaling activity and suppress COX-2 expression. COX-2 overexpression was found to inhibit PI3K activity and the Nrf2-mediated anti-oxidation reaction [30]. A study of healthy people and coronary artery disease (CAD) patients shows the activation of the Nrf2 pathway as an antioxidant response mechanism in monocyte-derived macrophages (MDMs) [66]. The activation of Nrf2 protects human coronary artery endothelial cells against oxidative challenge [67].

AP-1 family members include Jun and Fos. Jun subclasses include c-jun and JunB. Fos subclasses include c-fos and FosB. Different types of AP-1 transcription factor dimer combinations have different functions in gene expression regulation. Increased levels of c-Fos and c-Jun subunits negatively regulate the expression of the anti-oxidation genes NAD(P)H: quinone oxidoreductase 1 (NQO1), GSTα1, SOD1, and CAT. HO-1 is a rate-limiting enzyme that participates in anti-inflammatory reactions and is induced in response to oxidative stress. HO-1 gene activation has been demonstrated to inhibit AP-1 activity [68]. The activity of AP-1 was observed to increase in Nrf2 knockout cells throughout the JNK/c-Jun pathway, leading to the upregulation of HO-1 expression [69]. Our laboratory results indicate that Nrf2 upregulation, through the overexpression of TLR2 in transgenic goats, inhibited COX-2 expression and increased the expression of the c-Jun gene in monocytes/macrophages by TLR2 ligand stimulation. In addition, lower levels of anti-oxidation stress enzymes were observed in cells overexpressing TLR2 compared with wild-type cells. This, in turn, improved the activity of GSH, and the GSH consumed could be rapidly resynthesized [63]. TLR2 overexpression also upregulates PI3K and increases HO-1 gene expression. However, the concentrations of malondialdehyde (MDA), a marker of lipid peroxidation, and NO• in cells overexpressing TLR2 remained relatively low and were maintained at stable levels (Figure 2). In combination, these data indicate that tissue damage can be prevented through TLR2 overexpression.

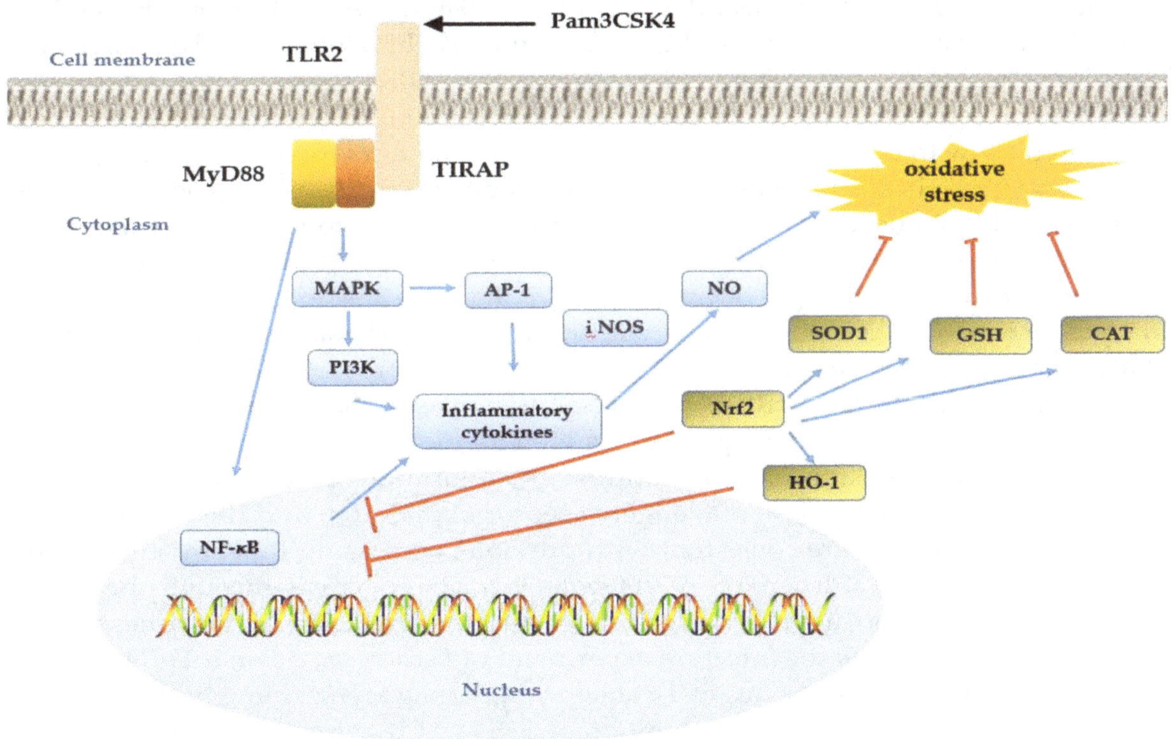

Figure 2. The TLR2 signal pathway involved in oxidative stress [48,55]. The TLR2 signaling pathway activates the MYD88 pathway to activate the NF-κB and MAPK pathways under the action of the ligand Pam3CSK4. Phosphorylation of MAPK then activates transcription factor activator protein-1 (AP-1) and PI3K/protein kinase signaling pathways to induce an immune response. AP-1 is a transcription factor that mediates pro-inflammatory factors, and NO• levels are regulated by iNOS. Nrf2 can induce the expression of pro-inflammatory factors by inhibiting the expression of pro-inflammatory NF-κB, which is a rate-limiting enzyme involved in the anti-inflammatory reaction and can induce oxidative stress. GSH, SOD, and CAT are all in a key antioxidant in the host organism. Activated NF-κB reduces cellular SOD activity.

4.2. *TLR4 and Nitroxidative Stress*

A specific ligand is mediated by the TLR4/MD2 complex, together with the co-receptor CD14, then recruits downstream adaptors to activate the MyD88- and TRIF-dependent pathways [58]. By activating these two pathways, TLR4 participates in innate immunity in the host's defense against Gram-negative bacterial infections, as well as in many inflammatory and autoimmune diseases. The TLR4 signaling pathway is activated upon the invasion of animal cells by pathogenic microorganisms. This triggers a cascade of reactions to promote the production and release of inflammatory cytokines, which induces the chemotactic aggregation of macrophages. However, recently, an interesting study has demonstrated palmitate-stimulated CD11b + F4/80 low hepatic infiltrating macrophage ROS generation by dynamin-mediated endocytosis of TLR4 and NOX2, independent from MyD88 and TRIF [14].

Several studies have demonstrated that TLR4-mutant enterocytes have a decreased sensitivity to LPS compared with wild-type cells [70]. Moreover, TLR4-mutant mice exhibit a suppressed inflammatory response [71]. This is because TLR4-deficient mice are unable to secrete IL-1 or IL-12 and express IL-6 at lower levels when stimulated with LPS [72]. Mice with targeted deletions of multiple inflammatory immune and antioxidant genes are susceptible to oxidative lung injuries [73]. Overexpression of TLR4 in mice amplifies the host response to LPS and provides a survival advantage of increased disease resistance [74]. LPS recognition stimulates TLR4 signaling pathways that lead to the activation of multiple downstream signaling pathways. In research on TLR4 overexpressing ovine macrophages, TLR4 initially promoted the production of proinflammatory cytokines TNFα and IL-6 by activating TLR4-mediated IRAK4-dependent NF-κB and MAPK (JNK and ERK1/2) signaling. This was later impaired due to the increased internalization of TLR4 into the endosomal compartment of macrophages. Then, the overexpression of TLR4 triggered TBK1-dependent interferon-regulatory factor-3 (IRF-3) expression, leading to the induction of IFN-β and IFN-inducible genes. The bacterial burden after infection with live *S. typhimurium* in these macrophages was decreased significantly [75].

TLR4 has also been shown to be involved in the phagocytosis of various bacterial species via its interaction with MAPK, Janus kinase 2 (Jak2), PI3K, and various receptors [76]. The expression of these factors was observed to increase following *S. typhimurium* infection of transgenic TLR4-overexpressing sheep. Ovine monocytes/macrophages overexpressing TLR4 were able to phagocytize higher numbers of bacteria and also showed a higher phagocytic ability at an early stage, even when infected with a lower bacterial dose. This suggests that TLR4 overexpression causes an increase in scavenger receptor expression. In addition, TLR4 overexpression increased the number of bacteria that could adhere to individual monocytes/macrophages, as well as the number of monocytes/macrophages able to participate in bacterial adhesion. In contrast, the inhibition of TLR4 reduced *S. typhimurium* internalization, actin polymerization, scavenger receptor expression, and the adhesive capacity of immunocytes. These findings are consistent with previous data using PI3K inhibitors, suggesting that TLR4 interacts with PI3K to enable *S. typhimurium* phagocytosis through the regulation of scavenger receptor expression, actin polymerization, and the alteration of the adhesive capacity of monocytes/macrophages [77]. In the infection experiment of *Escherichia coli* in a TLR4-overexpressing transgenic (Tg) sheep model, monocytes of Tg sheep could phagocytize more bacteria and exhibited higher adhesive capacity. Using specific inhibition of p38 MAPK, c-Jun N-terminal kinase (JNK), or extracellular signal-regulated kinases (ERKs), the TLR4-dependent *E. coli* internalization in sheep monocytes was reduced. p38, JNK, and ERK are all mitogen-activated protein kinases (MAPKs), which are important downstream signaling molecules of TLR4. These results provide valuable insight into the bacterial internalization mechanisms in sheep [78].

Our research laboratory generated transgenic sheep overexpressing TLR4 that had two TLR4 copies inserted into germ cells. About 90–95% of Gram-negative bacteria are considered to be harmful to their hosts. Many intracellular bacteria, such as *Brucella, Tubercle bacilli,* and *Salmonella,* are pathogenic to both animals and humans and they can be transmitted from animals to humans. As China has a large sheep breeding industry, the most harmful of these infections are brucellosis caused by *Brucella melitensis*. It not only brings huge economic losses but also causes security risks. The design of TLR4 expression sheep provides an important theoretical basis for transgenic disease-resistant breeding. LPS stimulation of these transgenic sheep enhanced the inflammatory response, which aids the clearance of pathogens. This could cause excessive oxidative stress that would cause tissue damage [79]. However, TLR4 is able to tightly regulate oxidative stress throughout these processes. Upon bacterial invasion, TLR4 triggers the activation of inflammatory factors such as NF-κB and AP-1. The mechanism by which TLR4 mediates the production of ROS involves the membrane-associated enzyme complex NADPH oxidase. Mouse peritoneal macrophages activated by LPS resulted in an increase in the functional activity of NADPH oxidase [80]. The NADPH oxidase inhibitor, apocynin, protected mice from LPS-induced lethality by decreasing the expression level of inflammatory cytokines in vivo [81]. Research using a yeast two-hybrid and glutathione S-transferase pull-down assay model has suggested that there may be a direct interaction between TLR4 and NADPH oxidase in mediating the LPS-induced production of ROS. The carboxy-terminal region of Nox4, a subunit of NADPH oxidase, interacted directly with the Toll/IL-1 receptor (TIR)-domain of TLR4 after LPS stimulation [82]. Intriguingly, another study on neutrophils shows that the synthesis of NADPH oxidase was also controlled by TLR4 through the interleukin-1 receptor-associated kinase 4 (IRAK4) pathways. Phosphorylation of the cytosolic factor p47 phox is essential for the activation of NADPH oxidase. The data shows that p47 phox is a substrate for IRAK-4 [83].

In addition, TLR4 can trigger the transcription of the iNOS gene, which promotes $NO^{\bullet}$ production. Furthermore, iNOS produces peroxide and $O_2^{\bullet-}$ radicals. $NO^{\bullet}$ synthesis not only requires L-arginine as a substrate but also several cofactors for its catalytic activity [84]. BH4 is an essential co-factor for all NOSs. BH4 bioavailability is a critical factor in regulating the balance between $NO^{\bullet}$ and $O_2^{\bullet-}$ production [85]. This has been demonstrated when diabetic mice underwent transgenic over-induction of BH4 synthesis to preserve $NO^{\bullet}$-mediated endothelial function [86]. ROS such as $O_2^{\bullet-}$ and $ONOO^-$ are able to rapidly oxidize BH4, leading to BH4 catabolism and depletion [87]. The oxidation of BH4 results in the formation of dihydrobiopterin (BH2), which binds to NOS and generates $O_2^{\bullet-}$, but not $NO^{\bullet}$ [88]. Guanosine triphosphate cyclohydrolase I (GCHI) is a rate-limiting enzyme in BH4 synthesis. A novel study investigating the relationship between activated GCHI with BH4 revealed that GCHI-transgenic mice stimulated with LPS showed a marked increase in the expression of renal iNOS and $NO^{\bullet}$ production (Figure 3) [89].

ROS can induce NF-κB activation and upregulate the cytokine-induced iNOS gene, resulting in the release of excessive amounts of $NO^{\bullet}$. Our research revealed that these inflammatory factors accelerate the inflammatory response by reducing SOD activity and increasing MDA production. SOD is depleted during the clearance of oxygen free radicals. Therefore, tissues can then be subsequently damaged by ROS accumulation [90]. Inflammatory events are often accompanied by oxidative stress, which generates lipid peroxidation products such as 4-hydroxy-2-nonenal (4-HNE). The study in primary neuronal cultures from TLR4 mutant mice and wild-type control mice show that TLR4 expression increases in neurons when exposed to the HNE, and TLR4 signaling increases the vulnerability of neurons to oxidative stress. This indicates that TLR4 signaling may play a role in Alzheimer's disease (AD) pathogenesis, possibly being activated by membrane-associated oxidative stress [91]. Another study found that 4-HNE blocks TLR4-mediated macrophage activation, gene expression, and phagocytic functions. This is done at least partly by suppressing receptor dimerization [92]. The expression of the anti-inflammatory enzyme HO-1 was found to increase in sheep overexpressing TLR4, which can directly regulate AP-1 expression. Increased HO-1 activity can suppress TLR4-induced signal transduction [93]. Our data showed that CAT activity was reduced and GSH-Px expression was increased following the LPS stimulation of sheep overexpressing TLR4. Another study suggested that GSH-Px acts in the place of CAT to eliminate H_2O_2 in certain tissues. We predict that GST may play a critical role in regulating antioxidative enzyme expression. GST transcription can be upregulated and downregulated with AP-1 and GSH, respectively. ROS-mediated inflammation can be reversed by adding exogenous GSH, which can be transferred to glutathione disulfide (GSSG) to eliminate free radicals and avoid oxidative damage to tissues. The overexpression of TLR4 in sheep resulted in a more rapid GSH consumption that dramatically increased GSSG and resulted in severe oxidative damage. Our laboratory showed that TLR4-induced oxidative stress occurred as a result of NO• synthesis.

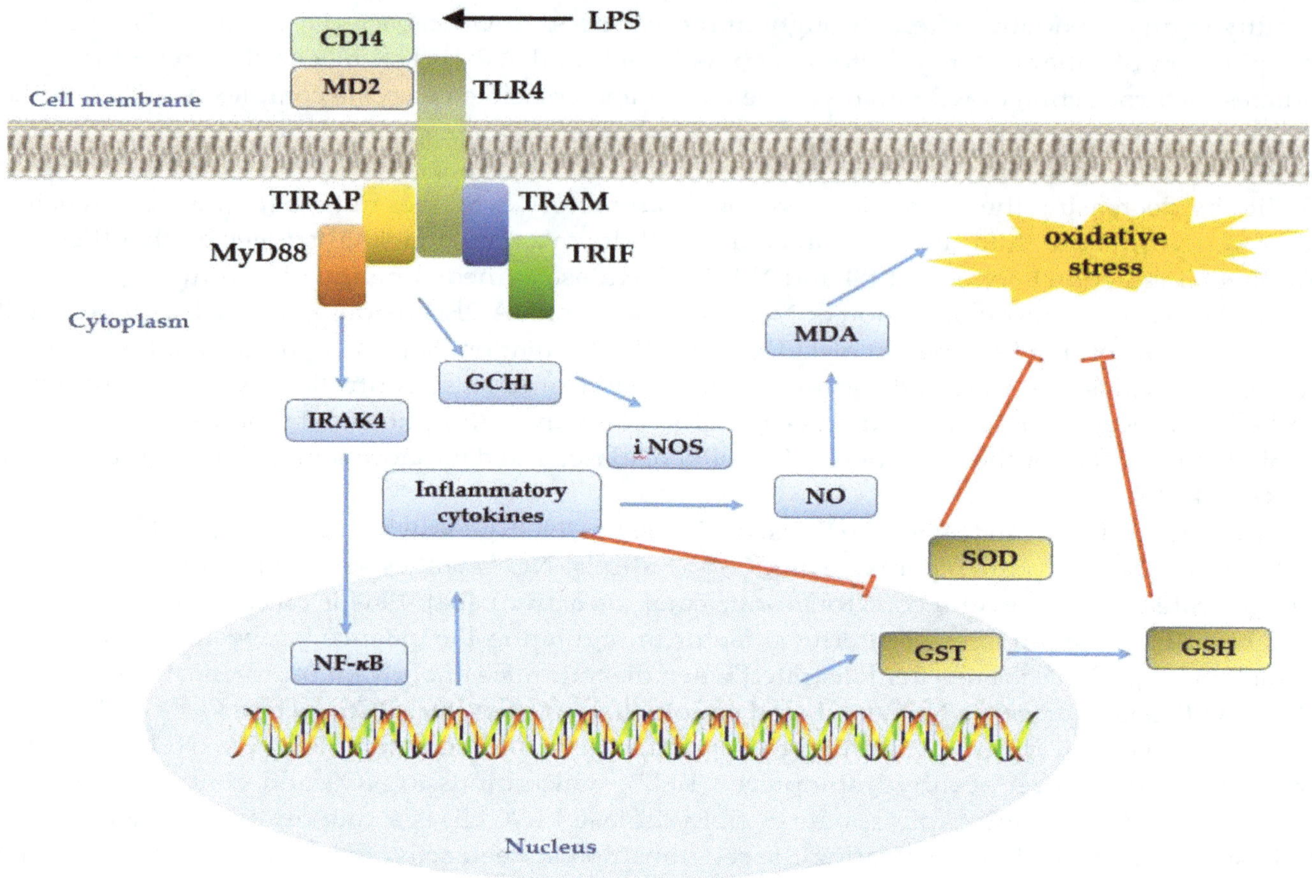

Figure 3. TLR4 signal pathway in oxidative stress [73,75]. TLR4 interacts with myeloid differentiation factor 2 (MD2), CD14, and specific ligand LPS receptors, then recruits downstream adaptors to activate MyD88– and TRIF-dependent pathway-mediated IRAK4-dependent NF-κB, which can trigger the transcription-promoting NO•. The GCHI-iNOS gene shows a significant increase in the expression of iNOS and NO, which accelerates the inflammatory response by reducing SOD activity and increasing MDA production. GSH eliminates free radicals and prevents oxidative damage.

TLR4 and its downstream signaling pathways are involved in the activation of GCHI expression [94]. In turn, GCHI plays an important role in the regulation of iNOS expression. The expression of NADPH oxidase and iNOS in a group of transgenic sheep overexpressing TLR4 was significantly greater compared with a WT group at 1 and 8 h, but these returned to normal levels at 48 h. This observation indicates that TLR4 can regulate the expression of both NADPH oxidase and iNOS. In addition, overexpressed TLR4 inhibited SOD activity and triggered AP-1 to activate downstream antioxidative genes that protect against oxidative stress [90].

5. Conclusions

Toll-like receptors (TLRs) act as immune receptors to initiate innate immunity and acquired immunity. They play an important role in the development of oxidative stress in the body. TLRs activate cells to produce pro-inflammatory factors in pathogen invasion, which can act as secondary messengers to regulate oxidative stress. Furthermore, TLRs can produce antioxidant mechanisms that interact to regulate oxidative stress. Our study shows that under the stimulation of LPS, transgenic overexpressing TLR4 sheep can rapidly trigger the TLR4 signaling pathway and upregulate the expression of cytokines in a short time, thus reducing the inflammatory reaction time, which is of great significance for improving the disease resistance of sheep. Positive individuals can significantly increase the adhesion of bacteria, which plays an important role in the timely removal of pathogens. Next, we will further evaluate the phagosome clearance ability of our transgenic sheep during an infection of phagocytic bacteria.

Author Contributions: S.-L.D., K.Y., Z.-X.L., and Y.L. wrote the paper.

References

1. Valko, M.; Leibfritz, D.; Moncol, J.; Cronin, M.T.D.; Mazur, M.; Telser, J. Free radicals and antioxidants in normal physiological functions and human disease. *Int. J. Biochem. Cell Biol.* **2007**, *39*, 44–84. [CrossRef] [PubMed]
2. Blokhina, O.; Virolainen, E.; Fagerstedt, K.V. Antioxidants, Oxidative Damage and Oxygen Deprivation Stress: A Review. *Ann. Bot.* **2003**, *91*, 179–194. [CrossRef] [PubMed]
3. Valko, M.; Rhodes, C.J.; Moncol, J.; Izakovic, M.; Mazur, M. Free radicals, metals and antioxidants in oxidative stress-induced cancer. *Chem. Biol. Interact.* **2006**, *160*, 1–40. [CrossRef] [PubMed]
4. Conner, E.M.; Grisham, M.B. Inflammation, free radicals, and antioxidants. *Nutrition* **1996**, *12*, 274–277. [CrossRef]
5. Ryan, K.A.; Smith, M.J.; Sanders, M.K.; Ernst, P.B. Reactive oxygen and nitrogen species differentially regulate Toll-like receptor 4-mediated activation of NF-kappa B and interleukin-8 expression. *Infect. Immun.* **2004**, *72*, 2123–2130. [CrossRef]
6. Janeway, C.A.; Medzhitov, R. Innate immune recognition. *Annu. Rev. Immunol.* **2002**, *20*, 197–216. [CrossRef] [PubMed]
7. Dröge, W. Free Radicals in the Physiological Control of Cell Function. *Physiol. Rev.* **2002**, *82*, 47–95. [CrossRef]
8. Stamler, J.S.; Singel, D.J.; Loscalzo, J. Biochemistry of nitric oxide and its redox-activated forms. *Science* **1992**, *258*, 1898–1902. [CrossRef]
9. Umezawa, K.; Akaike, T.; Fujii, S.; Suga, M.; Setoguchi, K.; Ozawa, A.; Maeda, H. Induction of nitric oxide synthesis and xanthine oxidase and their roles in the antimicrobial mechanism against Salmonella typhimurium infection in mice. *Infect. Immun.* **1997**, *65*, 2932–2940.
10. Zaki, M.H.; Akuta, T.; Akaike, T. Nitric Oxide-Induced Nitrative Stress Involved in Microbial Pathogenesis. *J. Pharmacol. Sci.* **2005**, *98*, 117–129. [CrossRef]
11. Umezawa, K.; Ohnishi, N.; Tanaka, K.; Kamiya, S.; Koga, Y.; Nakazawa, H.; Ozawa, A. Granulation in livers of mice infected with Salmonella typhimurium is caused by superoxide released from host phagocytes. *Infect. Immun.* **1995**, *63*, 4402–4408. [PubMed]
12. Barnes, P.J.; Shapiro, S.D.; Pauwels, R.A. Chronic obstructive pulmonary disease: Molecular and cellular mechanisms. *Eur. Respir. J.* **2003**, *22*, 672–688. [CrossRef] [PubMed]
13. Gaut, J.P.; Byun, J.; Tran, H.D.; Lauber, W.M.; Carroll, J.A.; Hotchkiss, R.S.; Belaaouaj, A.; Heinecke, J.W. Myeloperoxidase produces nitrating oxidants in vivo. *J. Clin. Invest.* **2002**, *109*, 1311–1319. [CrossRef] [PubMed]
14. Kim, S.Y.; Jeong, J.; Kim, S.J.; Seo, W.; Kim, M.; Choi, W.; Yoo, W.; Lee, J.; Shim, Y.; Yi, H.; et al. Pro-inflammatory hepatic macrophages generate ROS through NADPH oxidase 2 via endocytosis of monomeric TLR4–MD2 complex. *Nat. Commun.* **2017**, *8*, 2247. [CrossRef] [PubMed]
15. Singh, A.; Kukreti, R.; Saso, L.; Kukreti, S. Oxidative Stress: A Key Modulator in Neurodegenerative Diseases. *Molecules* **2019**. [CrossRef] [PubMed]
16. Halliwell, B. Antioxidants in human health and disease. *Annu. Rev. Nutr.* **1996**, *16*, 33–50. [CrossRef] [PubMed]
17. Sheppard, F.R.; Kelher, M.R.; Moore, E.E.; McLaughlin, N.J.; Banerjee, A.; Silliman, C.C. Structural organization of the neutrophil NADPH oxidase: Phosphorylation and translocation during priming and activation. *J. Leukoc. Biol.* **2005**, *78*, 1025–1042. [CrossRef] [PubMed]
18. Stumm, M.M.; D'Orazio, D.; Sumanovski, L.T.; Martin, P.Y.; Reichen, J.; Sieber, C.C. Endothelial, but not the inducible, nitric oxide synthase is detectable in normal and portal hypertensive rats. *Liver* **2002**, *22*, 441–450. [CrossRef]

19. Taylor, B.S.; de Vera, M.E.; Ganster, R.W.; Wang, Q.; Shapiro, R.A.; Morris, S.J.; Billiar, T.R.; Geller, D.A. Multiple NF-kappaB enhancer elements regulate cytokine induction of the human inducible nitric oxide synthase gene. *J. Biol. Chem.* **1998**, *273*, 15148–15156. [CrossRef]
20. Mosser, D.M.; Edwards, J.P. Exploring the full spectrum of macrophage activation. *Nat. Rev. Immunol.* **2008**, *8*, 958–969. [CrossRef]
21. Nairz, M.; Schleicher, U.; Schroll, A.; Sonnweber, T.; Theurl, I.; Ludwiczek, S.; Talasz, H.; Brandacher, G.; Moser, P.L.; Muckenthaler, M.U.; et al. Nitric oxide–mediated regulation of ferroportin-1 controls macrophage iron homeostasis and immune function inSalmonella infection. *J. Exp. Med.* **2013**, *210*, 855–873. [CrossRef] [PubMed]
22. Tkachev, V.O.; Menshchikova, E.B.; Zenkov, N.K. Mechanism of the Nrf2/Keap1/ARE signaling system. *Biochemistry* **2011**, *76*, 407–422. [CrossRef] [PubMed]
23. Bellezza, I.; Mierla, A.L.; Minelli, A. Nrf2 and NF-kappaB and Their Concerted Modulation in Cancer Pathogenesis and Progression. *Cancers* **2010**, *2*, 483–497. [CrossRef] [PubMed]
24. Thammavongsa, V.; Kim, H.K.; Missiakas, D.; Schneewind, O. Staphylococcal manipulation of host immune responses. *Nat. Rev. Microbiol.* **2015**, *13*, 529–543. [CrossRef] [PubMed]
25. Gutierrez, M.J.; McSherry, G.D.; Ishmael, F.T.; Horwitz, A.A.; Nino, G. Residual NADPH oxidase activity and isolated lung involvement in x-linked chronic granulomatous disease. *Case Rep. Pediatr.* **2012**. [CrossRef] [PubMed]
26. Ehrt, S.; Schnappinger, D. Mycobacterial survival strategies in the phagosome: Defence against host stresses. *Cell. Microbiol.* **2009**, *11*, 1170–1178. [CrossRef] [PubMed]
27. Koopman, W.J.; Nijtmans, L.G.; Dieteren, C.E.; Roestenberg, P.; Valsecchi, F.; Smeitink, J.A.; Willems, P.H. Mammalian mitochondrial complex I: Biogenesis, regulation, and reactive oxygen species generation. *Antioxid. Redox Signal.* **2010**, *12*, 1431–1470. [CrossRef]
28. West, A.P.; Brodsky, I.E.; Rahner, C.; Woo, D.K.; Erdjument-Bromage, H.; Tempst, P.; Walsh, M.C.; Choi, Y.; Shadel, G.S.; Ghosh, S. TLR signalling augments macrophage bactericidal activity through mitochondrial ROS. *Nature* **2011**, *472*, 476–480. [CrossRef]
29. Pandey, S.; Kawai, T.; Akira, S. Microbial sensing by Toll-like receptors and intracellular nucleic acid sensors. *Cold Spring Harb. Perspect. Biol.* **2014**. [CrossRef]
30. Janeway, C.J. Approaching the asymptote? Evolution and revolution in immunology. *Cold Spring Harb. Symp. Quant. Biol.* **1989**, *54*, 1–13. [CrossRef]
31. Vance, R.E.; Isberg, R.R.; Portnoy, D.A. Patterns of pathogenesis: Discrimination of pathogenic and nonpathogenic microbes by the innate immune system. *Cell Host Microbe.* **2009**, *6*, 10–21. [CrossRef] [PubMed]
32. Jault, C.; Pichon, L.; Chluba, J. Toll-like receptor gene family and TIR-domain adapters in Danio rerio. *Mol. Immunol.* **2004**, *40*, 759–771. [CrossRef] [PubMed]
33. Lin, Y.S.; Saputra, F.; Chen, Y.C.; Hu, S.Y. Dietary administration of Bacillus amyloliquefaciens R8 reduces hepatic oxidative stress and enhances nutrient metabolism and immunity against Aeromonas hydrophila and Streptococcus agalactiae in zebrafish (Danio rerio). *Fish Shellfish Immunol.* **2019**, *86*, 410–419. [CrossRef] [PubMed]
34. Wei, J.; Zhou, T.; Hu, Z.; Li, Y.; Yuan, H.; Zhao, K.; Zhang, H.; Liu, C. Effects of triclocarban on oxidative stress and innate immune response in zebrafish embryos. *Chemosphere* **2018**, *210*, 93–101. [CrossRef] [PubMed]
35. Takeuchi, O.; Akira, S. Pattern Recognition Receptors and Inflammation. *Cell* **2010**, *140*, 805–820. [CrossRef] [PubMed]
36. Rathinam, V.A.K.; Zhao, Y.; Shao, F. Innate immunity to intracellular LPS. *Nat. Immunol.* **2019**. [CrossRef] [PubMed]
37. Hoareau, L.; Bencharif, K.; Rondeau, P.; Murumalla, R.; Ravanan, P.; Tallet, F.; Delarue, P.; Cesari, M.; Roche, R.; Festy, F. Signaling pathways involved in LPS induced TNFalpha production in human adipocytes. *J. Inflamm.* **2010**. [CrossRef] [PubMed]
38. Beutler, B. Inferences, questions and possibilities in Toll-like receptor signalling. *Nature* **2004**, *430*, 257–263. [CrossRef]
39. Akira, S.; Uematsu, S.; Takeuchi, O. Pathogen Recognition and Innate Immunity. *Cell* **2006**, *124*, 783–801. [CrossRef]
40. Re, F.; Strominger, J.L. Toll-like receptor 2 (TLR2) and TLR4 differentially activate human dendritic cells. *J. Biol. Chem.* **2001**, *276*, 37692–37699. [CrossRef]

41. Akira, S.; Takeda, K. Toll-like receptor signalling. *Nat. Rev. Immunol.* **2004**, *4*, 499–511. [CrossRef] [PubMed]
42. O'Neill, L. TLRs: Professor Mechnikov, sit on your hat. *Trends Immunol.* **2004**, *25*, 687–693. [CrossRef] [PubMed]
43. Strunk, T.; Power, C.M.; Currie, A.J.; Richmond, P.; Golenbock, D.T.; Stoler-Barak, L.; Gallington, L.C.; Otto, M.; Burgner, D.; Levy, O. TLR2 mediates recognition of live Staphylococcus epidermidis and clearance of bacteremia. *PLoS ONE* **2010**. [CrossRef] [PubMed]
44. Triantafilou, M.; Triantafilou, K. Lipopolysaccharide recognition: CD14, TLRs and the LPS-activation cluster. *Trends Immunol.* **2002**, *23*, 301–304. [CrossRef]
45. Doyle, S.E.; O'Connell, R.; Vaidya, S.A.; Chow, E.K.; Yee, K.; Cheng, G. Toll-like receptor 3 mediates a more potent antiviral response than Toll-like receptor 4. *J. Immunol.* **2003**, *170*, 3565–3571. [CrossRef] [PubMed]
46. Hayashi, F.; Smith, K.D.; Ozinsky, A.; Hawn, T.R.; Yi, E.C.; Goodlett, D.R.; Eng, J.K.; Akira, S.; Underhill, D.M.; Aderem, A. The innate immune response to bacterial flagellin is mediated by Toll-like receptor 5. *Nature* **2001**, *410*, 1099–1103. [CrossRef] [PubMed]
47. Heil, F.; Ahmad-Nejad, P.; Hemmi, H.; Hochrein, H.; Ampenberger, F.; Gellert, T.; Dietrich, H.; Lipford, G.; Takeda, K.; Akira, S.; et al. The Toll-like receptor 7 (TLR7)-specific stimulus loxoribine uncovers a strong relationship within the TLR7, 8 and 9 subfamily. *Eur. J. Immunol.* **2003**, *33*, 2987–2997. [CrossRef] [PubMed]
48. Karin, M.; Delhase, M. The I kappa B kinase (IKK) and NF-kappa B: Key elements of proinflammatory signalling. *Semin. Immunol.* **2000**, *12*, 85–98. [CrossRef] [PubMed]
49. Kawai, T.; Akira, S. Toll-like receptors and their crosstalk with other innate receptors in infection and immunity. *Immunity* **2011**, *34*, 637–650. [CrossRef] [PubMed]
50. Kagan, J.C.; Medzhitov, R. Phosphoinositide-mediated adaptor recruitment controls Toll-like receptor signaling. *Cell* **2006**, *125*, 943–955. [CrossRef]
51. Yamashita, M.; Chattopadhyay, S.; Fensterl, V.; Saikia, P.; Wetzel, J.L.; Sen, G.C. Epidermal Growth Factor Receptor Is Essential for Toll-Like Receptor 3 Signaling. *Sci. Signal.* **2012**. [CrossRef] [PubMed]
52. Lu, Y.C.; Yeh, W.C.; Ohashi, P.S. LPS/TLR4 signal transduction pathway. *Cytokine* **2008**, *42*, 145–151. [CrossRef] [PubMed]
53. Gerold, G.; Zychlinsky, A.; de Diego, J.L. What is the role of Toll-like receptors in bacterial infections? *Semin. Immunol.* **2007**, *19*, 41–47. [CrossRef] [PubMed]
54. Gill, R.; Tsung, A.; Billiar, T. Linking oxidative stress to inflammation: Toll-like receptors. *Free Radical Bio. Med.* **2010**, *48*, 1121–1132. [CrossRef] [PubMed]
55. Lugrin, J.; Rosenblatt-Velin, N.; Parapanov, R.; Liaudet, L. The role of oxidative stress during inflammatory processes. *Biol. Chem.* **2014**, *395*, 203–230. [CrossRef] [PubMed]
56. Ruth, M.R.; Field, C.J. The immune modifying effects of amino acids on gut-associated lymphoid tissue. *J. Anim. Sci. Biotechnol.* **2013**. [CrossRef] [PubMed]
57. Shi, Q.; Le, X.; Abbruzzese, J.L.; Wang, B.; Mujaida, N.; Matsushima, K.; Huang, S.; Xiong, Q.; Xie, K. Cooperation between transcription factor AP-1 and NF-kappaB in the induction of interleukin-8 in human pancreatic adenocarcinoma cells by hypoxia. *J. Interferon Cytokine Res.* **1999**, *19*, 1363–1371. [CrossRef]
58. Imanishi, T.; Hara, H.; Suzuki, S.; Suzuki, N.; Akira, S.; Saito, T. Cutting edge: TLR2 directly triggers Th1 effector functions. *J. Immunol.* **2007**, *178*, 6715–6719. [CrossRef]
59. Qin, Q.; Laitinen, P.; Majamaa-Voltti, K.; Eriksson, S.; Kumpula, E.; Pettersson, K. Release Patterns of Pregnancy Associated Plasma Protein A (PAPP-A) in Patients with Acute Coronary Syndromes. *Scand. Cardiovasc. J.* **2002**, *36*, 358–361. [CrossRef]
60. Echchannaoui, H.; Frei, K.; Schnell, C.; Leib, S.L.; Zimmerli, W.; Landmann, R. Toll-like receptor 2-deficient mice are highly susceptible to Streptococcus pneumoniae meningitis because of reduced bacterial clearing and enhanced inflammation. *J. Infect. Dis.* **2002**, *186*, 798–806. [CrossRef]
61. Zou, L.; Feng, Y.; Chen, Y.; Si, R.; Shen, S.; Zhou, Q.; Ichinose, F.; Scherrer-Crosbie, M.; Chao, W. Toll-like receptor 2 plays a critical role in cardiac dysfunction during polymicrobial sepsis. *Crit. Care Med.* **2010**, *38*, 1335–1342. [CrossRef]
62. Mersmann, J.; Berkels, R.; Zacharowski, P.; Tran, N.; Koch, A.; Iekushi, K.; Dimmeler, S.; Granja, T.F.; Boehm, O.; Claycomb, W.C.; et al. Preconditioning by toll-like receptor 2 agonist Pam3CSK4 reduces CXCL1-dependent leukocyte recruitment in murine myocardial ischemia/reperfusion injury. *Crit. Care Med.* **2010**, *38*, 903–909. [CrossRef] [PubMed]

63. Deng, S.; Yu, K.; Jiang, W.; Li, Y.; Wang, S.; Deng, Z.; Yao, Y.; Zhang, B.; Liu, G.; Liu, Y.; et al. Over-expression of Toll-like receptor 2 up-regulates heme oxygenase-1 expression and decreases oxidative injury in dairy goats. *J. Anim. Sci. Biotechno.* **2017**. [CrossRef]
64. Connelly, L.; Palacios-Callender, M.; Ameixa, C.; Moncada, S.; Hobbs, A.J. Biphasic regulation of NF-kappa B activity underlies the pro- and anti-inflammatory actions of nitric oxide. *J. Immunol.* **2001**, *166*, 3873–3881. [CrossRef] [PubMed]
65. Ho, F.M.; Kang, H.C.; Lee, S.T.; Chao, Y.; Chen, Y.C.; Huang, L.J.; Lin, W.W. The anti-inflammatory actions of LCY-2-CHO, a carbazole analogue, in vascular smooth muscle cells. *Biochem. Pharmacol.* **2007**, *74*, 298–308. [CrossRef] [PubMed]
66. Fiorelli, S.; Porro, B.; Cosentino, N.; Di Minno, A.; Manega, C.M.; Fabbiocchi, F.; Niccoli, G.; Fracassi, F.; Barbieri, S.; Marenzi, G.; et al. Activation of Nrf2/HO-1 Pathway and Human Atherosclerotic Plaque Vulnerability: An In Vitro and In Vivo Study. *Cells* **2019**. [CrossRef] [PubMed]
67. Donovan, E.L.; McCord, J.M.; Reuland, D.J.; Miller, B.F.; Hamilton, K.L. Phytochemical activation of Nrf2 protects human coronary artery endothelial cells against an oxidative challenge. *Oxid. Med. Cell. Longev.* **2012**, *2012*, 132931. [CrossRef] [PubMed]
68. Calkins, M.J.; Johnson, D.A.; Townsend, J.A.; Vargas, M.R.; Dowell, J.A.; Williamson, T.P.; Kraft, A.D.; Lee, J.M.; Li, J.; Johnson, J.A. The Nrf2/ARE pathway as a potential therapeutic target in neurodegenerative disease. *Antioxid. Redox Signal.* **2009**, *11*, 497–508. [CrossRef] [PubMed]
69. Lee, I.T.; Wang, S.W.; Lee, C.W.; Chang, C.C.; Lin, C.C.; Luo, S.F.; Yang, C.M. Lipoteichoic acid induces HO-1 expression via the TLR2/MyD88/c-Src/NADPH oxidase pathway and Nrf2 in human tracheal smooth muscle cells. *J. Immunol.* **2008**, *181*, 5098–5110. [CrossRef] [PubMed]
70. Jilling, T.; Simon, D.; Lu, J.; Meng, F.J.; Li, D.; Schy, R.; Thomson, R.B.; Soliman, A.; Arditi, M.; Caplan, M.S. The roles of bacteria and TLR4 in rat and murine models of necrotizing enterocolitis. *J. Immunol.* **2006**, *177*, 3273–3282. [CrossRef]
71. Shi, H.; Kokoeva, M.V.; Inouye, K.; Tzameli, I.; Yin, H.; Flier, J.S. TLR4 links innate immunity and fatty acid-induced insulin resistance. *J. Clin. Invest.* **2006**, *116*, 3015–3025. [CrossRef] [PubMed]
72. Seki, E.; Tsutsui, H.; Nakano, H.; Tsuji, N.; Hoshino, K.; Adachi, O.; Adachi, K.; Futatsugi, S.; Kuida, K.; Takeuchi, O.; et al. Lipopolysaccharide-induced IL-18 secretion from murine Kupffer cells independently of myeloid differentiation factor 88 that is critically involved in induction of production of IL-12 and IL-1beta. *J. Immunol.* **2001**, *166*, 2651–2657. [CrossRef] [PubMed]
73. Nackiewicz, D.; Dan, M.; He, W.; Kim, R.; Salmi, A.; Rutti, S.; Westwell-Roper, C.; Cunningham, A.; Speck, M.; Schuster-Klein, C.; et al. TLR2/6 and TLR4-activated macrophages contribute to islet inflammation and impair beta cell insulin gene expression via IL-1 and IL-6. *Diabetologia* **2014**, *57*, 1645–1654. [CrossRef] [PubMed]
74. Roy, M.F.; Lariviere, L.; Wilkinson, R.; Tam, M.; Stevenson, M.M.; Malo, D. Incremental expression of Tlr4 correlates with mouse resistance to Salmonella infection and fine regulation of relevant immune genes. *Genes Immun.* **2006**, *7*, 372–383. [CrossRef] [PubMed]
75. Wei, S.; Yang, D.; Yang, J.; Zhang, X.; Zhang, J.; Fu, J.; Zhou, G.; Liu, H.; Lian, Z.; Han, H. Overexpression of Toll-like receptor 4 enhances LPS-induced inflammatory response and inhibits Salmonella Typhimurium growth in ovine macrophages. *Eur. J. Cell Biol.* **2019**, *98*, 36–50. [CrossRef] [PubMed]
76. Lee, J.J.; Kim, D.H.; Kim, D.G.; Lee, H.J.; Min, W.; Rhee, M.H.; Cho, J.Y.; Watarai, M.; Kim, S. Toll-like receptor 4-linked Janus kinase 2 signaling contributes to internalization of Brucella abortus by macrophages. *Infect. Immun.* **2013**, *81*, 2448–2458. [CrossRef] [PubMed]
77. Wang, S.; Deng, S.; Cao, Y.; Zhang, R.; Wang, Z.; Jiang, X.; Wang, J.; Zhang, X.; Zhang, J.; Liu, G.; et al. Overexpression of Toll-Like Receptor 4 Contributes to Phagocytosis of *Salmonella Enterica* Serovar Typhimurium via Phosphoinositide 3-Kinase Signaling in Sheep. *Cell. Physiol. Biochem.* **2018**, *49*, 662–677. [CrossRef] [PubMed]
78. Wang, S.; Cao, Y.; Deng, S.; Jiang, X.; Wang, J.; Zhang, X.; Zhang, J.; Liu, G.; Lian, Z. Overexpression of Toll-like Receptor 4-linked Mitogen-activated Protein Kinase Signaling Contributes to Internalization of Escherichia coli in Sheep. *Int. J. Biol. Sci.* **2018**, *14*, 1022–1032. [CrossRef]
79. Deng, S.; Wu, Q.; Yu, K.; Zhang, Y.; Yao, Y.; Li, W.; Deng, Z.; Liu, G.; Li, W.; Lian, Z. Changes in the relative inflammatory responses in sheep cells overexpressing of toll-like receptor 4 when stimulated with LPS. *PLoS ONE* **2012**. [CrossRef]
80. Lambeth, J.D. NOX enzymes and the biology of reactive oxygen. *Nat. Rev. Immunol.* **2004**, *4*, 181–189. [CrossRef]

81. Kim, J.H.; Na, H.J.; Kim, C.K.; Kim, J.Y.; Ha, K.S.; Lee, H.; Chung, H.T.; Kwon, H.J.; Kwon, Y.G.; Kim, Y.M. The non-provitamin A carotenoid, lutein, inhibits NF-kappaB-dependent gene expression through redox-based regulation of the phosphatidylinositol 3-kinase/PTEN/Akt and NF-kappaB-inducing kinase pathways: Role of H(2)O(2) in NF-kappaB activation. *Free Radic. Biol. Med.* **2008**, *45*, 885–896. [CrossRef] [PubMed]
82. Park, H.S.; Jung, H.Y.; Park, E.Y.; Kim, J.; Lee, W.J.; Bae, Y.S. Cutting edge: Direct interaction of TLR4 with NAD(P)H oxidase 4 isozyme is essential for lipopolysaccharide-induced production of reactive oxygen species and activation of NF-kappa B. *J. Immunol.* **2004**, *173*, 3589–3593. [CrossRef] [PubMed]
83. Pacquelet, S.; Johnson, J.L.; Ellis, B.A.; Brzezinska, A.A.; Lane, W.S.; Munafo, D.B.; Catz, S.D. Cross-talk between IRAK-4 and the NADPH oxidase. *Biochem. J.* **2007**, *403*, 451–461. [CrossRef]
84. Stuehr, D.J.; Santolini, J.; Wang, Z.; Wei, C.; Adak, S. Update on Mechanism and Catalytic Regulation in the NO Synthases. *J. Biol. Chem.* **2004**, *279*, 36167–36170. [CrossRef] [PubMed]
85. Higashi, Y.; Sasaki, S.; Nakagawa, K.; Kimura, M.; Noma, K.; Hara, K.; Jitsuiki, D.; Goto, C.; Oshima, T.; Chayama, K.; et al. Tetrahydrobiopterin improves aging-related impairment of endothelium-dependent vasodilation through increase in nitric oxide production. *Atherosclerosis* **2006**, *186*, 390–395. [CrossRef] [PubMed]
86. Alp, N.J.; Mussa, S.; Khoo, J.; Cai, S.; Guzik, T.; Jefferson, A.; Goh, N.; Rockett, K.A.; Channon, K.M. Tetrahydrobiopterin-dependent preservation of nitric oxide-mediated endothelial function in diabetes by targeted transgenic GTP-cyclohydrolase I overexpression. *J. Clin. Invest.* **2003**, *112*, 725–735. [CrossRef] [PubMed]
87. Vasquez-Vivar, J.; Kalyanaraman, B.; Martasek, P.; Hogg, N.; Masters, B.S.; Karoui, H.; Tordo, P.; Pritchard, K.J. Superoxide generation by endothelial nitric oxide synthase: The influence of cofactors. *Proc. Natl. Acad. Sci. USA* **1998**, *95*, 9220–9225. [CrossRef]
88. Xue, J.; Yu, C.; Sheng, W.; Zhu, W.; Luo, J.; Zhang, Q.; Yang, H.; Cao, H.; Wang, W.; Zhou, J.; et al. The Nrf2/GCH1/BH4 Axis Ameliorates Radiation-Induced Skin Injury by Modulating the ROS Cascade. *J. Invest. Dermatol.* **2017**, *137*, 2059–2068. [CrossRef]
89. Wang, W.; Zolty, E.; Falk, S.; Summer, S.; Zhou, Z.; Gengaro, P.; Faubel, S.; Alp, N.; Channon, K.; Schrier, R. Endotoxemia-related acute kidney injury in transgenic mice with endothelial overexpression of GTP cyclohydrolase-1. *Am. J. Physiol. Renal. Physiol.* **2008**. [CrossRef]
90. Deng, S.; Yu, K.; Wu, Q.; Li, Y.; Zhang, X.; Zhang, B.; Liu, G.; Liu, Y.; Lian, Z. Toll-Like Receptor 4 Reduces Oxidative Injury via Glutathione Activity in Sheep. *Oxid. Med. Cell. Longev.* **2016**, *2016*, 1–9. [CrossRef]
91. Tang, S.; Lathia, J.D.; Selvaraj, P.K.; Jo, D.; Mughal, M.R.; Cheng, A.; Siler, D.A.; Markesbery, W.R.; Arumugam, T.V.; Mattson, M.P. Toll-like receptor-4 mediates neuronal apoptosis induced by amyloid β-peptide and the membrane lipid peroxidation product 4-hydroxynonenal. *Exp. Neurol.* **2008**, *213*, 114–121. [CrossRef] [PubMed]
92. Kim, Y.S.; Park, Z.Y.; Kim, S.Y.; Jeong, E.; Lee, J.Y. Alteration of Toll-like receptor 4 activation by 4-hydroxy-2-nonenal mediated by the suppression of receptor homodimerization. *Chem. Biol. Interact.* **2009**, *182*, 59–66. [CrossRef] [PubMed]
93. Wang, X.M.; Kim, H.P.; Nakahira, K.; Ryter, S.W.; Choi, A.M. The heme oxygenase-1/carbon monoxide pathway suppresses TLR4 signaling by regulating the interaction of TLR4 with caveolin-1. *J. Immunol.* **2009**, *182*, 3809–3818. [CrossRef] [PubMed]
94. Deng, S.; Yu, K.; Zhang, B.; Yao, Y.; Wang, Z.; Zhang, J.; Zhang, X.; Liu, G.; Li, N.; Liu, Y.; et al. Toll-Like Receptor 4 Promotes NO Synthesis by Upregulating GCHI Expression under Oxidative Stress Conditions in Sheep Monocytes/Macrophages. *Oxid. Med. Cell. Longev.* **2015**, *2015*, 1–11. [CrossRef] [PubMed]

Permissions

All chapters in this book were first published by MDPI; hereby published with permission under the Creative Commons Attribution License or equivalent. Every chapter published in this book has been scrutinized by our experts. Their significance has been extensively debated. The topics covered herein carry significant findings which will fuel the growth of the discipline. They may even be implemented as practical applications or may be referred to as a beginning point for another development.

The contributors of this book come from diverse backgrounds, making this book a truly international effort. This book will bring forth new frontiers with its revolutionizing research information and detailed analysis of the nascent developments around the world.

We would like to thank all the contributing authors for lending their expertise to make the book truly unique. They have played a crucial role in the development of this book. Without their invaluable contributions this book wouldn't have been possible. They have made vital efforts to compile up to date information on the varied aspects of this subject to make this book a valuable addition to the collection of many professionals and students.

This book was conceptualized with the vision of imparting up-to-date information and advanced data in this field. To ensure the same, a matchless editorial board was set up. Every individual on the board went through rigorous rounds of assessment to prove their worth. After which they invested a large part of their time researching and compiling the most relevant data for our readers.

The editorial board has been involved in producing this book since its inception. They have spent rigorous hours researching and exploring the diverse topics which have resulted in the successful publishing of this book. They have passed on their knowledge of decades through this book. To expedite this challenging task, the publisher supported the team at every step. A small team of assistant editors was also appointed to further simplify the editing procedure and attain best results for the readers.

Apart from the editorial board, the designing team has also invested a significant amount of their time in understanding the subject and creating the most relevant covers. They scrutinized every image to scout for the most suitable representation of the subject and create an appropriate cover for the book.

The publishing team has been an ardent support to the editorial, designing and production team. Their endless efforts to recruit the best for this project, has resulted in the accomplishment of this book. They are a veteran in the field of academics and their pool of knowledge is as vast as their experience in printing. Their expertise and guidance has proved useful at every step. Their uncompromising quality standards have made this book an exceptional effort. Their encouragement from time to time has been an inspiration for everyone.

The publisher and the editorial board hope that this book will prove to be a valuable piece of knowledge for researchers, students, practitioners and scholars across the globe.

List of Contributors

Daniele Capitanio and Pietro Barbacini
Department of Biomedical Sciences for Health, University of Milan, 20090 Segrate (MI), Italy

Beatrice Arosio
Geriatric Unit, Fondazione IRCCS Ca' Granda Ospedale Maggiore Policlinico, 20122 Milan, Italy
Department of Clinical Sciences and Community Health, University of Milan, 20122 Milan, Italy

Fabio Trecate and Franca Rosa Guerini
IRCCS Fondazione Don Carlo Gnocchi ONLUS, 20148 Milan, Italy

Cecilia Gelfi
Department of Biomedical Sciences for Health, University of Milan, 20090 Segrate (MI), Italy
IRCCS Istituto Ortopedico Galeazzi, 20161 Milan, Italy

Enrica Torretta
IRCCS Istituto Ortopedico Galeazzi, 20161 Milan, Italy

Matteo Cesari
Department of Clinical Sciences and Community Health, University of Milan, 20122 Milan, Italy
Geriatric Unit, IRCCS Istituti Clinici Scientifici Maugeri, 20138 Milan, Italy

Daniela Mari
Laboratorio Sperimentale di Ricerche di Neuroendocrinologia Geriatrica ed Oncologica, IRCCS Istituto Auxologico Italiano, 20145 Milan, Italy

Mario Clerici
IRCCS Fondazione Don Carlo Gnocchi ONLUS, 20148 Milan, Italy
Department of Pathophysiology and Transplantation, University of Milan, 20122 Milan, Italy

Nehal Gupta
Department of Medicine, University of California San Francisco, 1600 Divisadero Street, San Francisco, CA 94115, USA

Kshitij Verma
Discovery Chemistry, Genentech, Inc., 1 DNA Way, South San Francisco, CA 94080, USA

Sarath Nalla, Alok Kulshreshtha, Rajiv Lall and Sahdeo Prasad
Discovery Chemistry, Genentech, Inc., 1 DNA Way, South San Francisco, CA 94080, USA
Noble Pharma, LLC, 4602 Domain Drive, Menomonie, WI 54751, USA

Anna Jastrząb and Agnieszka Gęgotek
Department of Analytical Chemistry, Medical University of Bialystok, 15-089 Bialystok, Poland

Elżbieta Skrzydlewska
Department of Analytical Chemistry, Medical University of Bialystok, 15-089 Bialystok, Poland
Department of Analytical Chemistry, Medical University of Bialystok, 15-222 Bialystok, Poland

Mariam Rado, Okobi Ekpo and David Fisher
Department of Medical Bioscience, University of the Western Cape, Bellville, Cape Town 7530, South Africa

Olufemi Alamu
Department of Medical Bioscience, University of the Western Cape, Bellville, Cape Town 7530, South Africa
Anatomy Department, Ladoke Akintola University of Technology, Ogbomoso 210241, Nigeria

Yu-Po Luo
Department of Surgery, Buddhist Tzu Chi General Hospital, Hualien 97002, Taiwan

Jui-Chih Chang
Department of Surgery, Buddhist Tzu Chi General Hospital, Hualien 97002, Taiwan
School of Medicine, Tzu Chi University, Hualien 97004, Taiwan

Huai-Ren Chang, Jing-Ren Jeng and Jen-Che Hsieh
School of Medicine, Tzu Chi University, Hualien 97004, Taiwan
Division of Cardiology, Department of Internal Medicine, Buddhist Tzu Chi General Hospital, Hualien 97002, Taiwan

Chih-Feng Lien
Institute of Medical Sciences, Tzu Chi University, Hualien 97004, Taiwan

Wen-Sen Lee
Graduate Institute of Medical Sciences, College of Medicine, Taipei Medical University, Taipei 11031, Taiwan

Yu-Cheng Hsu
Master Program in Medical Physiology, School of Medicine, Tzu Chi University, Hualien 97004, Taiwan

Kun-Ta Yang
Department of Physiology, School of Medicine, Tzu Chi University, Hualien 97004, Taiwan

Susanna Fiorelli, Benedetta Porro, Nicola Cosentino, Alessandro Di Minno, Chiara Maria Manega, Franco Fabbiocchi, Simone Barbieri, Giancarlo Marenzi, Viviana Cavalca, Elena Tremoli and Sonia Eligini
Centro Cardiologico Monzino, I.R.C.C.S., 20138 Milan, Italy

Giampaolo Niccoli, Francesco Fracassi and Filippo Crea
Department of Cardiovascular & Thoracic Sciences, Fondazione Policlinico Universitario A. Gemelli, I.R.C.C.S., Università Cattolica del Sacro Cuore, 00168 Rome, Italy

Carla Ferreri and Annalisa Masi
Istituto per la Sintesi Organica e la Fotoreattività, Consiglio Nazionale delle Ricerche, Via P. Gobetti 101, 40129 Bologna, Italy

Chryssostomos Chatgilialoglu
Istituto per la Sintesi Organica e la Fotoreattività, Consiglio Nazionale delle Ricerche, Via P. Gobetti 101, 40129 Bologna, Italy
Center for Advanced Technologies, Adam Mickiewicz University, 61-614 Poznań, Poland

Vladimir Shafirovich and Nicholas E. Geacintov
Department of Chemistry, New York University, 31 Washington Place, New York, NY 10003-5180, USA

Marios G. Krokidis
Institute of Nanoscience and Nanotechnology, N.C.S.R. "Demokritos", 15310 Agia Paraskevi Attikis, Greece

Yuan Liu
Biochemistry Ph.D. Program, Florida International University, Miami, FL 33199, USA
Biomolecular Sciences Institute, Florida International University, Miami, FL 33199, USA
Department of Chemistry and Biochemistry, Florida International University, 11200 SW 8th Street, Miami, FL 33199, USA

Pawlos S. Tsegay
Biochemistry Ph.D. Program, Florida International University, Miami, FL 33199, USA

Michael A. Terzidis
Laboratory of Chemical and Environmental Technology, Department of Chemistry, Aristotle University of Thessaloniki, 54124 Thessaloniki, Greece

Ewa Ambrożewicz and Marta Muszyńska
Department of Analytical Chemistry, Medical University of Bialystok, 15-222 Bialystok, Poland

Pavel V. Avdonin, Elena Yu. Rybakova, Piotr P. Avdonin, Sergei K. Trufanov, Galina Yu. Mironova and Alexandra A. Tsitrina
Koltsov Institute of Developmental Biology, Moscow 119334, Russia

Nikolay V. Goncharov
Sechenov Institute of Evolutional Institute of Evolutionary Physiology and Biochemistry, Saint Petersburg 194223, Russia
Research Institute of Hygiene, Occupational Pathology and Human Ecology, Leningrad Region 188663, Russia

Xinfang An, Zixing Fu, Chendi Mai, Weiming Wang, Linyu Wei, Dongliang Li and Chaokun Li
Sino-UK Joint Laboratory for Brain Function and Injury and Department of Physiology and Neurobiology, Xinxiang Medical University, Xinxiang 453003, China

Lin-Hua Jiang
Sino-UK Joint Laboratory for Brain Function and Injury and Department of Physiology and Neurobiology, Xinxiang Medical University, Xinxiang 453003, China
School of Biomedical Sciences, Faculty of Biological Sciences, University of Leeds, Leeds LS2 JT, UK

Grażyna Tokajuk
Department of Integrated Dentistry, Medical University of Bialystok, 15-230 Bialystok, Poland

Grzegorz Grynkiewicz
Pharmaceutical Research Institute, 01-793 Warsaw, Poland

Lidija Milković, Ana Čipak Gašparović and Neven Žarković
Laboratory for Oxidative Stress, Division of Molecular Medicine, Ruđer Bošković Institute, 10000 Zagreb, Croatia

Marko Tomljanović, Renata Novak Kujundžić, Dina Šimunić, Paško Konjevoda, Anamarija Mojzeš and Koraljka Gall Trošelj
Laboratory for Epigenomics, Division of Molecular Medicine, Ruđer Bošković Institute, 10000 Zagreb, Croatia

Shou-Long Deng
CAS Key Laboratory of Genome Sciences and Information, Beijing Institute of Genomics, Chinese Academy of Sciences, Beijing 100101, China

Nikola Đaković
University Hospital Centre Sisters of Charity, Institute for Clinical Medical Research and Education, 10000 Zagreb, Croatia
Department of Clinical Oncology, School of Medicine, University of Zagreb, 10000 Zagreb, Croatia

Yao Li, Zheng-Xing Lian and Kun Yu
Beijing Key Laboratory for Animal Genetic Improvement, National Engineering Laboratory for Animal Breeding, Key Laboratory of Animal Genetics and Breeding of the Ministry of Agriculture, College of Animal Science and Technology, China Agricultural University, Beijing 100193, China

Index

www.ingramcontent.com/pod-product-compliance
Lightning Source LLC
LaVergne TN
LVHW082216200926
848484LV00008B/2264